FINANCING THE ENERGY TRANSITION

TECHNOLOGIES AND CAPITAL FOR A SUSTAINABLE FUTURE

FINANCING THE ENERGY TRANSITION

TECHNOLOGIES AND CAPITAL FOR A SUSTAINABLE FUTURE

BRIAN SCOTT-QUINN
ICMA Centre for Financial Markets, Henley Business School, University of Reading, UK

DEYBER CANO
The New School, New York, USA

World Scientific

NEW JERSEY • LONDON • SINGAPORE • BEIJING • SHANGHAI • HONG KONG • TAIPEI • CHENNAI • TOKYO

Published by

World Scientific Publishing Europe Ltd
57 Shelton Street, Covent Garden, London WC2H 9HE
Head office: 5 Toh Tuck Link, Singapore 596224
USA office: 27 Warren Street, Suite 401-402, Hackensack, NJ 07601

Library of Congress Cataloging-in-Publication Data
Names: Scott-Quinn, Brian, 1945– author | Cano, Deyber author
Title: Financing the energy transition : technologies and capital for a sustainable future / Brian Scott-Quinn, ICMA Centre for Financial Markets, Henley Business School, University of Reading, UK, Deyber Cano, The New School, New York, USA.
Other titles: technologies and capital for a sustainable future
Description: London ; Hackensack, NJ : World Scientific Publishing Europe Ltd, [2027] | Includes bibliographical references and index.
Identifiers: LCCN 2026001704 | ISBN 9781800619012 hardcover | ISBN 9781800619029 ebook | ISBN 9781800619036 ebook other
Subjects: LCSH: Clean energy | Clean energy industries--Finance
Classification: LCC TJ808 .S36 2027
LC record available at https://lccn.loc.gov/2026001704

British Library Cataloguing-in-Publication Data
A catalogue record for this book is available from the British Library.

For any available supplementary material, please visit
https://www.worldscientific.com/worldscibooks/10.1142/Q0559#t=suppl

Desk Editors: Nambirajan Karuppiah/Gabriel Rawlinson

Typeset by Stallion Press
Email: enquiries@stallionpress.com

Preface

Without an understanding of the new underlying technologies of the energy transition, it is quite risky to undertake financial due diligence on any of the new types of asset which will be the backbone of the transition. For that reason this book is almost equally divided between financial asset financing issues and infrastructure asset technology issues. The need for finance professionals to invest on a much greater scale into carbon-free infrastructure assets has become clearer with each year of this decade. To enable this, a 'techno-economic' understanding of the new asset types is critical.

The first calendar year when global temperatures reached 1.5°C above the pre-industrial level was 2024. Already, 2025 looks like it will break 2024's record, specifically in relation to air and sea temperatures, particularly in Europe. On 25 October 2024, the UN reported that the world is on course for a temperature rise of more than 3°C – twice the 1.5°C Paris Agreement limit. In many regions, a warming world is supercharging the impact of wildfires, drought, and flooding and causing falling crop yields and falling labour productivity in agriculture. **Thus, increasing the speed of clean asset buildout and the consequent fall in emissions is a key requirement if global society is to minimise the impact of climate change on us and future generations.** Speed may require greater regulation rather than 'voluntarism'. On 23 July 2025, the Lex column in the *Financial Times* (FT) noted that 'companies that rushed to sign up to net zero initiatives [a few years ago] are stampeding for the exit'. Similarly, large banks are 'deserting the Net-Zero Banking Alliance'. The FT goes on to say that the obvious alternative is for 'governments to spearhead decarbonisation via legislation or capital allocation' as an alternative to voluntary coalitions that tend to fall apart over time.

There are other issues beyond the voluntarism/regulatory issue which must be overcome to speed up the transition. Scaling up the necessary technologies and the volume of financing while providing investors with a reasonable certainty of their project generating an appropriate risk-adjusted return will be key. The whole energy **system** needs a huge level of investment – it's not simply more solar, wind, and nuclear generation that is required. The spokesperson for the Dutch power grid operator has said that it is not only Holland that is having difficulties with the transmission grid system, but that 'Belgium is in trouble. The UK is in trouble. In Germany there's lots of trouble because in Germany all the wind is in the north and demand is in the south'. Holland is also short of 28,000 technicians necessary for the transition.

We focus on heat and heat networks as much as on electricity systems, as decarbonisation of an economy will require the decarbonisation of both industrial and space heating (buildings). While the energy transition in most countries will require around twice as much electricity as is being used at present, heat is currently a larger component of the total energy system than electricity, and thus we must also consider how heat can be decarbonised to eliminate the use of natural gas or coal. Because of our 'whole energy system' focus in this book, our most detailed chapter is on **nuclear**, as this is one of the few types of infrastructure asset which provides both **decarbonised *heat* and *power*** and, in addition, offers many advantages for desalination – something which will be increasingly required in Europe as well as globally (covered in Section 13.3). Will advanced nuclear technology (ANT) small modular reactors push out the use of natural gas in the next decade? We consider this issue in a key part of the text, Part V, *NUCLEAR – A New Approach to Decarbonisation: Innovation in Technology, Co-Location of Supply & Demand and Nuclear Financing.*

The 'New Energy System' requires a more complex network with many more types of distributed assets connected to it than before. Artificial intelligence (AI) technology is necessary to integrate all these assets into the 'system' to optimise individual projects and whole system performance. Such AI systems are needed to minimise the risk of brown-outs or blackouts, such as when around 30 GW of power disconnected from the Iberian Peninsula grid on 28 April 2025, causing a massive blackout. The digitalisation of electricity systems is covered in Chapter 12, which examines 'the other transition' – the **digital transition**. But the transition to **generative AI** and its use by customers has changed all forecasts of when global emissions will decline. It has led to a massive increase in the use of natural gas to power the new giga-scale data centres of the so-called 'hyperscalers'. One such facility, being constructed in Pennsylvania, is expected to require 4.5 gigawatts of power (from natural gas), which is enough to power several large cities. In Malaysia, data centres will account for 30% of total power demand by 2030, with gas-fired power capacity expected to rise by around 50% by 2030.

In the past, geopolitical risk has not been much considered by analysts. But what is now termed 'geoeconomics' has become critical. A European example of potential risk is that the cables, known as Offshore Transmission Owner (OFTO) electricity cables bringing offshore wind power to shore, can be severed by countries such as China, Russia, or Iran, causing severe disruptions to European power supplies.

We have aimed for a wide readership, not only finance industry professionals worldwide but also government ministers and their staff, energy regulators, and those employed in the power, energy, and transport sectors, who need to understand more about how their company is viewed by financiers when undertaking due diligence on a proposed project. In addition, it will be useful in business schools and executive education programmes in corporates and financial firms. We hope we have made a contribution to speeding the transition, which becomes ever more critical as each year goes by.

Acknowledgements

Dr James Martin is an energy sector and nuclear engineer. He was Mott MacDonald's director responsible for power transmission and distribution, and at SSE, he was the board member responsible for power generation.

Dr Michael Lai was a PhD student in the Department of Meteorology, University of Reading, whose research interest is climate variability. His personal interest is in climate finance and climate justice. He now works for the Met Office in Exeter.

Dr Lisa Schopohl is an associate professor of finance, focusing on socially responsible investing. Along with a colleague, Dr Ivan Sangiorgi, she contributed to data analysis for the Climate Bonds Green Bond Treasurer Survey.

Aditya Sharma is a civil engineer with an MBA from Henley Business School. At Rising Sun, he managed a 140 MW solar power project, then moved to ACME, India's largest solar power producer, and now works with Sterling and Wilson.

John Chessher was originally an investment banker in Europe and Asia and subsequently a fund manager in the City of London. He went on to found Cenkos Asia. He now teaches banking and sustainable finance at the ICMA Centre.

Morgan Qian Cheng is a PhD student at the ICMA Centre & Meteorology Department, researching the impact of weather on the power output, risk, and investment return of intermittent generation. He is jointly supervised by Professor Scott-Quinn. He now works for Alpha ESS, a global Chinese energy storage provider.

ICMA Centre for Financial Markets, Henley Business School, the University of Reading

We would like to acknowledge the huge contribution made by ICMA at the time of the Centre's founding in 1991–1992, its board of directors, CEO (John Langton), chairman (Jan Ekman), and deputy chairman (Osama Elansari), which included providing over £8 million of financing to enable the construction of the two buildings shown on the left-hand side of the photograph below. Furthermore, the contribution made by Reuters/Refinitiv/LSEG, which provided 20 data terminals for the first trading room in a European university, was significant. In addition, its founding was enabled by senior colleagues at the University of Reading who supported a new venture of a nature which, 30+ years ago, was an innovation that many other universities would have found difficult to support. Capital markets, derivatives markets, and trading were indeed activities of which most universities had no knowledge or interest. Without these organisations and individuals, there would be no ICMA Centre today (see the photo on the next page[1]). Like other finance schools, we also teach and research fintech, sustainable finance, and many other areas at the Centre today. The Centre is part of the world-renowned Henley Business School, one of the few schools in the world with accreditation from all three accrediting agencies, the US, the European, and the UK.

The University of Reading has been recognised as *The Times* and *The Sunday Times* Sustainable University of the Year 2025, a well-deserved celebration of its outstanding commitment to tackling climate change and driving sustainability. Since 2009, the university has slashed its carbon emissions by nearly 60% and reduced its waste by 35%. Whether its cutting emissions, saving energy, or finding smarter ways to manage resources, Reading has consistently shown that sustainability isn't just a goal, it's a way of life.

---

[1] Picture credit: https://www.icmacentre.ac.uk/about.

The ICMA Centre buildings (the two on the left), which are part of Henley Business School, were gifted by the International Capital Market Association.

International Capital Market Association (ICMA; www.icmagroup.org)

ICMA, with a head office in Zurich, promotes well-functioning cross-border capital markets, which are essential to fund sustainable economic growth. It is a not-for-profit membership association with other offices in London, Paris, Brussels, and Hong Kong, serving over 620 member firms in 70 jurisdictions. Among its members are private and official sector issuers, banks, broker-dealers, asset managers, pension funds, insurance companies, market infrastructure providers, central banks, and law firms. It provides industry-driven standards and recommendations, prioritising four core fixed income market areas: primary, secondary, repo and collateral, and sustainable finance. ICMA works with regulatory and governmental authorities, helping to ensure that financial regulation supports stable and efficient capital markets.

ICMA's Role in Sustainable Finance

ICMA plays a recognised leading role in supporting the development of sustainable finance in the bond and wider debt capital markets. It provides the Secretariat for the Green Bond Principles (GBP) and other principles, collectively known as 'the Principles' – which are the *de facto* global issuance standard for the international sustainable bond market.

The ICMA Centre and the University of Reading Meteorology Department

As well as undergraduate and MSc courses in sustainable finance, the Centre offers an **MSc Climate Change, Sustainable Business and Green Finance,** jointly with the **University's Meteorology Department**, which is the top-rated meteorology department in the UK and fifth highest-rated in the world. The Meteorology Department also offers an **MSc Climate Change and Artificial Intelligence**.[2] In addition, the European Centre for Medium-Range Weather Forecasts (ECMWF) will shortly relocate to the University of Reading campus.

[2]Brochure available at https://www.reading.ac.uk/ready-to-study/study/2025/meteorology-and-climate-pg/msc-climate-change-and-artificial-intelligence.

About the Authors

Brian Scott-Quinn, the principal author, is a specialist in capital markets and the financing of sustainable infrastructure. His work has increasingly focused on how financial markets, innovative financing structures, and emerging technologies can accelerate the transition to low-carbon infrastructure. This book reflects those interests. He spent many years working in international capital markets, holding senior investment banking positions in London with Kidder Peabody, Drexel Burnham Lambert, and Security Pacific Bank.

He founded the International Capital Market Association (ICMA) Centre at the University of Reading's Henley Business School with £8 million of support for the construction of two new buildings from the trade association for the international capital markets – the ICMA. Under his leadership, from small beginnings it became one of Europe's leading centres for education and research in capital markets. As a professor of sustainable finance, he teaches and undertakes research at the ICMA Centre.

He is a founding director of the fintech company Madiston Plc, which produces banking and capital markets software. He was also a founding director (2015–2018) and the Financial Conduct Authority (FCA) Senior Management Function 11 holder at the challenger bank Civilised Bank Ltd (now Allica Bank), where he was part of the team that enabled the bank to gain its banking licence from the FCA. He is a member of two Brussels-based EU High Level Groups on Policy Innovation in the Energy Transition. From 2014 to 2021, he was a member of the advisory board of

Penultimate Power UK Ltd, which focused on bringing the manufacture of advanced nuclear technology reactors to the UK.

He has undertaken many consultancies for banks, stock exchanges, investment managers, the Ministry of Finance, and the Capital Market Authority and was an advisor to Mrs Thatcher's Chancellor of the Exchequer, Sir Geoffrey Howe (later Lord Howe). Recently, he has been advising investors seeking an acquisition for a US clean energy Special Purpose Acquisition Company (SPAC), a company considering investing in a sustainable aircraft fuel project, and an entrepreneur planning a venture in the low-altitude economy market using battery electric and, at a future date, hydrogen-fuelled vertical and/or short take-off and landing air taxis and regional aircraft.

Deyber Cano, the associate author, is an infrastructure finance economist whose work centres on financing sustainable infrastructure, with particular emphasis on housing, decentralised energy systems, and climate finance. His research and professional interests explore developing financial mechanisms to mobilise private and institutional capital to deliver sustainable infrastructure, affordable housing, and carbon-free energy systems, particularly in emerging markets.

He holds an MSc with distinction in infrastructure investment and finance from University College London and an MSc with merit in international economics from the University of Essex. He is currently a fellow at The New School, New York, where his work explores the intersection of housing finance, sustainable urban development, and public policy. He contributed to the development and review of the book chapters covering decentralised energy systems, blended finance, project finance structures, and the application of environmental, social, and governance–aligned financing instruments, including green bonds and sustainability-linked loans.

His earlier joint publications with Brian Scott-Quinn include *A Guide to Infrastructure Financing* (published by the Association for Financial Markets in Europe, 2015) and the chapter on Financing the Energy Transition in *More on Risk and Energy Infrastructure* (published by Globe Law and Business, 2017).

Contents

Part I

Financing Established and Innovative Clean Energy Technologies

Introduction: The Pivotal Roles of Technologies and Financing in the Transition to Carbon-Free Electric Power, Heat, Hydrogen, and Desalination

Capital and loan markets provide the function of matching corporate **real asset** projects, such as wind and solar farms, nuclear power stations, geothermal energy sources, hydropower, energy storage, electricity, and transmission and distribution, and many other real tangible asset types with the buyers of the **financial assets** needed to finance them. But investors and lenders will only provide financing if the potential returns are at least equal to those of other financial assets of comparable risk available to them **globally**. It is the function of financial intermediaries, sometimes with the assistance of government agencies, to structure financial instruments in such a way that investors and lenders, through undertaking the necessary due diligence (DD) on a project, will be convinced that the expected returns will indeed likely meet their required return on investment for the level of risk they are taking. Only then will they supply the funds required for such projects.

However, neither side of the equation – the financing side or corporate decision-making regarding clean energy infrastructure asset investment – is working as well as it should. The problem appears to be that equity and debt investors are not used to projects which are not only large in terms of the capital required but also require many more years to see the initial investment returned – they are long-lived projects. In the case of nuclear, the expected life of a power station may be 60 years or more. In addition, many investors and lenders have not previously invested in this type of asset and are unwilling to undertake the necessary work to assess the level of risk involved and whether or not the expected return is at the required level.

The lack of understanding of new technologies, as well as the uncertainties inherent in the new business models of potential energy transition projects, creates **market failures in scaling up transition financing**. Overcoming this requires, *inter alia*, innovative public sector risk mitigation mechanisms and increased co-financing by both public and private sectors. The United States (US; until the second Trump administration), the European Union (EU), and China have all intervened in markets to support climate finance – the EU with the 'Green Deal' and China with support for solar, wind, energy storage, and automotive technologies, including batteries. However, much of the world, including those countries, has not come close to meeting the requirements for a 'pathway' to 'net-zero' emissions by 2050. Indeed, in the US, at the federal level, the new administration in 2025 seems to have chosen not only to withdraw from the global energy transition by abandoning the so-called 'Inflation Reduction Act', which, despite its name, had been designed to greatly speed up the energy transition, but also to withdraw from the Paris Agreement.

In all regions except China, scaling up the production and commissioning of the necessary assets has been much slower than expected. A key issue is regulatory and permitting problems. It can take 10 years to obtain permission for a project and as long to obtain the needed transmission cables to transport power from where it is produced to where it is needed. But the financing issues referred to above have also been critical in some cases. Many of these issues can be considered market failures.

Capital flows into clean energy technologies in developing economies have been particularly weak. Indeed, much of the benefit from reductions in global carbon emissions achieved by the actions of the US, EU, and China is being dissipated by the growth of emissions in developing economies. As the **global** concentration of greenhouse gases in the atmosphere determines the extent of global warming, the logic is that large economies would gain a net benefit by paying for a high proportion of the investment required by those countries to achieve their own energy transitions. Without this aid, it is unlikely the world will achieve the necessary transition before temperatures become unbearable for human and animal life in many parts of the globe. While superficially this is a financing problem, it is as much, or more, a political and social issue that can only be resolved through political will worldwide.

The need to scale up and speed up the energy transition worldwide is a key, indeed the key, existential issue that humankind faces at present. Much greater financing of established technologies such as wind and solar is a critical need, but so too is the scaling up of innovative and existing technologies, including energy storage, synthetic fuels, carbon capture, and advanced nuclear. In the case of new technologies, it is proving even harder than for existing technologies. While much of the problem arises from slow permitting of potential projects, many projects have been cancelled even after construction has commenced due to political uncertainty (such as the outcome of the November 2024 US election). This has made firms hesitate to invest.

One example of this is the cancellation in August 2024 by the large wind power developer Ørsted of a US project to make so-called e-fuels. These are manufactured using power from wind farms. Without the certainty of tax incentives from governments and mandatory offtake (purchases) by users through regulation of, for example, the proportion of clean fuel in aircraft fuel, investment has become simply too risky and too large for many investors. In 2024 and early 2025, share prices of many 'clean asset manufacturers', such as producers of wind energy equipment and manufacturers of electrolysers used in hydrogen generation, have declined sharply due to regulatory issues and a lack of offtaker demand. This and the cutting back of clean project investments, particularly into assets such as wind farms, to which the US President seems to have an emotional objection, have held back the transition.

There is no absolute global shortage of funds available for investment in financial assets of many different types, which are available to investors worldwide; however, in the case of 'clean tech', the needs of investors and lenders often do not align with the objectives of entrepreneurs, developers, and corporations, resulting in too slow a growth in clean energy assets and hence too slow an energy transition. Entrepreneurs and developers say that investors won't look at projects which have even a relatively small degree of uncertainty. This problem is often a **result of a lender's or investor's lack of knowledge and understanding of new technologies and, hence, of their risks and potential returns**. They therefore shy away from such investments in favour of asset types that they understand better and which have quicker returns. Competition for funds comes from investments in artificial intelligence (AI) and data centres, as well as from the health sector in companies such as Novo Nordisk.

There is a symbiosis between financing and the technologies that require scaling. That is why we have covered both sides of this equation in some detail. Without knowledge of technologies, many investors and lenders will continue to shun new types of assets and miss out on many potentially successful projects. This suggests, on the one side, the need for a better understanding of the techno-economics of new types of real assets by capital providers, regulatory agencies, planning authorities, official bodies, and governments. On the other side, there is a need for developers to better understand the approach lenders and investors take in conducting DD on potential projects and on structuring the financing of such projects.

Just as new technologies have to 'work' in the sense of achieving what they were designed to achieve technologically, so too must the economics of the transition to ensure that investors and lenders can expect comparable returns for the risks they take relative to other potential equity or debt investments that are available to them. This is what we mean by 'techno-economics'. Historically, energy transitions such as those we cover in Section 1.3 have been driven by favourable economics such as when animal power, water power, steam power, and electricity were introduced and supplanted earlier technologies. But even though the prices of solar and wind power and battery storage have all declined dramatically over the past 20 years, the economics are not yet in their favour. Despite the costs of solar and wind falling to record low levels, the costs of integrating these variable output technologies into a grid system that requires continuous generation 24/7/365 are high. In fact, power companies revert to using natural gas turbines on a large scale when power from renewables is in short supply.

In practice, there is a bifurcation in capital markets between '**infrastructure capital**', on the one hand, and '**venture financing**', on the other hand, which leaves a gap in the market – the so-called 'missing middle', which represents a **market failure**. Perhaps only government interventions could fill this gap, although many in the private sector have attempted to do so. This bifurcation misses out on the need for capital focused on companies that have matured out of the venture stage but have not yet scaled up and de-risked sufficiently to access the 'fully commercial asset' capital market.

In 2024, a number of companies and projects that looked as if they would soon scale up output to a mass-production level using capital market finance were, in practice, unable to do so. One reason is that investors have become more used to 'big tech' projects, as we have noted above, which are very different in nature compared to long-lived, asset-heavy projects. There is a big difference between digital projects, where human capital is central and returns may increase relatively rapidly, and 'hardware' projects, where the construction period can be long, as can the asset's operational life. In practice, it is clear that many potentially highly successful companies find it difficult to access the capital necessary to bridge the gap between **venture capital** and **infrastructure investment capital**.

The problem of the missing middle arises both in obtaining growth equity capital and the capital required for first-of-a-kind (FOAK) projects. Both are critical to enabling the technical de-risking of a project and the scale-up of the technology. This is necessary to enable companies to advance from the venture stage to becoming bankable by infrastructure debt providers.

In effect, the climate tech market consists of two distinct markets: the first is in relatively early-stage technologies, and the second is in well-established technologies. A successful and timely energy transition will require the use of both and their respective financing techniques. The first of these financing techniques is required to bring new technologies to the point where they can be commercialised and then scaled. The second is required not only to enable new technologies to move to a commercial scale but also to massively scale up existing technologies which are already in use. We cover both sectors in the following chapters. Where the problem still lies is in the 'missing middle' that we referred to above.

When considering the 'technology stage of development', the generally accepted approach to defining the stage a technology has reached is one developed by the US National Aeronautics and Space Administration (NASA) – the Technology Readiness Levels (TRLs), shown in Figure 1.

TRL 9 is the stage at which all companies are likely to be aiming, at which point they should certainly be able to access infrastructure finance, often by using project finance techniques, which we cover shortly. While some companies start life at TRL 1, most other companies would start at a later stage, as they are likely to be building on fundamental research (TRL1) frequently carried out by others, including universities, research institutes, and government laboratories.

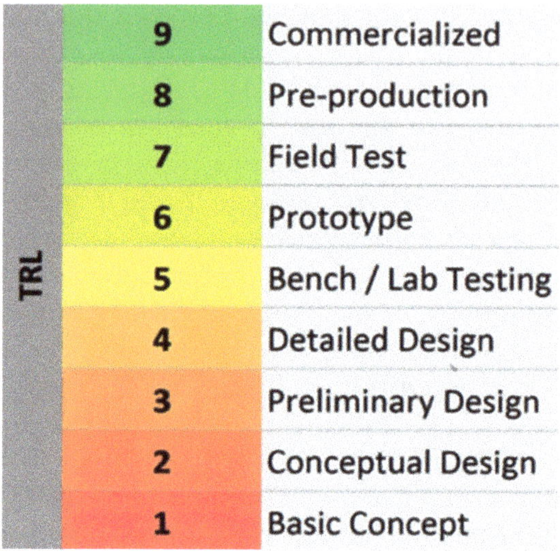

Figure 1. NASA Technology Readiness Levels (TRLs).

The winners in the finance industry of the future are likely to be those firms which develop a deep understanding of the science of climate change and relevant technological innovations to mitigate and adapt to climate change impacts and thus have a better insight into which technologies are likely to be most rewarding for corporates and, therefore, for asset owners, asset managers, and lenders. That insight will be provided by analysts who understand the new technologies. It is for this reason that we extensively cover innovative technologies as well as covering technologies.

Although the cost of solar and wind has fallen dramatically over time (though wind has risen in cost quite substantially in the years since COVID-19), it has been government regulations, taxes, and incentives – in particular, guarantees and revenue subsidies from governments for projects – that have driven forward solar, wind, and storage projects. But much investment is required beyond simply solar and wind farms. The cost of the necessary upgrade of, for example, the UK National Grid is estimated at £58 billion ($74 billion). This is due to an expected 64% increase in electricity demand by 2035, driven by sectors such as transportation and heating transitioning to electric power. The investment programme is also likely to involve the expansion of the offshore power grid and the construction of new north-to-south electrical 'spine' in a number of countries. Despite considerable support for new technologies, the UK government, in July 2025, declined to support the Xlinks project (see Section 10.2) on the grounds that it did not provide power generated domestically and also on 'security' grounds'. The Xlinks project cost was expected to require $22 billion of investment.

For lenders and investors to release their capital for investment in the energy transition and thereby divert it from being applied to other opportunities, clear signalling from governments and an absence of frequent changes in policy are required. It also requires actions to reduce risks to investors in both new and existing technologies. One risk arises from the possibility of long delays due to planning rules and resistance from local communities. These issues occur in most countries, but some countries are trying to tackle what is known as the 'NIMBY problem', or the 'not in my backyard' problem, which often delays projects by some years at the planning stage.

In the UK, the energy company SSE has noted that by the time communities have been consulted, ground conditions mapped, seabird flocks assessed, and grid connections secured, 10 years may have passed between the granting of a seabed lease by the Crown Estate and construction commencing. As a result, developers may abandon projects. Since the COVID-19 emergency ended, the costs of wind turbines and interest rates have risen so much as to invalidate the assumptions behind their original techno-economic assessment.

As well as covering projects which are at the stage of the final investment decision (FID), we cover the innovation and financing of start-ups, given the importance of research and development (R&D) in achieving the transition in time to avoid the ever-increasing hazards to life on Earth with each fractional rise in the Earth's temperature.

While at one time hydropower was a reliable source of dispatchable (continuous and always available) power, today in many regions it is no longer so reliable, and alternative sources of clean continuous electricity supply must be found. The reason is that droughts have caused very large reductions in power generation from hydro in many African countries, in Ecuador, in the US, and elsewhere. The Kariba Dam, constructed in the 1950s with a total capacity of 2 GW used to provide 70% of the electricity consumed in Zimbabwe. But its huge reservoir was only 3% full in 2022 due to repeated droughts. It continues to have a low or zero output due to drought, which may be a function of the changing climate in Africa because of global warming. In China, a severe drought started in 2022, forcing authorities to revert to the use of coal to meet demand for electricity. In China as a whole, hydro capacity rose 18% between 2019 and 2023, but the amount of hydro electricity generated fell 1%.

Two alternative non-fossil sources of continuous electricity generation are geothermal and nuclear. Geothermal energy comes from the heat generated during the formation of planet Earth and the nuclear decay of materials of which the Earth is composed and will be able to continue to provide 24/7 energy for thousands of years. It is likely that with more investment, geothermal could become a significant source of power in **some** geographic regions. Increasingly, technology originating in the oil and gas industry is being used, such as hydraulic fracturing of rock. Such techniques enable the construction of what are known as 'enhanced geothermal systems' (EGS).

Advanced nuclear technology (ANT) reactors, which we cover in some detail in Part V on nuclear and which will quite soon be available on the market, have few similarities to the megawatt reactors currently being installed worldwide. They introduce the concept of 'distributed' power, i.e. small (and also micro) power generation systems, similar in concept to solar and wind farms. But unlike wind and solar, they can also provide sources of heat and can be installed near, or within, built-up areas close to sources of demand. They can provide high-temperature heat which can substitute for fossil-generated heat in many industrial applications. Such a distributed power approach with 'micro' reactors installed very close to sources of demand would also greatly reduce the high cost of strengthening national grids, which is the investment urgently required to bring solar and wind power from often far-distant locations to sources of demand.

At this point, we must emphasise that when we write about the 'energy transition', we are not writing about a transition to renewables. **While renewables are critical for the energy transition, on their own they may not achieve the best balance between cost, security of power supply, and a fully decarbonised electricity system**. We use the term 'clean energy', 'carbon-free energy' (CFE), 'carbon-free electricity' (CFE), or 'carbon-free heat (CFH)', as the objective of the energy transition is to transition from a fossil-energy-based economy to a carbon-free energy system, not a renewable system, although renewables are likely to remain a key input in the transition.

Around the world, trade associations play an important role in helping companies contribute to a speedier climate transition. Internationally, the International Capital Market Association (ICMA), which sponsored the

ICMA Centre buildings, runs the 'Green Bond Principles'. Green bonds provide capital for new and existing projects with environmental benefits.

In the UK, a relevant trade association for financial companies focused on sustainability is the **UK Sustainable Investment and Finance Association (UKSIF)**. It has some 300 members managing over £19 trillion in assets under management (AUM), including investment managers, pension funds, banks, financial advisors, research providers, and NGOs. In turn, UKSIF is a member of Eurosif. Eurosif works as a partnership of Europe-based national Sustainable Investment Fora (SIFs), with the direct support of its network, which spans over 400 Europe-based organisations drawn from the sustainable investment industry value chain. These organisations include institutional investors, asset managers, financial services, index providers, and ESG research and analysis firms.

Chapter 1

Clean Energy Infrastructure and the Energy Transition

1.1 Infrastructure as a Financial Asset Class

It is not so long ago that infrastructure was not a capital market asset class, as most types of infrastructure in most countries such as electricity and gas production were state-owned. But during the late 1970s, in the global drive to privatise state-owned industries and companies, the shares of many such companies worldwide (after the conversion of the entity to a corporate form) were transferred from state to private ownership. The industries in which this happened include companies providing telecom services, electric power generation, water supply, coal gas, natural gas, railways, oil companies, airlines, bus companies, ports, iron and steel manufacturers, post offices, and mail delivery services. Institutional investors were keen buyers of the new financial assets being created, often simply because they believed they were being sold off too cheaply. They saw the opportunity to increase efficiency by reducing costs and so generate substantial increases in profitability and market value. Until 20 or 30 years ago, infrastructure investment referred mainly to investment in utilities such as telecoms, water, gas and electric companies, and energy pipelines, or 'economic' infrastructure as it is known. Today, we also have renewables, LNG terminals, public transport, toll roads, airports, ports, energy storage assets, transmission cables, mobile phone masts, data centres, 5G wireless for mobiles, and, by the early 2030s, possibly small modular reactors.

Also, in more recent times, infrastructure has transitioned from a relatively minor asset class in institutional investor portfolios to one that is increasingly important. Infrastructure equity is generally a 'private' asset class, i.e. not traded on a stock exchange. It has different performance characteristics than those of the listed equities, which are traded on a stock exchange. Infrastructure project finance debt also has characteristics which differ from those of corporate debt. These differences can make infrastructure equity and debt attractive as diversifying components of an equity, bond, or loan portfolio. As a result, investors have perceived considerable diversification and other benefits from adding these asset classes to their existing equity and debt portfolios.

Infrastructure assets are a component of the 'real' asset class, which comprises tangible assets such as real estate and commodities (gold, silver, copper, etc.). Retail investors cannot, in most instances, invest in a private infrastructure project for their own portfolios. However, there are many options for investing in **funds** which hold only infrastructure assets (including fossil-free funds) and are traded on markets such as the London Stock Exchange, giving both institutional and retail investors the ability to obtain the benefits that infrastructure equity and debt can provide. By spreading an investor's liquidity over a number of projects, such funds enable investors to overcome the lack of access to and illiquidity of individual assets, thus enabling them to benefit from stock market liquidity. We provide a case study of one of these funds in Section 9.4.

A fund manager may invest in **already existing** projects by purchasing a proportion of the total shares or debt of a project company in a bilateral transaction with the current holder of such financial assets. Alternatively, they may provide 100% of the necessary funding or a part thereof to a **developer** to enable a **new project** to be constructed. This latter type of investment creates '**additionality**', i.e. it increases the stock of such assets, whereas investment in existing shares does not. Infrastructure assets generally have a long expected lifetime – often up to 50 or more years. In any investment today, environmental, social, and governance (ESG) issues are important not only for ethical reasons but also for practical and regulatory reasons. It should be noted, however, that many 'clean energy' assets have low carbon emissions when operating but have a high carbon input in their manufacture. This includes the carbon content of the concrete in the foundations, the steel in the tower, and the mining of the copper required for the turbine itself. These embedded carbon costs of manufacture should be amortised over the expected lifetime of the asset when comparing GHG emissions of any two technologies. Thus, most generation technologies are, in practice, low-carbon assets rather than being truly 'carbon-free' assets.

We referred above to 'asset managers', who manage portfolios of financial assets. But there are also 'physical asset managers'. Their function involves sustaining physical assets so they continue to perform optimally over their expected life. It also includes optimising the use of such assets to generate maximum revenue without this causing a shortening of their economic lifespan. This is facilitated by the Internet of Things (IoT), which enables information on the often geographically scattered assets to be transmitted to artificial intelligence (AI) systems, allowing them to adjust parameters and increase the efficiency of the utilisation of such assets. The other type of asset manager is the financial, portfolio, or fund asset manager. While a good physical asset manager is critical to the success of physical assets in a portfolio, we will not be covering this function in any detail but will primarily focus on the management of financial assets in a portfolio.

Renewable energy is just one of many categories of economic infrastructure. The fact that renewable energy and other 'new energy assets' must compete with all the other options that investors have available to them in the wider arena of economic infrastructure means that project developers of clean energy assets must consider how they can offer equity and debt which provide the characteristics investors are seeking from private infrastructure assets and are competitive on the features that investors are seeking. These include:

1. a low correlation with other asset classes;
2. relatively low volatility;
3. in many cases, long 'offtake' contracts which guarantee revenue levels for many years;
4. returns which to some extent are linked to inflation; and
5. higher expected risk-adjusted returns than listed equities.

The fifth feature above, if manifested in practice, is certainly an attraction. It arises because private assets, such as infrastructure, have relatively low liquidity (i.e. cannot be sold quickly through a stock exchange). This means that investors will **expect** to receive a 'liquidity premium' and thus achieve a higher **expected** risk-adjusted return than from listed equities, though this premium may not materialise in practice.

While any asset, including listed and highly liquid equities, may be held for long periods, infrastructure is an asset class that is particularly appropriate for 'long-term' investors. In practice, what this means is that the transaction costs involved in buying and selling are higher than for most other assets and, hence, to achieve the expected return on the asset requires that the higher transaction cost be amortised over a lengthy period. Thus, while investors in listed assets may consider trading their assets on a regular basis in the belief that this may improve performance (which, in practice, it often does not) or to rebalance a portfolio, assets with high transaction costs cannot be traded frequently without substantially reducing expected returns. Thus, these are assets which are more relevant to investor classes that do not trade actively.

Infrastructure assets often have long-term contracts with entities possessing high credit ratings, such as governments, local authorities, and major companies. Contracts such as power purchase agreements (PPAs) 'guarantee' future revenues for many years ahead. As a result, such assets can be considered relatively low volatility. For some types of investors, the long duration of infrastructure assets **is** attractive, as it provides them with long periods of revenue generation and thus certainty. The types of investors to whom such assets should be attractive are, for example, pension funds and insurance companies that have long-term liabilities. In addition, infrastructure assets have a relatively low correlation to traditional asset classes, in part because their value does not rise or fall during business cycles as much as traditional assets since their revenue is not as volatile as that of other asset classes and is often inflation-linked. However, this low correlation is sometimes illusory, as it results from the absence of market prices for private assets and hence difficulty in valuing them, particularly when quoted market prices are falling.

In addition to sources of energy generation, the energy transition will require significant investment in energy storage (energy which can be quickly converted to electricity) to 'balance' supply and demand at every moment, as renewable energy has a highly variable output. Such storage needs to cope with short-term variations in supply and demand and also seasonal variation (such as winter, when there is much less sun) and times when there is a 'wind drought', as has happened in Europe for periods of weeks at a time. Also required will be facilities that manufacture hydrogen from water and electric power using electrolysers for use in so-called 'hard-to-decarbonise sectors' and that manufacture synthetic liquid hydrocarbon fuels from hydrogen and captured carbon.

The total amount of financing required to achieve the Paris Agreement goals will also need a much wider range of assets than just those we have referred to above, **not all of which will be held in the financial portfolios of infrastructure funds**. Bloomberg New Energy Finance (BNEF) produces a report published annually entitled 'New Energy Outlook' (NEO). In its NEO21 and subsequent issues, BNEF has used a new, broad measure of '**energy transition investment**', which, as well as including wind and solar generation, now includes investment in a wider scope of transition areas and end-uses, **including the purchase of battery electric vehicles by consumers and new heating systems by households, as well as energy storage, hydrogen production, and carbon capture and storage**. They estimate that 'the investment required globally in energy supply and infrastructure between 2020 and 2050 to achieve net zero by 2050 is between $92 trillion and $173 trillion'. If this range is correct, then between $3.1 trillion and $5.8 trillion per year on average will need to be invested over the next three decades, which is greater than twice today's investment of around $1.7 trillion per year, as we noted earlier.

The European Commission estimates that the 'twin transitions', the green energy transition and the digital transition (which is also highly relevant to the energy transition), require €620 billion and €125 billion each year, respectively, through 2030 if Europe is to achieve its committed contribution to the global climate mitigation effort. For other countries and regions, similar sums are, proportionately, likely to be necessary. Unfortunately, in Europe, unlike the US, capital markets do not seem to be structured appropriately to enable them to channel these amounts of funding into these two types of infrastructure – energy and digital. The grand plan to create a 'Capital Market Union' in the European Union has come to nothing in the past decade, making large-scale financing of projects either impossible or feasible only at a considerably higher expected rate of return for equity investors and higher interest returns to lenders. The regulatory requirements of what is known as Solvency II also seem to inhibit capital investment in infrastructure by institutional investors – though there are considerable pressures for these rules to be modified. In addition, Europe's venture capital sector is unlikely to be able to fund sufficient early-stage ventures to enable them to reach commercialisation. This is due in part to 'the missing middle', as it has been called, to which we referred in the previous section; the term refers to finance for growth-stage climate tech companies, particularly those at the Series B+ investment stage, having achieved initial venture financing from 'seed' finance and Series A equity issues (which we cover in Section 2.3).

Other countries have managed to resolve such issues far more effectively than those in Europe and are also better able to mobilise savings for investment in the transitions (both green energy and digital) – particularly the US with its single national capital market on Wall Street and its so-called Inflation Reduction Act (IRA) designed to subsidise green infrastructure (though President Trump may phase this out). China, it should be noted, has created more clean energy assets than any other country and transitioned to electric vehicles more quickly than most, although its use of fossil fuels has also increased substantially.

1.2 What Is Energy, What Is the Energy Transition, Sources of Energy, and Decarbonisation

Clearly, 'the Energy Transition' is about changing the energy system from one that has been in use for many decades and is based on low-cost oil, gas, and coal to one based on 'clean assets'. To create a CFE economy, project developers and innovators must understand how lenders and equity investors think and ensure that their projects are likely to meet their needs. On the other side of the equation, lenders and investors need to have an almost equally deep understanding of the new CFE assets and other relevant assets in order to undertake the necessary due diligence (DD) on potential opportunities. The first step in this process is understanding the energy system as a whole and the different ways in which power (by which we mean electricity) and green hydrogen can be produced, as these two energy carriers (along with derivatives of hydrogen and clean heat energy) will be the basis for almost all decarbonisation efforts.

Energy is the potential to do work and comes in many forms – chemical, atomic, thermal, human power, animal power, etc. It has been the availability of fossil fuels since the start of the Industrial Revolution which commenced in the United Kingdom around 1760 that has driven economic growth and productivity dramatically upwards in the major economies. Figure 1.1[1] shows energy in two forms: first kinetic energy, and second potential energy (that has yet to be released).

Under the 'Kinetic Energy' heading, the first type is energy is in the form of motion, such as cycling on a bicycle (human energy). Electrical energy is the energy that can be carried through a cable. Thermal energy arises from combusting wood, coal, or natural gas and can then be utilised by, for example, using the heat generated from combustion to turn water into steam and passing the steam through a turbine to create mechanical energy. This energy can then be used to drive an electric generator, which converts mechanical energy into electrical power.

[1] University of Central Arkansas https://faculty.uca.edu/njaustin/PHYS1400/LabSims/sim05Energy.html.

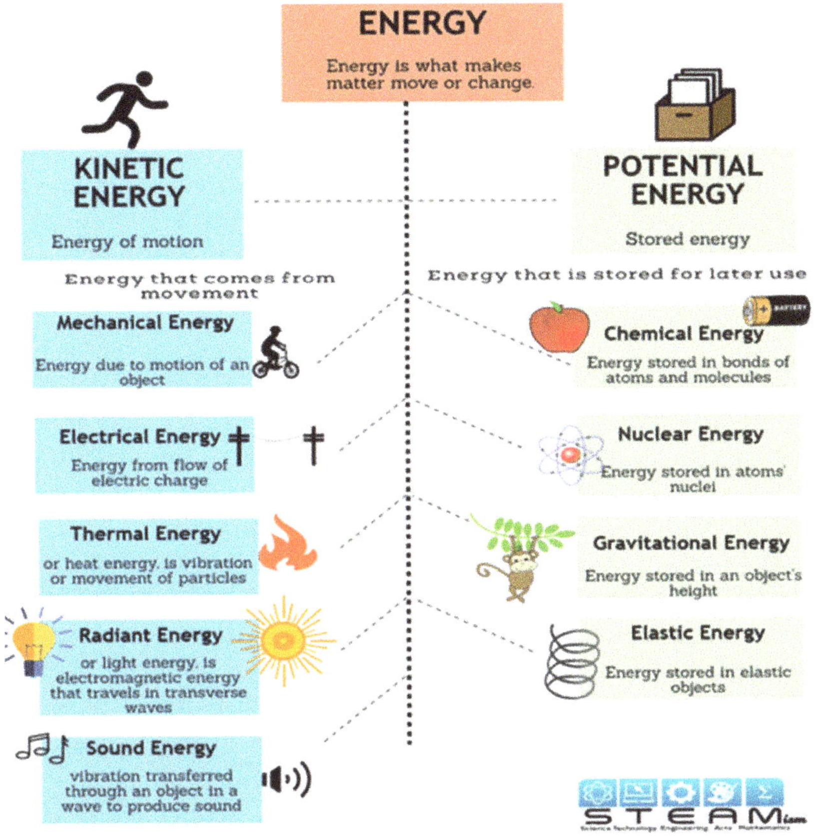

Figure 1.1. Types of energy.

Potential energy (on the right-hand side of the diagram) is not, in itself, energy but is energy stored in some form. Gasoline/petrol/essence is chemical energy in the form of a liquid. But as soon as it is oxidised by igniting it in air, it explodes, releasing a huge amount of heat energy. In an internal combustion engine (ICE), this explosive force is controlled and turned into mechanical energy. Likewise, hydrogen, which is a chemical energy source, is not energy itself, but when oxidised, releases its stored energy as heat.

Nuclear energy requires a fission process to release its potential energy content. This is a process whereby an atom splits into parts, which releases a large amount of energy, also in the form of heat. This heat can then be used to generate steam to drive an electrical (power) generator, which converts heat energy into electrical energy (power).

1.2.1 *Sources of Energy: Conversion of Primary Energy Sources to Secondary Energy (Energy Carriers) and Decarbonisation of the Economy*

To understand the transition better and the opportunities it offers developers and financiers, it is helpful to start with an explanation of the original source of energy (primary energy), as shown in Figure 1.2. For example, primary energy in the form of crude oil cannot be used directly but has to be converted to a usable form, such as petrol, diesel, kerosene (for aircraft), or bunker fuel (for shipping). Nuclear energy is also primary energy but produced initially in the form of heat, which, if not required directly as heat, can be converted to electricity.

Solar photovoltaic (PV) produces electricity **directly** (but not 24/7) by converting the light component of the energy in sunlight (primary energy) into electricity (an energy carrier). Concentrating solar power (CSP), which we cover later, converts the heat energy of the sun (infrared rays) into usable heat at a high temperature. Wind turbines convert the kinetic energy of wind (produced by the sun's heat energy generating air currents) directly to electricity by means of an electric generator. Hydropower is kinetic energy driven by the heat energy of the sun (infrared) raising water into the sky as moisture, which then falls on higher ground as rain. Reservoirs therefore store kinetic energy, which can be used to drive a generator and produce power.

In Figure 1.2,[2] fossil fuels (at the top left of the graphic under the heading 'Primary Energy') have been the mainstay of the global economy for many decades – initially, coal, then oil, then natural gas.

A major part of the transition involves ceasing to use these sources of primary energy and moving to nuclear, hydro, geothermal, and renewables (note that batteries do not **produce** energy or electric power and only store energy). Primary energy is energy in a form found in nature, e.g. crude oil, wind energy, solar energy, and nuclear energy from uranium or thorium. It then goes through a conversion process. In contrast, primary energy from CSP provides heat energy directly for its many final demand uses.

An energy carrier (secondary or useful energy) is a means of providing energy in a more usable form. The uses of energy, namely 'Final Demand', are shown in the last column. Examples of final energy use as heat can be found in industries such as steel, aluminium, and glass

[2]Hori (2010).

Primary Energy	→	Energy Carrier (Secondary ~ Final Energy)	→	Final Demand
Fossil Fuels (Petroleum, Natural gas, Coal, etc.) Nuclear Renewables (Hydro, Solar, Wind, Biomass, etc.)	Conversion	Hydrocarbons (Fossil fuel products, such as Gasoline, Kerosene, Diesel oil, City gas, etc. In the Future, Synthetic fuels, including Bio fuels) Electricity Hydrogen	Utilization	Heat Light Power Transportation Electronics Communication

Figure 1.2. From primary energy through energy carrier to final demand.

manufacture and, of course, domestic and commercial space heating. Transport is one of the largest uses of energy for achieving motion. Domestic lighting was originally created by heating a metal filament by passing electricity through it, which then glowed with a warm white colour. It provided light but at the cost of also mainly producing a high (wasteful) output of heat energy. Today, most light is produced by passing electricity through light-emitting diodes (LEDs), which produce much less waste heat and are therefore much lower in cost to run. It also means that less generating capacity is required for this particular use of power.

Another way of looking at energy carriers is to consider the two forms in which they exist: **electrons or molecules**. Electrons are what provide electric power. Molecules are gases or liquids (usually hydrocarbons). Examples of molecules include hydrogen itself and its derivatives (e.g. methane, methanol, ammonia, and synthetic liquid hydrocarbon fuels). In the first instance molecules produce heat energy through oxidation (combining with oxygen normally from the air), otherwise known as combustion. This energy is then converted to either mechanical energy (gasoline or petrol energy to propel a motor car) or into electrical power (by using the heat when oxidised to generate steam and turn a generator). The problem of the energy transition lies in generating electrical power, motive power for transport and heat for industry and space heating (homes and offices) without using fossil fuels and in the most cost-effective manner.

The key objection to fossil fuels is that their production and use create emissions of carbon dioxide (CO_2), methane (CH_4), and nitrous oxide (N_2O), which are, collectively, the principal greenhouse gases (GHGs). While we tend to consider mainly CO_2, in fact methane is 30–80 times more potent per kilogram in trapping heat in the atmosphere than CO_2. In addition to the global heating impact of GHGs, the local impact of fumes emitted by combusting fossil fuels in cities and homes can also have severe impacts on human health, including through the emission of very small particles called PM2 from the combustion of diesel fuel in trucks and cars.

While the key intention of the energy transition is to convert as many end uses of energy as possible to electrical power, some uses of fossil fuels are difficult to convert. Some industrial processes which use coal, oil, or natural gas as sources of heat, and air travel which cannot yet (and may never be able to) use electrical power to replace the high energy density of aviation fuel are examples. This is where hydrogen (as a molecule) may play a role, at least for short-haul planes. We cover hydrogen in some detail in Chapter 16, as well as its derivatives, including methanol, methane, and ammonia, and also e-fuels such as petrol, kerosene, and diesel (synthetic fuels in the above graphic). Nuclear energy, although not referred to above, is another important source of carbon-free heat, and this heat energy enables the production of CFE power that is required by society, which we cover in Chapter 13.

1.3 Energy Transitions Through Time: Innovation in Energy Technologies

With respect to the successive industrial transitions that the world has experienced, Figure 1.3[3] shows the radical changes in technologies employed since the first wave of innovation in the world, which started with the UK's Industrial Revolution. The Industrial Revolution is considered by some historians to have been the most important event in the

[3]Right to reproduce this graphic kindly provided by Professor T. Clark of University of Technology, Sydney, Australia.

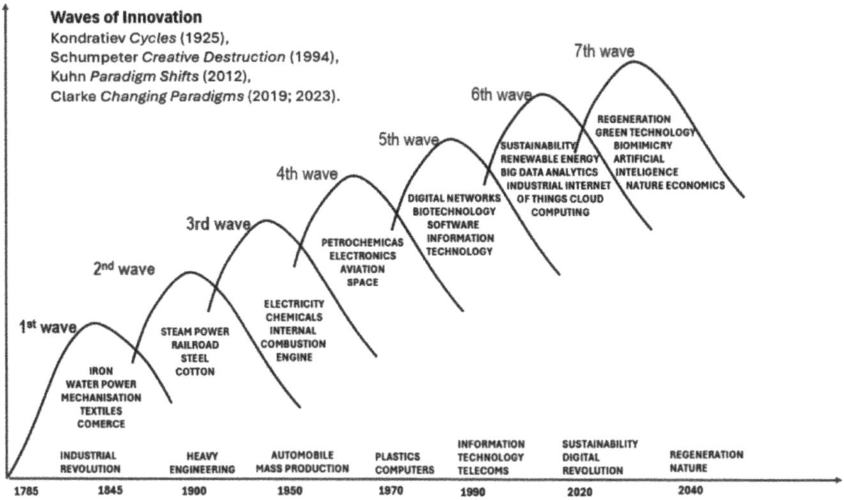

Figure 1.3. The five waves of innovation over 250 years.

history of humanity since the earlier era when humankind started to domesticate animals (horses, cows, etc.) and plants (the agricultural revolution). It is these industrial technologies, along with the massive increase in farming – in particular of beef cows which emit methane and whose pasture land requires large amounts of nitrogenous fertiliser made from natural gas – which has resulted in huge volumes of greenhouse gases being emitted into the atmosphere. Growth of beef cow farming has also resulted in large areas of land being deforested to provide pasture land for cows. Such land use change is a particular issue in the Amazon.

The first wave of the Industrial Revolution was a transition from the use of energy in the form of human power to animal power (horses and oxen) to drive the economy and, subsequently, to the use of energy from flowing rivers (waterpower). Waterpower technology enabled the development of a large-scale textile industry in the UK (mainly cotton mills) and water-powered flour mills to make flour for bread. This source of power was limited by the need to be near a river with sufficient flow of water to provide the necessary energy. So, just as solar and wind power are dependent on location (and weather), so too was water power. Wind power has also been used over the ages and, in particular, in Holland (the Netherlands), where it was, and still is, used to pump water to prevent reclaimed land from flooding. Indeed, it is believed that windmills were used in Holland as early as 1,200 CE (AD) for grinding grain.

In the second wave, fossil fuel (coal) was used to create heat (to make iron and subsequently steel) and to provide mechanical power through the invention of the steam engine. This development enabled the creation of the iron industry and subsequently the steel industry, followed by the development of steam-powered iron ships (rather than wooden ships with sails) and railways, which required iron and steel for the tracks, engines, and coaches and used coal as the source of energy for propulsion. It was in this second wave that global warming emissions started to increase dramatically. Later in the industrial revolution, coal was used to produce steam, which was fed to a steam engine (an external combustion engine) used to drive an electric generator.

The third wave involved the use of fossil fuels other than coal. Crude oil, refined into vehicle fuels, could be used in internal combustion engines (ICEs), which enabled a massive increase in efficiency compared with the external combustion engine (the steam engine). The fourth wave was electronics and aviation. The fifth wave was digitisation. At the present point in the 21st century, humankind is in the sixth wave, and we are seeing yet again 'creative destruction', which refers to the incessant technological (product, process, and business model) innovation by which new production units replace outdated ones. In particular, as shown in the graphic, this wave includes most of what we cover in this book. Today, we are probably entering the seventh wave (the graphic is from 2019), which involves new clean technologies, generative AI, the IoT, and robots and drones (including drones for weaponised attacks in war). The expression '**creative destruction**' was coined by Joseph Schumpeter, a political economist and a one-time finance minister who considered it 'the essential fact about capitalism' (along with the need for competition and innovation). In today's case, some of the destruction will be of formerly valuable fossil fuel assets, which, in the new fossil-free world, will be valueless, so-called *stranded assets*.

Heat energy, supplied almost completely by fossil fuels, is as important as electrical energy, as it fuels much of industry as well as its use for heating domestic and commercial premises. Transport is a huge user of (fossil) energy to generate heat, which the ICE converts to motion. When 'peak' oil and gas usage is reached, many of the assets in the oil and gas industry will start to become stranded assets.

A key reason for focusing mainly on electric power production and use in this book is that electrification of end uses, such as road vehicles, home heating, and industrial processes, will require a large increase in the supply of CFE power. In addition, almost all the green hydrogen that may become necessary for the energy transition (and its derivative products such as synthetic liquid hydrocarbon fuels) is manufactured in electrolysers which require large amounts of electric power. AI large language models are also much more energy intensive than traditional AI and may consume 5–8% of total power generation by the next decade.

Chapter 2

Financing Infrastructure Projects: Corporate Finance, Project Finance, and Financing Climate Tech Innovation

2.1 The Financing Choices: Corporate Finance or Project Finance

Financing the capital expenditure of *any* of a company involves the exchange of a cash inflow to the company now (a loan or share investment) against an expectation by lenders or shareholders that they will receive a greater amount of cash in the future generated by the assets financed by the cash inflow. In the case of debt financing, the inflow may be in the form of a fixed or floating-rate loan, a floating-rate security, or a fixed-rate bond. In fact, there is little legal difference between loans and bonds – the once common distinction that loans cannot be traded while bonds can is no longer the case. But the key issues from a financing point of view are, firstly, whether the venture will be financed using corporate finance techniques or project finance techniques. Following that decision, the next decision is the proportion of debt to equity that is optimal. This is determined mainly by focusing on what structure is likely to achieve the lowest cost of capital.

There are two ways of 'housing' a new project:

1. using the **corporate structure** of an existing large company, which is the corporate finance structure also known as **on-balance sheet financing**;

2. using the **project structure** by creating a new company just for the purpose of 'housing' the new project using project finance techniques, also known as **off-balance sheet financing**.

Corporate finance structures are relatively well known, as corporate finance is a standard topic in a finance degree programme. Project finance is much less well known, so we cover it in much more detail. In any financing, there are two aspects to which lenders pay attention:

> *First*, we assume that the economic unit will be a going concern for the life of the loan and therefore that the project will generate sufficient cash to service its debt;
>
> *Second*, if the entity ceases to be a going concern (becomes a gone concern), there must be a mechanism to enable investors to recover as much principal and unpaid interest as possible.

The two financing techniques – corporate finance and project finance – are shown in Figures 2.1 and 2.2.[1]

Corporate finance (1) is where we have an existing company running an ongoing business.

The average cost of each type of capital is its Weighted Average Cost of Capital (WACC). The company has decided to start a new project and **house it within the existing corporate structure**. The project may be financed by existing resources obtained from retained earnings on the balance sheet, additional debt and/or the issue of additional shares by the corporate. The key point is that if the new project fails, existing shareholders will take any loss. If the project is thought by investors to be risky, the WACC of the company will increase, meaning that the share price is likely to fall and the cost of any future debt raise will be higher. To avoid this risk, the company may opt instead to use project financing to minimise any direct impact on its own cost of capital and share price.

Project financing (2), as shown in Figure 2.2, is where there are now **two companies**, with the lower one being the project company. This company is called a special purpose vehicle (SPV) or special purpose company, which is 'special' only in the sense that it was created only for the

---

[1]Project Finance 101 Bryan C. Associate Director @ Buro Four | Data-driven Project Management & Engineering Professional | Mission Critical https://www.linkedin.com/pulse/project-finance-101-bryan-chan-ceng-mba-pmp/.

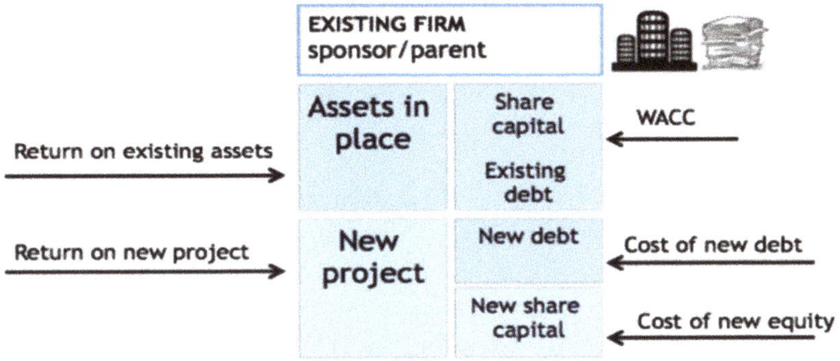

Figure 2.1. Corporate finance.

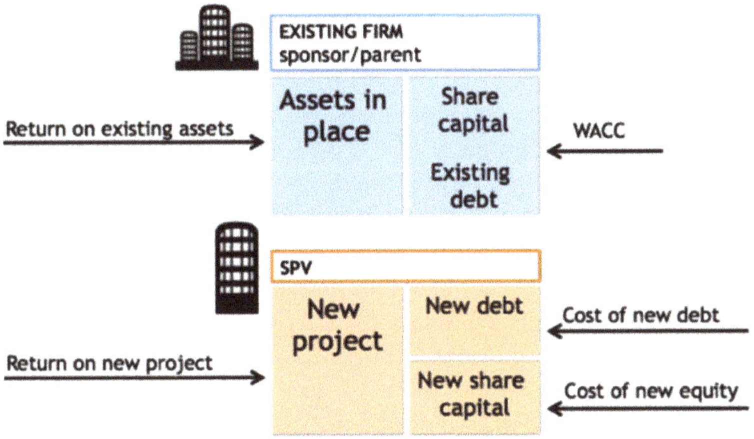

Figure 2.2. Project finance.

purpose of this one project. If the project fails and lenders are not repaid, this does not directly impact the sponsor (which may be a corporate or a developer), as the debt is not on its balance sheet. As a result, it should not have much if any impact on the corporate's cost of capital. However, the drawback from the point of view of lenders is that they cannot demand repayment of the debt from the corporate sponsor or any other equity investor since the venture is quite separate legally from the sponsor and any other investors. As a result of the potentially greater risk to lenders in consequence, to obtain debt funding steps must be taken to mitigate the risk of default, though even with such steps the venture may still remain riskier for lenders compared with corporate financing. As a consequence, the spread (cost above the interest rate on loans to government of the same maturity (tenor) on any loan is likely to be higher than if corporate finance were employed, and thus the WACC is likely to be higher than if financed on the balance sheet.

It is frequently said that the main difference between these two structures is that in the first, lenders have recourse, i.e. have a claim on all the *assets of the corporation* if the debt is not serviced, i.e. interest and any capital repayments are not made on time; while in the second, they only have recourse to the cash flows of the *project* and to its assets but not to the sponsor(s). That is, however, rather too simplified an explanation. In fact, most project financings involve only 'limited' recourse, not non-recourse, at least until a project is operational.

In addition to the fact that the spread on a project loan or bond is likely to be higher than on a corporate loan or bond, leading to a higher cost of debt, the *structuring* of a project financing is much more complex than a corporate financing and can be costly in fees. The structuring must overcome (mitigate) the problem of the potentially highly volatile nature of cash flows from a single new project, i.e. undiversified idiosyncratic cash flows. The structuring also must also try to overcome the fact that if a project fails due to problems in completing it, then it is very difficult to realise much value from the assets of an incomplete infrastructure project. In project finance, the continuation of the project after restructuring is almost always preferable to liquidation, which may not be the case if the project was in a corporate structure.

Because project companies are almost always much more highly leveraged (high proportion of debt to equity) than corporates, any rise in the cost of debt over the life of the project (which could result if the initial financing has a relatively short term and requires refinancing) or any reduction in revenues less expenses can have potentially disastrous effects on the Cash Flow Available for Debt Service (CFADS).

Lenders also gain risk mitigation in project financing by requiring that, even when it is a going concern (an accounting term for a business assumed to meet its financial obligations when they become due), **no dividend payments can be made to shareholders before all debt service has been covered**. If debts cannot be repaid from the project as a going concern, then the financial structuring of a project company should have been designed to ensure either that debt holders could take steps (such as finding new managers) as they consider appropriate to allow it to revert to being a going concern; or, if this is not possible, the structuring should ensure that lenders could recover the maximum percentage of money owed to them.

An additional key difference between corporate and project financing lies in the structure of loan repayments, i.e. principal (P) repayments. In a corporate, which is expected, at least in theory, to have an ongoing infinite life, investors are willing to accept that loans will not be repaid from surplus cash generated by operations but from new loans being issued (rolling over a loan). Thus, corporate bonds will often have 'bullet' repayment terms, i.e. no principal payments will be made during the life of the bond, and redemption will occur in a single payment when the bond reaches its final maturity date. The cash required for repayment will be raised by issuing a fresh bond (often bought by the same investors who then receive the proceeds from the redemption of the maturing bond).

In contrast to this, in most project financings a bond issue, as with many bank financings, will have amortisation payments, i.e. capital repayments at regular intervals. This is because a project company will generally have a fixed and predetermined life – normally the period of any concession agreement to which it is a counterparty or a few years short of this. If a project company will be dissolved in, say, 20 years, lenders would require it to have amortisation payments made to them on regular payment dates – thus taking cash out of reach of management and enhancing the probability of full repayment. Generally, for additional security, the loan would also have a final redemption date a few years prior to the project's termination date. This so-called *tail* provides the borrower with the additional security of a few years' more revenue flowing into the project company should the borrower default on the scheduled final redemption payment.

Some bond issues and many bank loans have structures that try to match amortisation payments with expected cash flows over the life of the project. Sometimes, indeed, capital payments for redemption will be very much 'back-ended', i.e. highly biased towards the end of the project life in order that the cash requirements in the early years are minimised.

A project company has no existing projects generating revenue and no revenue will be generated until the completion of the construction phase (which will be measured in years). It will also only ever have one project, which means that *without risk mitigation measures, both revenue and costs may be highly volatile*. This could result in a project company missing debt servicing payments, even though in a subsequent year it may be easily able to meet such payments.

Project finance is undertaken through an SPV, which, as we noted above, is a company with only one clearly defined objective: the development and operation of a single project. The sponsors who are promoting the project will often be those companies which will benefit from contracts for construction, equipment (such as solar installations and wind turbines), and operation and maintenance (O&M). Those who intend to buy the power (offtakers) who are intending to sign power purchase agreements (PPAs) with the SPV, which guarantees the purchase price of power for a period of years, may also invest in the equity. Tech companies such as Microsoft and Amazon and others have an enormous, and rapidly growing, need for power, particularly as large language models (LLMs) such as ChatGPT consume much more power than traditional AI. Such companies are likely to assist financing by investing in equity as well as being offtakers.

The directors and officers of the SPV are constrained by **covenants** imposed by lenders, which may restrict management discretion. A covenant is a formal agreement or promise between two or more parties that specifies what actions may or may not be taken by the parties. A key feature of project finance covenants is that lenders demand significant control over how the company is operated, particularly in terms of the use of cash flow, compared to corporate lending. Covenants lay down activities that may or may not (through negative covenants) be undertaken. These are **designed to protect lenders by minimising management discretion particularly over the use of cash flow. This is to avoid cash outflows unrelated to the project.**

Four covenants which most project company SPVs will be required to consent to in the credit agreement are:

Negative pledge (the borrower agrees not to pledge any of its assets if that would reduce the security of the existing lenders);
Restrictions on additional borrowing;
Restriction on activities that the borrower may undertake;
Restrictions on capital spending and acquisitions.

The effect of these restrictions is that the SPV management may not take assets (collateral) away from the control of the lenders, may not borrow additional amounts (which would make the position of existing lenders less secure), may not diversify (which would introduce different business risks), and may not acquire another company or undertake capital spending.

Another somewhat different but key restriction is a *covenant to restrict the distribution of dividends*. This can be achieved by laying down uses of cash flow which must be prioritised prior to any dividends (distributions) being paid to shareholders. These are outlined as follows.

The revenue and cost stabilising measures that are taken should reduce cash flow volatility, and the restrictions on management discretion should reduce opportunities for cash to leak from the company. But the lenders will also wish to ensure that the leverage of the company is such that it is almost certain that more cash will be generated than the minimum required to service the debt. To achieve this, lenders will set certain 'cover ratios'.

Cover Ratios

Cash Flow Available for Debt Service (CFADS)
This is the starting point for calculating ratios. The lenders will require, for example, that in each period the Debt Service Coverage Ratio (DSCR), which is CFADS divided by debt service cost (P&I payment), is greater than 1. The level they will require for the DSCR is determined by the riskiness of the project, measured as the expected variability of operating revenue.

For example, if the payment mechanism is designed so that the company does not take demand risk (meaning revenue risk), lenders might be satisfied with a projected debt service cover ratio of 1.3×. But if the company does bear substantial risk (such as vehicle numbers on a toll bridge), then lenders may insist on a minimum DSCR as high as 2.0×. Lenders use detailed forward-looking financial models to estimate future cash flows and cover ratios.

There are other cash flow coverage ratios in addition to the annual measure, which are intended to provide an even greater assurance of future ability to service debt, such as the following:

Loan Life Cover Ratio (LLCR)
This is a comparable measure of ability to service debt but measured over the whole life of the loan. At any point in time, this is the discounted value of future CFADS over the life of the project plus any cash

in the Debt Service Reserve Account (DSRA) discounted back to a present value divided by the outstanding amount of debt at the time the calculation is undertaken. A further calculation that is sometimes made, particularly if there is some doubt about the project being able to service its debt before maturity, is the Project Life Cover Ratio (PLCR). This adds cash flow after the point when the debt should have been redeemed, i.e. during the *tail* period of the project, before a concession expires, to test whether this would enable full debt service.

Use of Cash Reserves to Minimise Risk of Temporary Debt Service Shortfalls

If, despite the measures mentioned above, the actual cash flow, rather than the expected cash flow from the financial model, in a particular period is not adequate to service the senior debt, perhaps due to a temporary fall in revenue, the lenders still want to receive the expected P&I payments. If the actual DSCR has fallen below 1, then the company will not have generated sufficient cash in that period to meet debt service. To protect themselves against this possibility, there is a backup measure included in the security documentation for the project. The lender will require that *debt service cash reserves* to be built up after operating costs have been met and senior debt has been serviced. This reserve might cover a period of six months or more, but it cannot overcome a continuation of the shortfall in cash flow once it has been used. It can also be applied to subordinated debt.

The Cash Flow Waterfall and Distribution Covenants

The successive uses of project revenue during only the operational stage are shown in Figure 2.3[2] (a toll road bridge project). Distribution *covenants* are designed to ensure that there is a clear priority in the use of cash to meet upcoming liabilities, with senior debt service obviously having first priority and distributions to shareholders having the lowest priority. Operating expenses are paid first (since if the project does not operate, there will be no revenue). Then, reserve funds are built up to enable maintenance and debt service for some months, even if the bridge is closed with no revenue. What remains can be distributed to shareholders if the proposed amount does not exceed the year's profit plus any reserves from previous years (restricted under company law).

[2]US Department of Transportation Center for Innovative Finance Support https://www.fhwa.dot.gov/ipd/p3/toolkit/toolkit_webinars/financing_highways/.

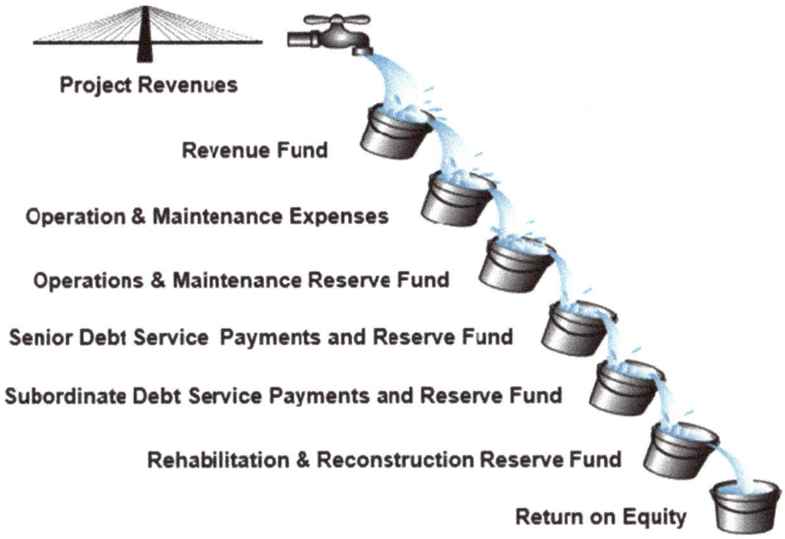

Figure 2.3. The cash flow 'waterfall' in project finance.

The distribution covenants are designed to ensure that there is a clear priority in the use of cash to meet upcoming liabilities, with senior debt service having first priority and distributions to shareholders having the lowest priority. This is often referred to as the *cash flow waterfall.* In the waterfall graphic in Figure 2.3, these various outflows can be seen. Operating expenses are paid first (since if the project does not operate, there will be no revenue). We assume that there are no more construction costs at this point when operations start. The first call on the available cash after that is the payment of P&I on the senior debt, followed by P&I on the subordinated debt. If there is cash remaining after these withdrawals, then in the waterfall in the graphic, the lenders have the right to undertake a *cash sweep.*

A cash sweep refers to cash available in the company in excess of that required to provide debt service on all debt. One-third to one-half of the 'surplus' cash can then be 'swept' out of the company and used to further pay down the senior debt beyond any mandatory amortisation payment. The remaining cash flow will then be used to top up the major maintenance reserve account (though it may already be at the required level and not require topping-up) and can also then be used to repay the subordinated debt if required by the loan agreement. Finally, what is left can be **distributed** to shareholders (**this is not a dividend payment**) provided the amount proposed to be distributed does not exceed the year's profit plus any reserves from previous years (restricted under company law). Thus, the distribution account enables any remaining cash to be distributed to shareholders provided there have been no defaults and all financial tests area met.

After the construction period and once the project is operating satisfactorily and physical and revenue risks are low, the owners of the SPV may choose to refinance its debt using either bank finance once again or by means of a bond issue. With risks greatly reduced at this point, the credit risk spread will fall sharply. At this point, the original sponsors may also sell their shares to an infrastructure corporate or to institutional investors, such as pension funds or clean energy funds, and take the capital gains resulting from the fall in the level of risk compared to the point in time at which they invested. This then enables them to 'recycle' their capital into another project.

While project finance, as covered above, is a common way to finance projects, the 'big oil' corporates which have large cash flows when oil prices are high will often finance a new project, whether it is oil and gas exploration and production (E&P) or a solar farm, from internal resources, i.e. cash on hand (from retained earnings), corporate borrowing, and, if necessary, a new equity issue using corporate finance, i.e. *on-balance sheet financing*. Alternatively, if undertaking a joint venture project, oil companies may choose to use project finance and create a new, separate joint venture company just to undertake a single, jointly owned project using project finance, i.e. *off-balance sheet financing*.

2.1.1 *Deeply Subordinated Debt*

Project equity, unlike corporate equity, is not permanent capital in the normal meaning of the term for the simple reason that project companies have a finite life. Project equity is available to the company for the duration of the project and must be repaid at the latest at the time the project company is liquidated (in the case of concessions, when the concession expires).

Ideally, project equity would be redeemed over the planned life of the company along with debt redemption. This is due to the fact that, over time, the company needs less permanent capital simply because its assets are falling in value over time as recognised by the depreciation charges, and its liabilities are also likely to be falling as bonds or loans amortise. However, company law in most countries only allows dividends to be paid out of current and retained profit. Dividends cannot be financed from 'paid-in capital', i.e. the original equity capital put into the project cannot be paid out. This leaves project companies with the problem that paid-in equity (the original owners' capital contribution) cannot be paid back to them except when the project company is liquidated and all the assets sold.

In fact, 'dividends' paid out by project companies to investors are not dividends as they are understood in corporate finance. In corporate finance, a dividend is paid out as a proportion of annual profit after taxation as a result of a decision by the directors – often around a 50% payout. In a project company, what is paid out at intervals is a *distribution* of cash generated beyond that required by the project company, which in practice means what the lenders agreed with the company in the covenants and in the **cover ratios** (see later). What is 'required' by the company and what may be paid out is not at the discretion of the directors.

However, even when a given level of cash distribution is allowed under the terms of the loan agreements, this might exceed what is permitted by company law, i.e. be more than the annual profit and any retained profit. This can easily happen in a project company, as it is not allowed to use surplus cash to undertake new investments. As the depreciation charge on the profit and loss account is a non-cash charge, the company may have low profits for the year yet have a high level of cash generation. Thus, cash can accumulate in the company on the asset side of the balance sheet as a counterpart to the original paid-in equity on the opposite side of the balance sheet. In effect, the assets are being liquidated, but the surplus cash is trapped within the project company.

To overcome this problem, many project companies use *deeply subordinated debt* to act as quasi-equity, along with a small proportion of actual ordinary share capital (common stock). As debt can be repaid to investors, it is possible to use such debt as a means of repaying this 'quasi-equity', which will normally have been provided by sponsors plus any financial investors in the project. It also has a tax advantage in that any interest paid on such debt is deductible against corporate taxation.

When considering what the word *deeply* means, we are actually asking what is the difference between equity and debt, and thus the real question is whether we can develop financing instruments that fall somewhere in between. If there is such an instrument, this would be valuable as part of the financing in the 'capital structure' of a project company. In fact, there is such an instrument, if properly structured.

Deeply subordinated debt **will** act as 'equity' provided it is seen as a 'cushion' for the senior debt by having almost all the features of 'real' equity without conferring voting or ownership rights. Provided a rating agency accepts that the debt is truly, deeply subordinated, the objective of providing virtually full support to the senior debt will be achieved.

Deeply subordinated debt, because it is considered virtually equity by rating agencies, allows for a higher proportion of debt in the capital structure without reducing the credit rating (or credit quality if unrated) on the senior debt. However, its use in the capital structure gives another benefit, as we noted above. It provides a means to return to investors the initial 'equity' put into the business by the sponsors or financial investors. By paying back this debt using any surplus cash available after meeting all the requirements of the cash flow waterfall a project company can achieve a higher IRR on its equity than it could if return ***of*** equity had to be postponed to a time closer to when the project company was being wound up.

2.1.2 *Summary of the Difference Between Corporate and Project Finance*

Corporate finance, in contrast to project finance, involves financing new projects by a medium or large company which has **an existing revenue and earnings stream** and has generated retained earnings which are now available for funding the project. In addition, funds raised from new on-balance sheet corporate borrowing or a secondary equity issue can be used. Energy majors such as large oil and gas companies can raise finance from banks and from bond markets but raised against **the strength of their own balance sheet and cash flow (corporate finance) and not on the strength of the expected future cash flows from the project as in project finance**.

Borrowing to finance such a new project using corporate finance will often be described in a prospectus or loan documentation as being ***for general corporate purposes***, meaning that it is not described as being for the purpose of financing a specific new asset. Consequently, the lender to the corporate has security from a call (ability to take control of) on all corporate assets on the balance sheet in the event of default. To sum up the differences, we have listed the main ones as follows:

- **Corporate finance is about financing a company which currently has multiple existing activities and new projects**
  - and is permitted to undertake future acquisition, expansion, and diversification plans using retained earnings.

- **Project finance is about financing a company which has no existing activities and only one new project**
  - and is not permitted to make future acquisition, expansion, or diversification plans;
  - will return all surplus cash to shareholders and not reinvest it;
  - uses a special purpose entity (SPE) or SPV and is a 'limited recourse company';
  - and the debt issued by the SPV has support only from the cash flows from the project, not from any sponsors (unlike an on-balance sheet corporate finance project).

Additional differences include the following:

- **A corporation (a corporate)** is:
  - **a multi-purpose vehicle** not a special purpose vehicle;
  - generally has a very diverse range of activities and intended projects;
  - is expected to have an **infinite life.**

- **A project company (a special purpose vehicle or SPV)** is:
  - a **single purpose vehicle**;
  - has only one project and no growth projects;
  - has a pre-determined **finite life.**

- **Project finance equity and debt will have been repaid by the time the SPV is liquidated. Corporates, on the other hand, are assumed (usually wrongly) to have an infinite lifespan (unlike the humans who manage them) and therefore do not repay their equity financiers (unless they choose, for some reason, to buy back their own shares in the stock market.**

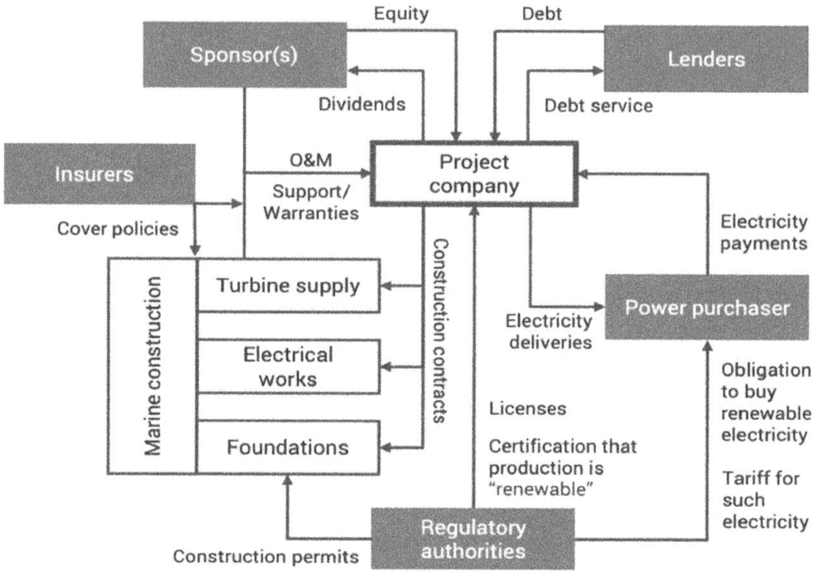

Figure 2.4. Contractual structure of a project financing for offshore wind.

Figure 2.4[3] shows a simplified graphical representation of the contractual structures involved in a project, in this case an offshore wind project. At the centre is the project company (the SPV), which is sometimes known as the nexus of contracts (the connection of links between things). The boxes show the main parties involved in these contracts and what the contracts are for. The construction contracts are between the SPV and the parties involved in constructing the offshore wind farm.

There are likely to be different specialist contractors for each type of work, though some contractors will provide a 'turnkey' service, i.e. many or all aspects of the construction – engineering, procurement, and construction (EPC). A link between construction and sponsors is shown, and the reason for this is that some or all of the sponsors are likely to be the firms which have bid for the contracting work and will generally also have invested in some of the equity of the SPV.

There is also likely to be an advisory firm which will work alongside project developers, contractors, investors, and asset managers. There are

[3]Green Giraffe. 'Recent trends in offshore wind finance' April 2019 S7 https://green-giraffe.com/wp-content/uploads/2019/04/190411_green_giraffe_bilbao_-_trends_in_ow_finance_final.pdf.

a number of firms which provide such advice, including large consulting practices. One of the independent firms is Operis, whose advisory teams specialize in infrastructure, the energy transition, and digital infrastructure projects. This is a good combination of skills, as the energy transition requires not only an understanding of generating and transmission infrastructure but also digital networks and artificial intelligence systems which control projects, networks, and indeed the whole electrical power system in a country. It has a particular expertise in the financial modelling of project finance transactions.

An example of the type of services an advisory firm might provide for a greenfield project, i.e. the development of a new clean energy infrastructure asset, would include:

- advice on bid strategy,
- contractual negotiation,
- transaction model development,
- running funding competitions from multiple potential sources of debt,
- sensitivity analysis based on due diligence.

One of the objectives would be to optimise the SPV's financial structure and advise the sponsors on debt raising. Advice on contractual negotiations of project documents, such as CfDs and PPA agreements, can also be critical to the ongoing success of a project.

There will be a number of regulatory authorities involved, which will provide the necessary licenses, certifications, and permits that are required. Not shown in the graphic is the agreement with the owner of the seabed for offshore wind, which, in the case of the UK, is the Crown Estate. If it is a land-based wind or solar project, then there will be an agreement for rental payments for the use of the land during the life of the project. Insurers are necessary for any project of this type, both third-party cover in case of damage to others and commercial and industrial (C&I) cover for risks to the assets. Lenders require such insurance cover so that, in the event of damage to the facility or a claim by a third party, this will not affect their cash flows.

Following the design and construction stages, the operational phase requires mechanisms to stabilise *operational revenue and operational costs* over a major part of the life of the project in order to stabilise the cash flow (revenues minus cash costs) of the project company. This is to maximise the probability of having a stable cash flow and hence the highest probability of being able to meet P&I payments when they are due. Mitigating revenue and cost variability will reduce much of the perceived credit risk at the operational stage of a project and hence reduce the credit risk premium that a lender would charge at that point.

> *On the operational* ***revenue*** *side*, mechanisms such as a PPA or long-term take-or-pay contracts (where the buyer of the output, such as electric power, must pay for an agreed amount of output even if it does not accept delivery) can be used. To reduce credit risk **on the SPV**, the counterparty to the PPA must have a strong credit rating of its own.
>
> *On the operational* ***cost*** *side*, solar and wind are not subject to the high cost variability of fossil fuel sources of energy. Other types of clean energy assets which have electric power as an input can clearly benefit from a PPA from the point of view of fixing input costs. An example would be fixed-price, long-term supply contracts between a wind farm owner and an electrolyser facility to manufacture hydrogen. These would be offtaker agreements with a generator (it would be the counterparty side of a wind farm PPA). Such contracts are designed to achieve stability of the major cost input.
>
> The operation and maintenance (O&M) of an asset once it is operational is one of the ongoing costs in any project. An example of a contractor which provides both EPC and O&M services is Belelectric, a large multinational Germany based contractor. It is a 'turnkey' provider which offers all the necessary services to enable it to hand over a completed project. Subsequent to handover, it offers plant maintenance generally through multiyear contracts in order to fix the cost over that period.

Project Phases

Figure 2.5[4] (an onshore wind power project) summarises the four stages of a project's life with the final investment decision (FID) and financial close (the point at which construction can commence) in the middle.

Before financial close (prior to the point at which contracts are signed for the debt financing), at which point the sponsors are committed to the project going ahead and construction can start, the first stage is the pre-planning development stage, where, critically, the fundamental techno-economics of the project must be evaluated (conceptual design or feasibility study) and all necessary permits obtained. After that is the

---

[4]Author's image, based on Platina Partners research: Javier Herrero y Saenz de Cabezon at http://jherrerosdc.typepad.com/.

Figure 2.5. How equity value increases at each stage of a project development.

Note: *Figures for development value increase are a generic, rough European wide onshore wind 'index', pre-financial crisis in 2008, not including debt.

post-planning development, including Front-End Engineering Design (FEED). The FEED focuses on the technical requirements, as well as estimating an approximate investment cost for the project and expected rates of return on debt and equity capital invested after risk mitigation. Without these steps, it is not possible to present a proposal to potential funders. After this stage is the construction period, which for the wind-farm in this example is estimated to be about two years. During this time, the project does not generate revenue. Once construction is completed, the turbines will be brought up to full operational output and the whole system tested and certified. Finally, the operational phase, which is estimated to be around 20 years, commences and revenues start to flow.

What can be seen in the graphic above is that **an investor coming into a project at a very risky early stage** (on the left-hand side of the graphic), where cash is being spent on project planning and design with only a 1 in 3 chance of getting planning permission, **is able to 'buy into' a potential 20-year revenue stream from selling the power output at a cost of no more than €5,000 per MW**. It would normally be the project sponsor(s)/developers who would invest at this stage. Once permits have been granted, the next stage is agreeing all the contracts necessary for construction, operation, maintenance, power purchase agreements, etc. As the project is now much less risky, since the 2 in 3 probability of the project not receiving planning permission has not materialised, the price at which further equity shares in the project can be sold shoots up to

€150,000–€200,000 per MW. Then a high-risk-taking private equity fund might be interesting in investing.

Once this stage is completed, the project company moves to financial close. At this point, an investor buying into the project will pay €300,000–€400,000 to buy-in since risks have been very substantially reduced (the higher the price paid, the lower the subsequent return to reflect the lower risk). At this stage, other private equity funds may be interested in investing, but there is likely to still be substantial 'construction risk', i.e. project over-budget and completion delay. Finally, once the project is up and running, pension funds and insurance companies may want to invest in the equity to generate relatively safe, long-term returns without any construction risk to match their need for safe, relatively low-return but long-maturity investments. These investors will pay as much as €600,000–€800,000 per MW, which means, as noted above, that their returns on a successful project will be much lower than for those who bought into the equity earlier, but equally their returns are more certain as major risks have been mitigated by completing construction, testing, and signing of guaranteed revenue contracts. Note that these costs per MW are hypothetical and change year by year.

On the debt side, while the project is likely to have been financed initially by a bank or syndicate of banks using a relatively short-term loan (probably three years maximum), once the project has been shown to be operating as designed and is generating revenue, it is likely that bond investors (relatively low-risk investors which once again are mainly pension funds and insurance companies) will wish to invest. Thus, at this point, the bank lending may be *re-financed* by a long-term bond issue. As bonds have a fixed rate of interest, this avoids the risk of an interest rate rise for the remainder of the project life (20 years) and thus contributes to stabilising the margin between costs and revenue.

2.2 Bond Financing of Infrastructure Projects

The transition to greater use of capital market (bond) financing requires a substantial shift in mindset from *sponsors* and *developers* seeking to finance projects using project bonds. Investment bankers with whom developers have to deal when arranging a bond issue take a transactional approach to clients rather than the relational approach with which the industry is more familiar. This arises in part from the fact that a debt capital market (DCM) banker is not providing other services to their client such as foreign exchange, treasury, transaction services, interest rate hedging, working capital lending, revolving credit, and invoice discounting, which a commercial banker may be providing and on which the bank makes a good return. In contrast, the investment banker makes a return only on a single bond issue, as a project company is unlikely to ever issue another bond or equity for the same company.

In addition, a commercial banker will also be the investor in the debt, at least initially, and will therefore serve as both the arranger (originator) and the investor. In contrast, in the debt markets, the arranging (originating) banker is only an intermediary between the developers/sponsors and investors, and consequently, a second key to a successful bond issue is an understanding of the needs of bond investors. Frequently, institutional investors seek long-term assets to match their long-term liabilities, such as pension policies, which is in direct contrast to bank lenders for whom short-term assets are more attractive and require less regulatory capital.

A bond issued at the time of financial close (rather than at a subsequent refinancing of a bank loan) offers the advantage that it will not require such subsequent refinancing because it can have as long a maturity as required by the project company. This therefore mitigates the perceived risk of a future refinancing raising the cost of debt service or not being available at all at any reasonable price.

One of the key advantages of using longer-term debt finance at the time of the initial financing, if it is possible, is that debt service is spread out over a longer period. This results in a smaller principal repayment each year (amortisation) and hence a greater amount of free cash flow is available to the project company in the early years, which helps the project get off to a good start. In this respect, project financing, which involves the annual amortisation of the bond principal, is different from corporate financing, where 'bullet' bonds with no or limited amortisation are issued, and in effect, the old bond is 'rolled over' by issuing a new bond to refinance the old one at maturity.

If, after meeting all prior claims in the cash flow waterfall, there is a cash surplus available to make an early equity distribution as a result of the lower amortisation payments, then the IRR of the project will be higher. This obviously makes such financing more attractive to equity investors. The abbreviation 'RET' stands for renewable energy technology.

Corporate Bond Financing and the Green Bond Market

A large energy *corporate* such as Exxon Mobil is regularly in the market with new corporate bond issues (both to raise new finance and to refinance maturing issues) in which the 'Use of Proceeds' paragraph in the prospectus would simply state 'For General Corporate Purposes'.

To the extent that such corporates choose to finance new projects on balance sheet using 'green bonds', increasingly, this *Use of Proceeds* clause is **not** going to satisfy investors. They will wish to know if any of the proceeds from the issue might leak into carbon-emitting investment projects. If so, many will not buy such securities, or the corporate will have to offer a substantial premium. The reason is simple: many of the underlying beneficiaries of institutional investment funds will not permit their funds to make investments which increase global CO_2 emissions.

One (partial) solution to the problem of the *Use of Proceeds* clause in a corporate bond is to provide the necessary disclosures on use of proceeds to satisfy the Green Bond criteria. To help with this process, and in order to make this growing asset class more appealing to institutional investors, the International Capital Market Association (ICMA), the sponsor of the ICMA Centre, developed the 'Green Bond Principles'. These principles are supported by over 115 institutions participating in the Green Bond market. Traditionally, green bonds have been issued by corporates. Increasingly, they are also being issued by project companies. Indeed, it is likely that carbon-emitting projects (including projects such as combined cycle gas turbine (CCGT) power stations) will find that new financing but also refinancing in the bond markets becomes increasingly difficult as investors choose to hold only green bonds and thus invest only in green project bonds.

Project Bond Financing and Solving Potential Drawbacks of Such Financing

While institutional investors find long bonds attractive, whether corporate or project, there are a number of problems which arise when trying to fit project bonds into the capital structure of a greenfield project. The 'sticking point' for bond investors is that project bonds without risk mitigation features may have some characteristics which are not attractive to them.

Credit Risk Mitigation from Sponsor Support (Construction Performance Support Packages)

Institutional investors generally require a higher credit rating (at least investment grade) than bank investors and are thus likely to require some additional type of credit enhancement if they are to be attracted to investment at the start of the construction phase.

One of the ways investment bankers have tried to overcome this issue is by using a construction performance support package (which would include support from credit worthy sponsors), which limits the exposure of bond investors to construction cost overruns or extended construction time and thereby may enhance the bond's rating.

If investors are persuaded that the risk of the construction phase has thereby been mitigated adequately, they may be persuaded to finance this phase. Some of these mechanisms are considered in the following sections.

Construction Phase – Risk Mitigation by Limited Recourse

Project finance is not always *non*-recourse. In fact, non-recourse finance is relatively unusual. There is usually some form of recourse back to sponsors. If additional funds to meet cost overruns come from the sponsors, this would be provided in the form of additional equity.

The World Bank views this type of limited recourse as 'Sponsor Support'.

Sponsor support may include[5]:

- shortfall guarantees, where the banks, after enforcing all other security rights, experience a shortfall;
- buy-down undertakings: a promise to prepay project debt to ensure specified ratios, in certain circumstances;
- price guarantees, to ensure the price of off-take;
- market price purchase guarantees, to purchase a minimum quantity of product at market price over a set period;
- tax loss purchases, where a shareholder agrees to purchase certain tax losses from the project company;
- technical support, extended warranties and maintenance arrangements; and
- contingent equity or subordinated debt commitments to cover construction or other price overruns.

It should be noted that during the initial periods (permitting, contracting, and obtaining financing), all cash inflows are likely to be in the form of equity from the developer and other equity investors, with no dividend payments or cash outflows to investors. This is because many projects do not manage to reach the 'start construction' phase and fail at an earlier stage. Thus, until all permitting, etc. are in place and the lenders have agreed that its conditions have been met and that the project can proceed to the point of the **final investment decision (FID)** and the **financial close**, the risk to providers of capital is much greater than after the financial close.

[5] World Bank Group PPPIRC. http://ppp.worldbank.org/public-private-partnership/financing/sources.

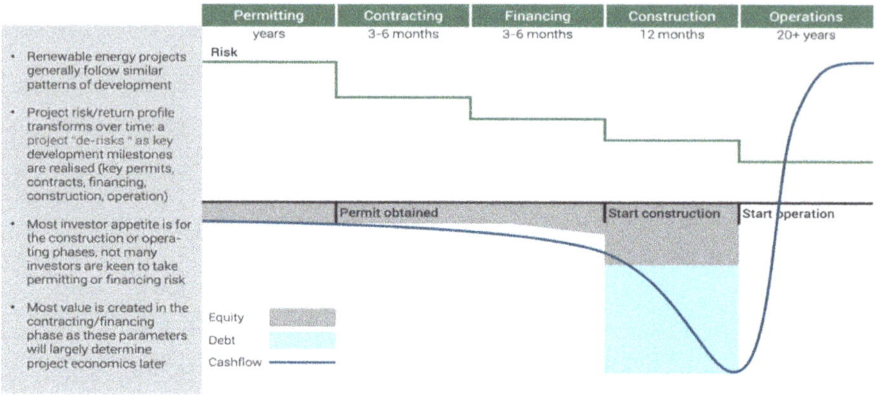

Figure 2.6. Most value created during the development and contracting phases.

The green line in the graphic shows the level of risk and how it changes over time, while the blue line shows cash flow becoming ever more negative over time, and as construction begins, debt funding comes in to use and enables a very negative cash flow until production commences.

Project stages, risks, and cash flow in graphical form are presented in Figure 2.6 from Green Giraffe,[6] a specialist advisory firm focused on renewable energy. The graphic shows risks declining over the period up to the end of construction (as we examined above) and also the pattern of cash flow. Large amounts of capital are not required prior to the construction stage and are met by sponsor (and other investor) equity.

Examples of Actual Projects

The first two are road projects. The third is for an offshore transmission operator (OFTO), which means the cable which connects an offshore wind farm to the national grid. The fourth is also an OFTO, while the fifth (Castor) was intended to be for natural gas storage as a reserve for use in winter in Spain. We provide a case study of Castor below. Table 2.1 is from the AFME/ICMA Guide to Infrastructure Financing,[7] on which Brian Scott-Quinn and Deyber Cano were the principal authors with valuable input from a committee of AFME/ICMA member firms and staff members from ICMA.

---

[6] https://green-giraffe.eu/sites/green-giraffe.eu/files/190411_green_giraffe_bilbao_-_trends_in_ow_finance_final.pdf.

[7] https://www.afme.eu/globalassets/downloads/publications/afme_guide_to_infrastructure_financing2.pdf.

Table 2.1. Examples of actual project financings.

	CAV Roads Project	N25 Road Project	WoDS OFTO	Greater Gabbard OFTO	Castor Gas Storage Project
Country	Italy (Baa2 stable)	Ireland (Baa1 positive)	UK (Aa1 stable)	UK (Aa1 stable)	Spain (Baa2 stable)
Project Co	Concessioni Autostradali Venete S.p.A.	New Ross N25 By-Pass Designated Activity Company	WoDS Transmission Plc	Greater Gabbard OFTO Plc	Escal UGS S.L.
Issuer	Concessioni Autostradali Venete S.p.A.	New Ross N25 By-Pass Designated Activity Company	WoDS Transmission Plc	Greater Gabbard OFTO Plc	Watercraft Capital S.A.
Bonds	EUR 830 million 2.115% due Dec 2030	EUR 145.4 million 2.684% due Jan 2042	GBP 254.8 million 3.446% due Aug 2034	GBP 305.1 million 4.137% due Nov 2032	EUR 1,400 million 5.756% due Dec 2034
Issue Date	12-Apr-16	2-Feb-16	25-Aug-15	29-Nov-13	2-Aug-13
Project Summary	Operation and maintenance of a motorway network comprising four sections with an aggregate length of 74 km, in the region of Venice, north-east Italy	Construction, operation & maintenance of approximately 13.6 km of new dual carriageway, 1.2 km of new/upgraded single carriageway and various structures including a 0.9 km bridge	Operation and maintenance of electricity transmission assets that connect the 389 MW West of Duddon Sands wind farm, some 14 km offshore north-west England, to the UK grid	Operation and maintenance of electricity transmission assets that connect the 504 MW Greater Gabbard offshore wind farm, some 40 km east of the port of Harwich, to the UK grid	Construction and operation of an underground gas storage facility utilising a depleted oil field located some 20 km offshore, near the port of Vinoraz, Spain
Project Life	16 years	25 years post completion	20 years	20 years	24 years
Term of Debt	14 years	Construction + 23 years	19 years	19 years	21 years

2.3 Venture Capital and Private Equity: Financing Climate Tech Innovation

> *For new technologies to reach TRL 9 (scaling-up and commercial production levels) they need to be financed through all the stages prior to this. Venture capital and private equity are involved in these early stages as well as financing through grants from governments and foundations.*

In this chapter, we examine the techniques for financing risky start-ups and financing their growth to enable them to reach TRL 9.

To achieve a much more effective and rapid energy transition from 2030 to 2050 using new technologies currently being developed, companies with well-thought-out concepts must find sources of risk capital, initially 100% equity (the risk capital). The technologies that are being developed, trialled, and demonstrated will have to be trialled in the market and then, if successful, put into large-scale production. Having covered the financing of projects at TRL 9, we are now looking at those from TRL 1–8.

The issue with financing 'post-2030 technologies' is that each stage requires a different type of investor and a different scale of capital. In addition, there are relatively few 'analysts' able to assess these new technologies and choose companies for a fund to invest in from the hundreds that may approach them. Figure 2.7[8] provides a good picture of the problem and, in particular, of the so-called **'valley of death'** that companies need to traverse before even potentially valuable technologies become mainstream. The valley of death has two phases, but both result from the difficulty of finding sources for the additional capital that needs to be continually provided for such ventures, which consume capital but at this stage have no revenue and indeed may never generate revenue. Thus, cash flow remains negative for some years.

Initially, many projects at the R&D stage (pre-seed capital, as illustrated in the lifecycle diagram above) are financed by research institutes, government agency grants (for example, through the US Department of Energy, universities, or/and with funding from entrepreneurs' savings or borrowing or 'friends, family, and fools!!'). That's what the 3F means in

---

[8]Graphic from Business Development Bank of Canada (BDC) *BDC's Industrial, Clean and Energy Technology (ICE) Venture Fund* https://www.bdc.ca/en/bdc-capital/venture-capital/funds/energy-cleantech-venture-fund.

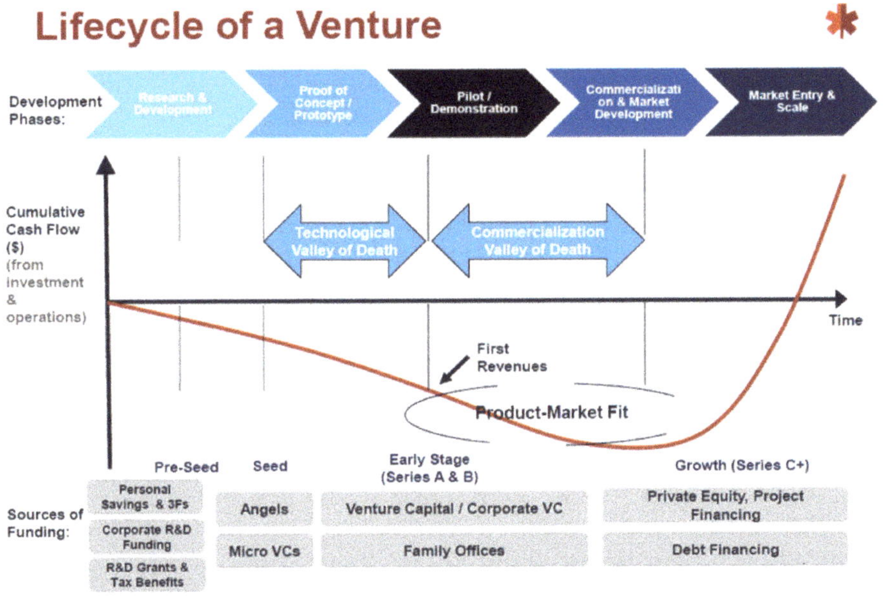

Figure 2.7. Lifecycle of a new venture.

the 'sources of funding' above, described this way since putting money into a venture at this stage which has only a small chance of being successful is only for those who are foolish (hoping to make a quick buck) or are able to bear the high probability of loss.

On the other hand, venture capitalists operate by putting relatively small amounts of money into a large number of ventures. While 80% of their investments could fail to make a high return or fail completely, the 20% which succeed may provide very high returns. Thus, this diversification across many projects protects professional investors from the most serious losses in their portfolios.

In the graphic, the terms Series A and B are used. While a Series A funding round is intended to develop the product, a Series B funding round is when a business is being taken to the next level, past the development stage. Series A is about getting the product to the point where there actually is a product. Series B is designed to enable the business to start selling the product, while Series C is about scaling up operations and possibly preparing for an exit, such as an IPO or acquisition, to enable earlier investors to realise their capital gains (or losses).

Entrepreneurs can also work within large corporates when they are known as **intrapreneurs**. Indeed, the first personal computer, the IBM PC, was funded internally by IBM (corporate R&D funding) but in a separate facility far from head office (known as the 'skunk works'). Many of those in head office thought the venture was almost certain to fail. The PC was introduced to the market in August 1981, only a year after the company gave the team of intrapreneurs the go-ahead. Initially, it was thought by IBM management that there were not enough applications for this device to justify the idea of large-scale production. But as we now know, it was a roaring success against all odds, particularly the odds against incubating a completely novel idea for a desktop computer being produced by the manufacturer of 'big iron' (mainframes).

Following the pre-seed stage, when basic R&D has shown the possibility of a saleable product, we have the seed capital stage as the next stage in the timeline. At this point, some capital may come from 'Angel Investors' (those outside family and friends but who are willing to help what seem like promising ventures, such as Henley Business Ventures). At this stage, some small VC firms may also make a contribution. This funding is entirely in the form of equity, i.e. risk capital, as it is unlikely that any financial institution would lend to a venture at this stage.

The technological 'valley of death' is the period when the technology is being tested, and almost certainly, there will be technology failures along the way. If investors maintain faith and the entrepreneurs and scientists seem able to overcome the technological problems, investors may put in more capital. If not, the company will collapse (we give two ongoing examples of this problem at the end of this chapter). Those investors would be the larger venture capital companies and the venture capital arms of corporates. In addition, family offices, which are the private wealth management companies of ultra-high-net-worth (UHNW) individuals, may contribute capital. They would be subscribing to what is known as Early-Stage Series A and B funding rounds prior to the hoped-for Series C funding round. A key distinction between these successive funding rounds is the **valuation of the business, which is likely to rise at each stage as risk is reduced and hence value in increases**. The different stages also reflect its maturity level and growth prospects; the greater these are, the higher the valuation.

While there may be some early sales of the product or service to 'early adopters', this does not guarantee a market large enough to ensure its survival. This is the 'commercialisation valley of death' point in the timeline, which is partly during the early stages (which may not go as well as hoped) or during the early growth stage when expectations are unmet and promised funding is discontinued due to the company not meeting certain targets. This is still a period in which cash flow is negative (the red line) and revenues are still well behind costs. But once the cash flow graph turns up as it does in the graphic, the company 'is on its way', and at that point, but generally not before, it may (as the graphic shows) be able to obtain some debt financing, including project finance.

Financing a project becomes increasingly difficult as its capital intensity (in this case defined as the absolute amount of capital) rises and also as technology risk rises. While venture capital firms may be able to finance a high-risk project if the absolute capital requirement is not too large (and they take only a proportion) if the project also has a high capital requirement that is not within the business model of most such firms. The reason for this is that a VC firm, while only investing in firms where they have a high expectation of success, in practice, as they also know, many of these firms will fail. Thus, they need to make many investments (a large, diversified portfolio as we noted before) to ensure that, on average, they generate their expected rate of return for the risks they take and consequently cannot put too large an investment into any one firm.

Unfortunately, most clean energy infrastructure assets involve a high absolute amount of capital and a long period of time to recover the initial investment. We are looking here at the 'bankability' of a project. Simply put, this means structuring a project such that the type of financier being approached is satisfied by the expected return in relation to risk. In practice, this means designing an optimal risk-sharing protocol during the project development phase. This will frequently mean a need for government involvement in risk sharing. It is likely to apply to, for example, nuclear reactors, whether these are giga-scale power generation projects or small modular reactors.

2.3.1 *Case Study: British Volt and Northvolt*

While we have not covered innovation in the transport transition in detail, it is important to note here that the transition from fossil-fuel powered vehicles to battery electric is also proceeding much more slowly in Europe in particular than expected by some. Converting factories from internal combustion propulsion to electric (EV or, though less so, hybrid) has not been going well. This is also a result of a dramatic change in technology from traditional vehicles to so-called 'software-defined vehicles' (SDVs), which has been developing concurrently. These are vehicles in which software has a key role to play in managing their operations and features, rather than, as traditionally, employing hardware. SDVs are also a key part of the automotive industry's shift toward self-driving and connected cars. Additionally, batteries are a key component (and comprise up to 40% of its cost) of such a vehicle and European companies have been less successful in making such batteries at the required scale.

Most batteries come from South Korea and China. The UK's attempt at creating a start-up battery factory, called BritishVolt, ended in failure within around three years when the company filed for bankruptcy. In Continental Europe, Northvolt was a brave attempt to create an operation able to provide batteries on a 'gigawatt-hour' scale to power vehicles made in Europe. But by September 2024, it was in danger of going the same way as BritishVolt and has now filed for Chapter 11 reorganisation. Northvolt raised more capital than any other private group in Europe – more than $15 billion in equity, debt, and government support. Its problems signify **the difficulty in 'scaling up' production from initial prototypes or small-scale production to the huge volumes of battery cells required by road vehicle manufacturers**. That then leads to difficulties in obtaining scale-up financing and breaking through to mass production. Without this breakthrough, sales are low, revenues are low, costs are high, and investors are very cautious. Indeed, this stage of financing, as we noted above, is known as 'the missing-middle' – between venture capital/private equity and infrastructure finance. Meanwhile, such a company is 'burning cash', and investors become increasingly nervous. This is a fate that governments need to ensure that Gen IV AMR manufacturers do not approach.

Chapter 3

Green and Sustainable Finance

3.1 Financing the Climate Transition: ICMA Green, Social, and Sustainability-Linked Bond Principles and Sustainability Guidelines (Collectively GSS Bonds), and the Transition Finance Market

For many investors, there is a requirement, either from a regulatory perspective or from the conditions in an investor mandate, to invest in 'sustainable' financial assets. Here, we cover mainly investment in bonds, as infrastructure assets are generally financed with a high proportion of debt relative to equity. However, there is often confusion regarding terminology related to the word 'sustainable'. There are many terms in use, making it necessary to define them more precisely. The International Capital Market Association (ICMA), a Zurich-based body which originally financially sponsored the ICMA Centre, has taken a particular interest in sustainability. ICMA serves as secretariat to the Green Bond Principles (GBPs), Social Bond Principles (SBPs), Sustainability Bond Guidelines (SBGs), and Sustainability-Linked Bond Principles (SLBPs) – collectively known as 'the Principles'.[1] ICMA has also produced a

[1] https://www.icmagroup.org/sustainable-finance/ and https://www.icmagroup.org/assets/documents/Regulatory/Green-Bonds/Sustainable-Finance-High-Level-Definitions-May-2020-051020.pdf.

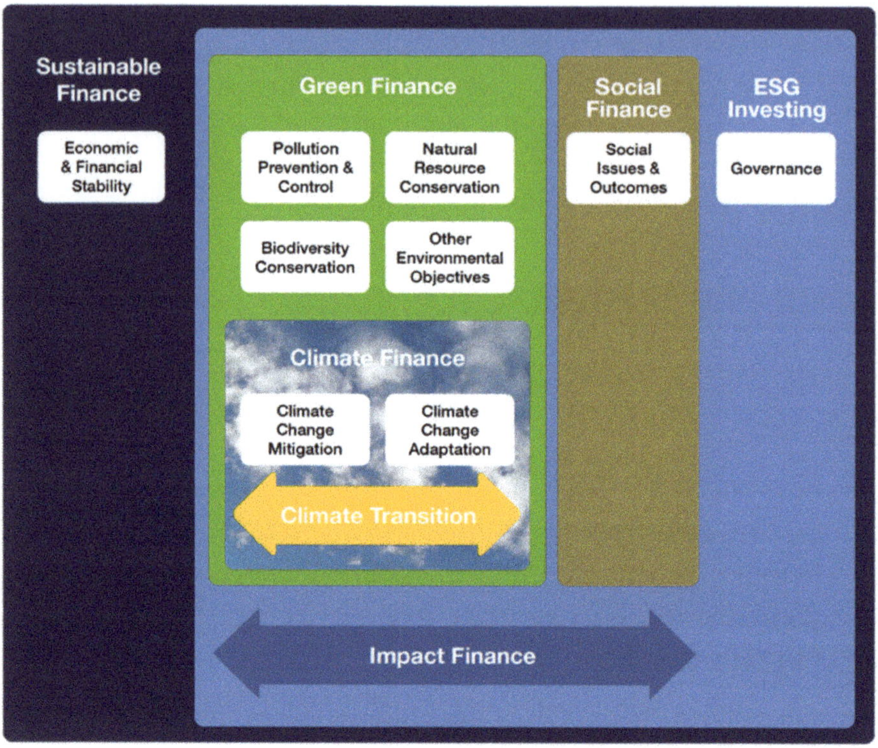

Figure 3.1. Wide and narrow definitions within sustainable finance.

document entitled 'Sustainable Finance – High-Level Definitions',[2] which provides a good graphical presentation (Figure 3.1) of how various terms – sustainable, green, social, and ESG – relate to each other.

The graphic illustrates the interaction among key high-level definitions and, especially, how wider definitions incorporate narrower ones. For example, 'Sustainable Finance' (which, if achieved, should lead to greater economic and financial stability) is the widest definition, incorporating all the others – 'Green Finance', 'Social Finance', 'Impact Finance', 'ESG Investing', and 'Climate Finance'. The text in white boxes in the figure aim to identify key differentiating themes between the definitions,

[2] https://www.icmagroup.org/assets/documents/Regulatory/Green-Bonds/Sustainable-Finance-High-Level-Definitions-May-2020-051020.pdf.

e.g. social issues and outcomes in relation to social finance. The double-arrowed boxes identify definitions that interact dynamically with themes or other definitions, e.g. 'Impact Finance', in relation to green finance and social finance.

Achieving a net-zero carbon economy by 2050 requires transforming business models, as well as wider economic and social transformations, by companies, banks, investors, insurers, and households. In this context, sustainability reporting has a key role to play, meaning reporting by companies on the extent to which they have set and achieved sustainability objectives and targets. While not all sustainability issues such as rising temperatures and water droughts and shortages may be deemed immediately existential by some companies, they can nonetheless have severe impacts on companies and their stock market value.

Sustainable Finance

Sustainable finance incorporates climate, green, and social finance while also considering broader, long-term economic sustainability of the organisations that are being funded, as well as the role and stability of the overall financial system in which they operate.

Climate Finance

Climate finance is the financing of changes in the real asset structure of society required to transition the world from carbon-based energy to carbon-free energy and to finance adaptation of the world (households, corporates, cities, and countries) through, for example, protective sea walls (such as those required in Venice), air conditioning to cope with increased temperatures, and modifications to road surfaces to minimise tar melting. Climate finance is about rewiring the financial system, which we consider in the following chapter.

Climate finance seeks to support mitigation, adaptation, and other actions that will address climate change. While climate change is clearly the existential issue right on the immediate horizon (the 'Tragedy of the Horizon' that Mark Carney has talked about) and is the focus of this book, there are many other environmental issues that may impact the sustainability of the planet and, therefore, individual companies and individuals, and even countries.

Air pollution in the home (from natural gas or wood cooking), in city air (particularly in cities with vehicle smog problems), water pollution in rivers (from the dumping of dangerous/carcinogenic materials), and plastic in the ocean (from plastic being dumped into rivers and flowing into the oceans) are all environmental issues. Today, however, maintaining biodiversity of plants and animals, keeping the air fresh, and ensuring the quality of water, soils, and minerals – which, in aggregate, means preserving and enhancing natural capital – are aspects of sustainable finance, an all-encompassing term.

Green Finance

Green finance has been defined as any financial activity that has been designed to ensure a better environmental outcome. Thus, in addition to climate finance, we would be concerned with financing that helps maintain the biodiversity of plants and animals, keeping the air fresh and ensuring the quality of water, soils, and minerals, which, in aggregate, means preserving and enhancing natural capital.

The OECD definition of 'Green Finance' is finance for 'achieving economic growth while reducing pollution and greenhouse gas emissions, minimising waste and improving efficiency in the use of natural resources'.

People's Bank of China's (PBoC) definition of 'Green Finance Policy' refers to a series of policy and institutional arrangements to attract private capital into green industries, such as environmental protection, energy conservation, and clean energy, through financial services—including lending, private equity funds, bonds, shares, and insurance.

Impact Finance

Impact finance refers to investments in corporates or funds and loans to corporates, made with the intention of generating a beneficial social or environmental impact as well as a financial return.

The Transition Finance Market

This can be described as financing that supports an organisation to credibly decarbonise its activities, respond to any climate-related risks and opportunities, and contribute to a global and economy-wide net-zero transition. This includes financing of hard-to-abate and high-emitting areas of the global economy.

Net Zero

'Net zero' refers to the balance between the amount of greenhouse gas produced and the amount removed from the atmosphere. We reach net zero when the amount we add is no more than the amount taken away. How to achieve this is the focus of this guide.

Net zero in financial asset portfolios is highly contentious in practice, even though the general definition above would probably be widely accepted for corporates. One area of contention centres around the concept of 'offsetting' by, for example, buying carbon credits or planting tree saplings. Also, what we need is a maximum temperature rise of 1.5°C, with net zero being only one possible mechanism for achieving it.

The first observation we have to make regarding green finance is that any type of financing can be considered green. However, particularly for investors with a 'green mandate', a means of knowing that a particular security is 'green' is vital. For this reason, the Principles Community described as follows set out to develop Principles and Guidelines for 'Green Bonds'. This was several years prior to the EU's Taxonomy and Green Bond Standard (EU GBS) being launched.

The second observation is that in the future, all financing – whether bond, bank, or equity – *should* become 'green', i.e. there should be no financing of fossil fuel assets and the so-called financed emissions should cease. If this is not achieved, then it is unlikely that the Paris Agreement's temperature objective will be achieved.

Most 'green' financing, i.e. financing which follows principles laid out by the 'Principles Community', is bond financing and is large in size. Although there are also green loans, they would not generally be for financing the development of new technologies but for project or corporate finance. Therefore, green bonds are much more likely to be used as a financing instrument in scaling existing proven technologies.

Many investors have preferred to purchase green bonds (as defined under ICMA's GBPs or EU GBS) to help achieve one aspect of their ESG mandates and may even accept a slightly lower return on them than on other bonds, although green bonds currently comprise only a very small proportion of the global debt market. As the EU GBS becomes established, bonds meeting this standard's requirements are likely to predominate. However, ICMA has also developed principles and guidelines for other types of bonds, such as social bonds, sustainability bonds, and sustainability-linked bonds. More information on all these bond types can be found at https://www.icmagroup.org/sustainable-finance/.

There are two other private sector sustainable finance initiatives. The first, known as the ICMA Sustainable Finance Committee (SFC), is part of ICMA, though it is separate from the GBPs initiative. We consider its role first. Launched in September 2019, the ICMA SFC brings together various ICMA committees and constituencies (i.e. the Asset Management and Investors Council, the Corporate Issuer Forum, the Financial Institutions Issuer Forum, the Legal & Documentation Committee, the Primary Market Practices Committee, and the Regulatory Policy Committee), all of which have a growing interest in sustainable finance. Included in this committee is the Steering Group of the GBP SBP SLBP Executive Committee.

The SFC is designed to:

- provide a forum for discussing sustainable finance developments;
- provide a wide platform for ICMA regulatory responses and dialogue on sustainability-related topics;
- liaise with the Global Sustainable Finance Council (GSFC) to further leverage cooperation among leading financial industry associations;
- provide input into specific cross-cutting sustainability-related deliverables.

The other initiative is the GSFC, which we cover as follows:

> The GSFC was created in 2017 (originally as the Global Green Finance Council) with the objective of bringing together key global and regional associations and other stakeholders involved in green and sustainable financing. The industry joined forces to coordinate efforts to promote sustainable finance and facilitate cross-fertilisation between related markets and asset classes, with the ambition to act as a representative counterparty to the official sector on sustainable finance policy matters.
>
> In 2017, GSFC published a directory to provide policymakers and global and regional market participants with a simple, easy-to-use reference guide on international and regional governments and industry bodies that have implemented or are implementing major initiatives related to green, sustainability, and climate change.
>
> The GSFC also served as the forum where the need for guidance on green loans was identified, which led to the publication in 2018 of the Green Loan Principles (GLPs) by the LMA, together with the APLMA and LSTA, with support from ICMA.

The Principles Community

The Principles Community consists of more than 400 members and observers. Members are institutions that have issued, underwritten, placed, or invested in a green, social, sustainability, or sustainability-linked bonds. Members elect the executive committee, decide on changes to the governance of the Principles, and provide input into their updating. The ICMA operates the secretariat for the Principles and provides advice on governance and other issues.

3.2 Green Loan Principles

In the international loan market, the use of similar principles to those developed for international bond markets would clearly benefit debt markets as a whole. For this reason, the other debt market associations have produced principles in line with the ICMA Principles. These associations are: The Loan Market Association (LMA) in Europe, the Asian Loan Market Association (ALMA) and the Loan Syndication and Trading Association in the US. They have jointly produced two sets of principles: one covering green loans and one covering sustainability-linked loans.

The GLPs build on and refer to the GBPs of the ICMA, with a view to promoting consistency across financial markets. The GBPs are the internationally recognised voluntary issuance guidelines that promote transparency, disclosure, and reporting in the green bond market. The GLPs are intended for broad use by the market, providing a framework within which the flexibility of the loan product can be maintained and will be reviewed on a regular basis in light of the development and growth of the global green loan market.

Chapter 4

Financing Case Studies: Innovative Structures

4.1 Castor Gas Storage Project[1]

Table 4.1 provides a description of the Castor Gas Storage Project. The refinancing of the Castor gas storage project off the coast of eastern Spain was one of the most significant deals in European energy finance in 2013. It marked the debut of the **project bond credit enhancement product from the European Investment Bank (EIB)**, a risk mitigation product designed to make private sector investment in the energy transition more attractive.

Note that the currency of the transaction shown in Table 4.1 is euro.

We can learn a great deal from the above information. First, Watercraft Capital, which is what the SPV is called, is quite a highly leveraged company, with most of the financing being debt. Part of the enhancement of the credit rating of the bond is the result of the new EIB project bond initiative, which encourages such bond issues. There is a wide range of advisors and lawyers involved in the transaction. The following information, provided by IJ Global, is very helpful in understanding some of the issues in project financing:

> Special purpose vehicle Watercraft Capital issued €1.43 billion in bonds, and lent the proceeds on to project company Escal UGS, the concession company for the construction and operation of an underground gas

[1]Case study based on material from IJ Global website https://www.ijglobal.com/articles/127334/european-refinancing-deal-of-the-year-2013-castor-ugs#comments.

Table 4.1. A description of the castor gas storage project.

Watercraft Capital SA/Escal UGS	
STATUS Priced 25 July 2013	PROJECT BOND ENHANCEMENT EIB
SIZE 1.7 billion	MONITORING ADVISER Trifinium
DESCRIPTION 1.9 billion cubic feet gas storage facility located 15 km off the coast of Spain	SPONSOR LEGAL COUNSEL Watson Farley & Williams
SPONSORS ACS (67%); Castor UGS (33%, with Dundee Energy representing 74% of that shareholding)	BOOKRUNNER LEGAL COUNSEL Allen & Overy (Luxembourg, English and Spanish law)
EQUITY 252 million	EIB LEGAL COUNSEL Clifford Chance
DEBT 1.43 billion	TRIFINIUM LEGAL COUNSEL Ashurst
BOOKRUNNERS Bankia, BNP Paribas, La Caixa, Crédit Agricole, Natixis, Santander, SG	INDEPENDENT ENGINEER Gaffney Cline
MATURITY 2034	INSURANCE ADVISER Willis
COUPON 5.76%	MODEL, ACCOUNTING AND TAX ADVISER Deloitte

storage facility. The concession company used the proceeds of that loan to refinance the bank loans that funded Castor's construction. The original financing closed in 2010 and stretched the bank market in the aftermath of the 2008 financial crisis. Banesto Santander, Caja Madrid, Crédit Agricole, SG, WestLB, BayernLB and Natixis headed up a 12-strong group of mandated lead arrangers.

The bond issue priced at the end of July of 2013 and settled on 2 August 2013. The deal was the first fruit of the European Commission and EIBs project bond initiative. The initiative is meant to attract additional long-term lending to infrastructure from institutional investors such as insurance companies and pension funds.

Spain has a troubled economic climate and its energy sector faces severe structural weaknesses. But Castor featured minimal remaining construction, and a regulatory framework that allowed the project company to recover its costs and a return on equity from gas network users.The EIB provides a €200 million liquidity line that can be drawn in a variety of scenarios that reduce the project's ability to service its debt.

The global coordinator was Santander. The structuring banks were Bankia, Credit Agricole, Caixabank, Natixis, Santander and SG. Bookrunners were BNP, Bankia, Crédit Agricole, Caixabank, Natixis, Santander and SG.

The coupon on the amortising bonds, which mature in 2034, is 5.756%, equivalent at pricing to around 100bp over the 20-year Spanish government bond. The bonds have an average weighted life of 12 years, and a rating from Fitch of BBB+, a notch above the Spanish sovereign rating.

The order book comprised institutional investors (61%), agencies (including the EIB, with a €300 million ticket, 25%), fund managers (10.1%) and banks (3.9%). The investors' geographical split was Germany (28%), Benelux (including the EIB, 23%), Spain (18%), France (11%), the UK (10%) and Italy (10%). The EIB was pleased, in particular, at the level of response from insurance companies, which had been following the project bond initiative closely.

After the bonds closed, the sponsors detected seismic activity in the region. The Spanish government again delayed final cushion gas injection and acceptance, while additional seismic surveys took place.

The vagaries of the Mediterranean's geology, then, could test the termination provisions of Castor's concession contract sooner rather than later. But these delays should not put investors off the entire bond initiative. Since then, the EIBs bond enhancement has featured on the Greater Gabbard offshore transmission operator project, which runs under the more seismically stable North Sea. If the product can help refinance a project with Castor's technical challenges and history, it should be able to enhance a variety of other unusual credits.

In December 2017 (four years after its original analysis above), IJ Global published the following information[2]:

The Earth Moved!

WHAT WE SAID THEN: 'Construction and geological risk, particularly in the latter part when the cushion gas is being inserted, is difficult to quantify'.

---

[2] IJ Global *The Earth Moved: Castor Gas Storage* 03 October 2018 https://www.ijglobal.com/articles/135619/the-earth-moved-castor-gas-storage?home=1.

Possibly the most dramatic instance over the last 20 years of event risk, the Castor underground gas storage project in Spain failed because it unexpectedly caused earthquakes. But while the Castor facility was closed down, bond holders and sponsors were not left out of pocket, making it a fine example of good structuring.

The original financing for Castor closed in July 2010. It funded construction of the underground gas storage facility which included both onshore and offshore elements. Wells were located 21 km off the east coast of Spain providing around 1.9 billion cubic metres of storage capacity. The gas was compressed 15km inland from the coastal town of Vinaros for injection into the gas reservoir, ready to be pumped back into the Enagas-operated pipeline network.

The project carried no market or volume risk, but as we said at the time, pretty high geographical risks which were not compensated by high returns as in oil exploration development. To provide some protection, the sponsors agreed a five-year window in which they could walk away (on reasonable grounds) and receive compensation.

Construction began in 2008 and was completed in 2012, although when the debt was refinanced in July 2013, the project was still waiting for Spanish government approval after running into trouble. The lingering impacts of the financial crisis led the government to increase the debt tenors on the project from 10 to 20 years, in order to reduce the burden of charges on Spanish gas and power users.

Then in January 2013, the Spanish authorities approved a ministerial order allowing the project to use gas from a third party. Enagas, the operator of Spain's gas grid, and Castor's largest customer, had already agreed to buy half of ACS' stake at completion. It now took over from the project the role of buying cushion gas.

The bond refinancing represented the first ever use of the European Investment Bank's project bond credit enhancement (PBCE) product. This attracted criticism due to the lack of demand risk or much completion risk on the transaction. But the EIB seem very keen to test its new toy, which saw it take a €200 million unfunded commitment that starts out as equivalent to 14% of senior debt, eventually decreasing as the bonds amortise, and staying equivalent to 20% of outstanding principal.

With the PCBE, Castor achieved a rating from Fitch one notch above Spain's sovereign rating a raised a total of €1.43 billion. Then Castor was shut down in 2014 after the detection of more than 200 minor earthquakes the previous year. Full repayment of the bonds happened in November 2014 following a Royal Decree which terminated the Castor concession. Escal was forced to relinquish the project, but the decree ordered Enagas to arrange around €1.3 billion in bank debt from a club of lenders to repay the concessionaire.

The government got the banks to compensate Escal and in exchange for payment rights from the Spanish gas system. The banks agreed to receive payments for 30 years (around €80 million per year) from January 2016. Spain was stuck with a significant bill, but the deal participants were compensated quickly and in line with legislation. In May 2017 the Spanish government confirmed that it has no plans to reopen Castor.

4.2 Case Study of the Financing Structure for a Typical Offshore Wind Farm

> *Press announcement 18 July 2017*: British Wind Energy GmbH a wholly owned subsidiary of Northland Power Inc., has today successfully achieved Financial Close for the approximately EUR 1 (one) billion project financing of its offshore wind farm project Deutsche Bucht ("DeBu"). The non-recourse project financing has been entered into with a consortium of 10 commercial banks. It has a capacity of 252 MW and comprises 31 MHI-Vestas V164 wind turbines. Green Giraffe[3] was the sole financial advisor to DeBu, which was initially owned by Highland Group Holdings and is now 100% owned by Northland Power Inc.

Figure 4.1 from the Green Giraffe shows British Wind Energy GmbH (the SPV) at the centre of a 'nexus of contracts'. The SPV organises contracts with counterparties to achieve its objectives. It has no cash or revenue of its own but raises cash from equity investors (including Northland Power) and 10 banks.

The SPV does not construct the wind farm nor operate it (it does not have the staff or the skills to undertake either of these tasks) but writes contracts with others to undertake this work. It also writes contracts to provide risk *mitigation*, which is necessary to make a project 'bankable', i.e. to lower risk to a level at which banks feel comfortable lending. Therefore, the construction contract will include a penalty clause under

[3] Courtesy of Green Giraffe https://green-giraffe.eu/.

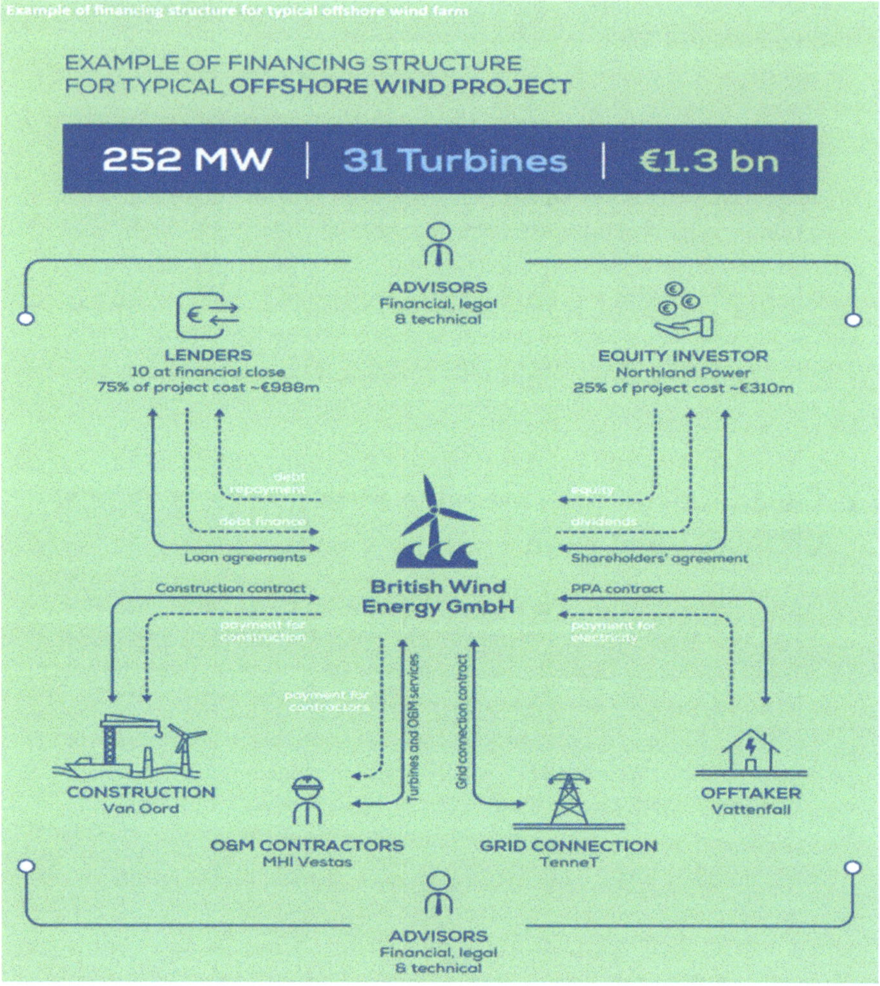

Figure 4.1. Financing structure for an offshore wind project.

which Van Oord will have to cover some of the additional costs or delays during construction. This means that the costs involved if there is an overrun or delay in completion do not all hit the P&L or balance sheet of the SPV (and hence impact lender credit risk) but instead hit the P&L and balance sheet of the construction company. Similarly, the offtaker contract with Vattenfall compels them to take the amount of power they have contracted for, even if they don't need it. If they refuse, they still have to pay for it (under a take-or-pay contract). Such risk mitigation by, in effect, paying more than otherwise to the construction company and offering power to the offtaker for less than without the take-or-pay contract means that the lenders to the project have greater protection than otherwise (effectively, the equity shareholders have paid for insurance).

July 2017 marked financial closure for this project, which commenced operations in 2020. In 2021, it was re-financed with Green Giraffe acting as the sole financial advisor on the non-recourse refinancing. The re-financing transaction closed in March 2021, with the sponsors successfully arranging an €886 million senior debt refinancing package. This will reduce the interest rate on the project's senior debt to 2.3% all-in (from approximately 2.6% all-in) and accordingly improve cash distributions from the project over the remaining 13-year senior debt term. The refinancing also included the addition of a debt service reserve facility, which released €50 million in cash, thereby enhancing Northland's corporate liquidity. The refinancing won the 2021 Inspiratia award for Top Refinancing Deal of 2021.

In terms of the timing of different types of financing from different types of investors, as shown in Figure 4.2,[4] it can be seen that the

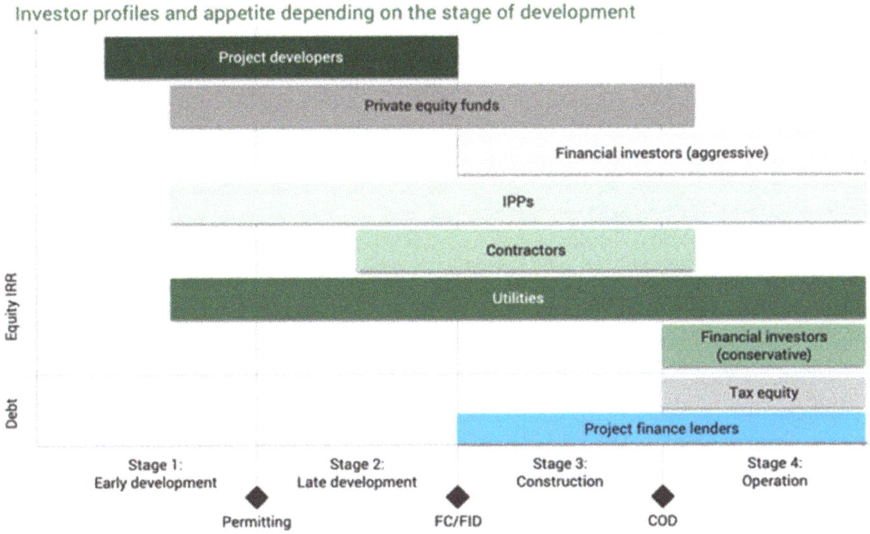

Figure 4.2. Investor types in the financing of a project at different stages up to the operational stage.

[4] https://green-giraffe.eu/.

developers are the first to put cash into the project (in the form of equity). A little later, PE funds may come in, and later still, some risk-taking investors (perhaps hedge funds). Some independent power providers (IPPs) may come in as contractors on the project, and utilities, which, once operational, may wish to buy power from the SPV. The lenders come in at the bottom of the graphic after the decision is made to go ahead (FC/FID). Finally, once commercial operations commence (COD), financially conservative investors, i.e. pension funds and insurers, may enter (FC/FID is financial close/final investment decision; COD is commercial operation date).

4.3 Ørsted: Case Study of Corporate Financing of Wind Power

We have not written much on corporate financing, as traditional corporate finance is well covered in university finance courses and many textbooks. Instead, we look at a few particular features that were utilised by Ørsted before the wind industry as a whole faced significant changes in the environment in which it operated after 2022, primarily due to COVID-19 and rising energy costs resulting from Russia's invasion of Ukraine. These significant challenges included above-general inflation rates of increase in their input costs. Not only were material costs much higher, but high interest rates also suddenly became an issue, as did supply chain bottlenecks. This has led many developers to reduce their investment targets and postpone or, in some cases, abandon large projects while taking substantial financial losses or impairments.

Ørsted has been a leader in the decarbonisation of the power industry, mainly through the development of North Sea wind farms, and has made a high level of investment in *new real (green) assets*. Ørsted was originally a state-owned oil and gas company set up in 1972 to exploit the new oil and gas fields which were being discovered in the North Sea. In 2018, it changed its name from DONG (Danish Oil and Natural Gas) to Ørsted after it sold its upstream oil and gas activities to Ineos, a private UK company whose main products are now chemicals, polymers, and oil and gas. The new name for Dong honours the Danish scientist Ørsted, who discovered electromagnetism (on which principles power generation is based). By 2018, they had shifted the company to their 85/15 vision, i.e. 85% of their power production was green compared to only 15% in 2009. Ørsted serves as a holding company for almost 300 subsidiaries (effectively projects), with all major business activities conducted through those subsidiaries, which is a common practice in most corporates but with far fewer such subsidiaries.

Corporates frequently use project finance, where the assets of a project are initially held on the balance sheet of a separate company, an SPV, which is also the issuer of equity, bonds, and the counterparty in bank lending. In Ørsted's case, borrowing was mostly on the balance sheet of the parent company (on-balance sheet corporate borrowing), and thus there *was* recourse to the parent. It is unusual because it finances each project through its corporate financing group, but each project is structured within a separate subsidiary, allowing it to sell shares in a project to its partners to reduce its total capital commitment and risk.

> Its financing strategy[5] is to concentrate group borrowings at the group parent level to simplify its liability profile while avoiding any structural subordination issues.

Thus, for Ørsted, its financial structure protected those who lent at the group parent level and thus reduced their risk, and hence the credit spread on loans as a result of this technique raised credit ratings compared with a structure with structural subordination. This is because if all financing for a subsidiary comes from its parent, then lenders to the parent will know they do not need to concern themselves with debt raised by other lenders within the group through its subsidiaries, which, given that there are 300 subsidiaries, would involve a great deal of work by a lender to the group.

They suggested at one point that by avoiding the use of off-balance sheet financing (project finance structures), they could reduce their cost of debt by 100 basis points (1%). Financing comes from retained earnings, a range of debt issue types, joint ventures, and so-called farm-downs. This last expression means selling equity stakes in a project pre-construction to outside investors to free up their own equity capital for further projects. This is more commonly known as asset rotation or build, sell, operate, which enables Ørsted to take out profits from developing and operating facilities without requiring to have 100% ownership in them. Under their

---

[5] https://orsted.com/en/investors/debt/funding-strategy.

partnership model for wind projects, they average 50% ownership of future *offshore* wind farms with Ørsted. On the other hand, they intend to sell down more *onshore* projects in order to invest more in offshore and hydrogen manufacturing.

Their approach of not financing each project separately, as they would have to if they used a project finance approach, also means that the parent company of the subsidiaries holds a diversified (real asset) portfolio, which enables risk reduction for lenders without some of the costs of normal project finance risk mitigation measures. This, in turn, results in both a lower cost of equity and lower project costs. They also chose to reduce risk further compared with project financing structures by avoiding the high levels of leverage seen in project finance, which can be of the order of 75%. With lower leverage, the variability of profits is much reduced. Thus, the overall weighted average cost of capital (WACC) is reduced.

They reduce risk further by having a natural hedge for much of their currency risk which means that rather than using derivatives, they achieve it by holding borrowings in the currency of denomination of the funds from operations (FFO) for each project. Thus, the currency of denomination of the asset side of the balance sheet is approximately matched with the currency of denomination of its debt (liability side of the balance sheet). This reduces the variability in total costs relative to revenues compared with unmatched financing.

Before 2022, following an assessment of their balance sheet and financing strategies by Moody's and S&P and Ørsted having chosen a BBB+/Baa1 as the optimal one for their company given its substantial debt requirements, they have then chosen not to try to raise their rating to the next highest level – single A. As their rating was heading up towards single A, they have therefore reduced their Funds from Operations to Debt Ratio (FFO/adjusted net debt) from 30% to 25%, thus enabling them to raise more debt for a given level of cash flow while maintaining their BBB+/ Baa1 rating. However, in February 2025, Ørsted was downgraded by Moody's to Baa2 (close to 'junk' status). In August 2024, it was reported that the company had abandoned an e-fuel project. But the final straw for the company's downrating may have been US President Trump's view of wind power, to which he seems to have some emotional objection and which led to a large project having to be discontinued. In consequence, the share price has suffered, having peaked in 2021 and declined since. Nonetheless, Ørsted remains a great and successful company.

4.4 A Model for Nuclear Financing: Thames Tideway Tunnel Regulated Asset Base (RAB) Financing and Risk Mitigation Through a 'Contingent Financial Support Package'

The traditional proven solution for financing capital-intensive infrastructure in many countries is the regulated asset base (RAB) model, which is utilised in electricity, water, gas, and telecommunications worldwide. The UK's Department for Business, Energy and Industrial Strategy (BEIS, but since 2023, split into two departments – the Department for Energy Security and Net Zero (DESNZ) and the Department for Science, Innovation and Technology (DSIT)) has been having an ongoing consultation on the use of the RAB model for nuclear. In a document the UK government published in 2019, there is a key paragraph which states that 'there will still be a crucial role for low carbon "firm" power i.e. always available power, in 2050'. The UK Committee on Climate Change includes 38% firm low carbon in their further ambition scenario – 'to meet net zero while maintaining security of supply and keeping costs low'. As there are only very limited technologies for the provision of firm low-carbon power apart from nuclear, this confirms that nuclear remains a key component of the UK's energy policy.

BEIS has also stated: **'Our assessment has concluded that, by providing regulated i.e. virtually guaranteed, returns to investors, a RAB model has the potential to reduce the cost of raising private finance for new nuclear projects, thereby reducing consumer bills and maximising value for money for consumers and taxpayers'. As economic regulators are required by statute to ensure that investors obtain the agreed rates of return, this financing mechanism has already been successful in enabling private sector investment in other infrastructure sectors such as the National Grid's UK investments in gas and electricity, and in the US, it is used for nuclear, coal, and gas power stations.**

There are many types of risks in new nuclear, and if these can be successfully mitigated, the overall cost of financing (including the cost of risk transfer, which is in the nature of insurance) will be lower than it would be without such mitigation. A 'megaproject' in the UK which achieved a record low financing cost was the construction of a tunnel under the Thames to transport sewage and excess surface drainage water from places where it flowed directly into the River Thames into the tunnel, which would be constructed under the Thames to take the effluent directly to a sewage works further down the river. The project construction commenced in 2014 and was successfully completed 10 years later, in late 2024. Beyond the use of RAB financing and specifically for the Thames Tideway Tunnel project, the UK government developed a new type of UK Treasury risk mitigation model for certain risks, which could serve as an archetype for an SMR/AMR project.

The so-called 'Contingent Financial Support Package', known as the GSP, was designed to cover certain risks on the Thames Tideway Tunnel (TTT) project. We cover it in detail because of its applicability in nuclear and hence to governments aiming to achieve their legally mandated greenhouse gas reduction targets, for example, through the use of nuclear power. The government's provision of the GSP to Bazalgette (the SPV) is aimed at mitigating some risks and transferring liability to the taxpayer if those risks materialise. The government concluded that private delivery of the project would not be financially viable without some form of government support, due to the project's scale, associated risks, and the implications for financing costs that customers would face. Under the terms of the GSP, which expires when the three-year post-construction project testing period is expected to be complete in February 2027, the department has agreed to provide financial support in certain limited circumstances:

- *Cost overruns*. The project's target price is £3.2 billion, but if Bazalgette's costs exceed this threshold, the department agrees to either provide equity to Bazalgette or to discontinue the project and pay compensation.
- *Disruption to capital markets.* A 'Market Disruption Facility' commits the department to providing short-term liquidity but on appropriate commercial terms.
- *Indemnity.* The department agrees to provide an indemnity to Bazalgette to cover for liability claims exceeding the company's insurance limits.
- *Compensation for discontinuation.* In the event that the department decides to discontinue the project, compensation will be paid.

The regulatory licence from Ofwat (the UK regulator) has, according to Ofwat, 'strong incentives in the form of clear rewards and penalties to complete the project within time, to quality and at minimum cost'. This is not a typical incentive for a project company to minimise risk to investors.

A further interesting aspect of the Thames Tideway project is that a mechanism was proposed and implemented to enable the project to pay dividends to the equity investors during construction, i.e. through the ten-year construction period, even though no revenue would be generated by the tunnel during that time. This was achieved through an agreement that Thames Water customers would pay an additional charge on their water and sewerage bills during this period. This additional charge started at £7 on annual water bills and rose to between £20 and £25 in the period up to the mid-2020s. As Thames Water is the monopoly supplier to most of its customers, it was able to impose this additional charge. In a competitive retail and wholesale power market, on the other hand, it would be more difficult. But just as interest on loans is capitalised during construction by adding it to the loan principal, so too could the equity dividend be capitalised into the total loan amount on this project or on a nuclear power project. A dividend could thus be provided to investors during the construction phase, with the cash coming from the increase in the loan principal. This whole structure – RAB, GSP, and dividend payments – resulted in the WACC for the whole project being well below expectations at only 2.497%. Given that for many such infrastructure projects, including a nuclear power plant, the cost of debt service is likely to be the largest annual cost element, a low cost of capital is critical. In addition, in 2025, interest rates are well above those that pertained at the time the financing package for TTT was set up. In a nuclear project, a structure such as that

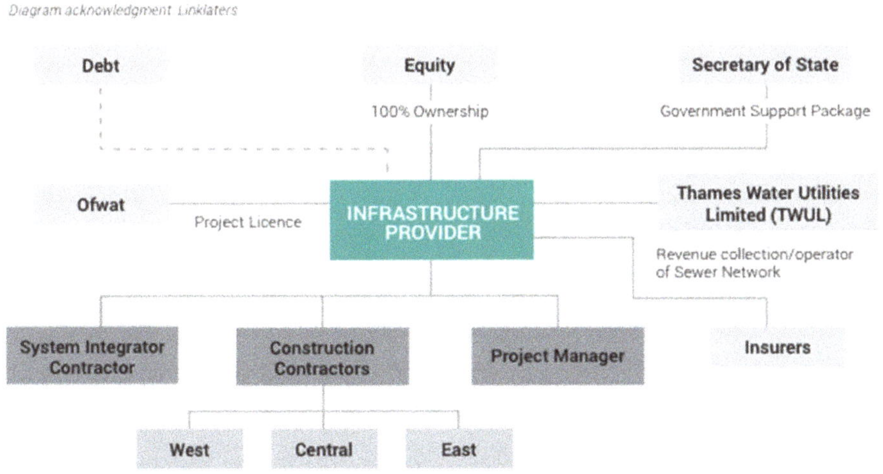

Figure 4.3. The project structure for the Thames Tideway Tunnel project.

used for TTT would almost certainly reduce the cost of capital substantially by enabling a fall in the investor's required return on equity. The package as a whole would almost certainly also reduce the cost of debt.

Given the success of the TTT financing, as shown in Figure 4.3, and the lack of proven low-carbon solutions apart from nuclear to meet the need for firm power (except for storage, which still requires investment in power generation), it is clear why the UK is seeking to improve financing options for nuclear.

Chapter 13 covers nuclear and nuclear financing in detail.

Part II

Carbon Accounting, Disclosure, and Risk Reporting Frameworks for Environmental and Social Factors

Chapter 5

Task Force on Climate-Related Financial Disclosures: Strategy, Scenarios, and Risks

For those readers who are not well acquainted with the science of climate change, we suggest that it would be very useful to read Appendix 1 on the science of climate change prior to covering the remainder of the book.

5.1 The Carbon Questions and Financial Risk: What Does Net Zero Mean? What Are Science-Based Targets? Net Zero or 1.5°C; Carbon Budgets, Externalities, Social Cost of Carbon, Carbon Taxes, and Alternative Routes to Paris

Before we can consider the legal and other frameworks/directives/regulations/accounting rules and societal trends within which corporates and financial firms must operate and the risks of not compliance, we must consider climate science and our understanding of what needs to be done to achieve the desired objectives of society in relation to climate change. We cover the science of climate change in Appendix 1. It is also difficult to ensure effective decision-making in corporates or in financial firms without understanding the carbon background, which we call the 'carbon questions', before moving on to frameworks.

In relation to the finance industry, climate change – in particular, rising temperatures and water shortages – presents risks to banks, investors, and insurers, not only due to its direct impacts on companies but also from the potential **impact of a changing climate on mass conflict and mass migration** from areas of Earth which become uninhabitable.

Efforts to mitigate rising temperatures on Earth did, in fact, begin many years ago. One of the first formal efforts was in 1994 when the United Nations Framework Convention on Climate Change (UNFCCC) entered into force. Today, it has near-universal global membership. The 197 countries that have ratified the Convention are called 'Parties to the Convention'.

At the Conference of the Parties (COP 21) in Paris, on 12 December 2015, the 197 Parties to the UNFCCC reached a landmark agreement to combat climate change and to accelerate and intensify the actions and investments needed for a sustainable low-carbon future. The so-called 'Paris Agreement' builds on the Convention and unties all nations in a common cause: to undertake ambitious efforts to combat climate change, to adapt to its effects, and to assist developing countries in their efforts. In 2025, for the second time, the US withdrew from the Paris Climate Agreement under an executive order from President Trump.

The Paris Agreement's central aim is to strengthen the global response to the threat of climate change by limiting global temperature rise. The agreement has three aims, of which the third is most relevant to those in the finance industry:

(a) Holding the increase in the global average temperature to well below 2°C above pre-industrial levels and to pursue efforts to limit the temperature increase to 1.5°C above pre-industrial levels, recognizing that this would significantly reduce the risks and impacts of climate change;

(b) Increasing the ability to adapt to the adverse impacts of climate change and foster climate resilience and low greenhouse gas emissions development, in a manner that does not threaten food production;

(c) **Making finance flows consistent with a pathway towards low greenhouse gas emissions and climate-resilient development.**

To enable the world to achieve the target, the Paris Agreement developed the concept of nationally determined contributions:

> Nationally determined contributions (NDCs) are the core of the Paris Agreement and the achievement of these long-term goals. NDCs

embody efforts by each country to reduce national emissions and adapt to the impacts of climate change.

5.1.1 *What Does 'Net Zero' Mean? Would It Achieve the '1.5°C' Target? What Is the Finance Industry's Role in Contributing to Climate Change Mitigation?*

What does the finance industry have to do to achieve its full potential contribution to climate change mitigation (based on the above targets), thereby ensuring the realisation of the third of the Paris Agreement's three objectives – *Article 2.1c* 'Making finance flows consistent with a pathway towards low greenhouse gas emissions and climate-resilient development'?

When we talk about net-zero targets of the world's governments, such targets for the real sector of the economy are not yet embedded in the world's financial systems as an objective that shapes their decision-making. Instead, **the financial sector is much more concerned with mitigating risk in financial asset portfolios as distinct from mitigating climate change risk to the Earth. They are thus more concerned with the management of climate risk to portfolios, regardless of whether or not it actually mitigates climate change risk for the world.** A key debate is whether or not the actions and decisions being taken by the finance sector will, in practice, drive the decisions that need to be taken on reducing emissions in the real economy, which requires a massive increase in 'climate finance'. Climate finance is new financing provided by financing institutions for the creation of new clean energy assets. Instead, financial institutions are still providing a high volume of funds to enable the creation of new carbon-emitting assets by corporations. Such financing is known as 'financed emissions', i.e. additional emissions enabled by the provision of financing from the issuance of new equity, bond issues, and bank loans for large global emitters.

This last paragraph is emphasising that net zero is about the **real economy** – ensuring decarbonisation and hence reducing the extent to which the Earth is warming and the risk to humankind. On the other hand, management of portfolio climate risk by financial institutions is, at present, mostly about minimising the risk of loss to asset owners on existing portfolios of financial assets from physical and transition risks. More particularly, it is about trying to avoid the risk of finding that a portfolio holds a high proportion of 'assets at risk', the so-called stranded assets.

To the extent that demand for oil and gas may peak by 2026–2027, some assets owned by oil and gas companies will cease to have little economic value.

The Paris Agreement Target is 1.5°C, which is a different objective from achieving net zero by a particular date. Net zero was simply a proposed, and agreed, means to achieve the 1.5°C target. In practice, there can be **'many a slip between net zero and 1.5°C'.** Net zero is a good target in some ways since achieving it is the only way to stop a continually rising temperature on Earth – which would eventually wipe out most life on Earth, including humans. But as the net zero target is not on track, there is no reason to believe that, by the time we do achieve it (which we are now unlikely to achieve by 2050), we will have in fact limited the global temperature rise to 1.5°C or even to 2 or 2.5°C. Even achieving it by 2050 does not guarantee success with the temperature target for the world. Thus, net zero is not by itself an objective but simply a possible means if companies and households had kept on track with decarbonisation targets.

5.1.2 *Zero or 'Net Zero'?*

'Zero emissions' is different from 'net zero'. Net zero accepts (whether rightly or wrongly) that it is not possible to eliminate all emissions. Instead, it is about 'offsetting' or cancelling out any carbon emissions that a company or country produces. Net zero is achieved when the amount of greenhouse gas produced in a country is no more than the amount removed. Countries, cities, and companies need to have a plan as to how they intend to achieve net zero and to have set targets such as planned and actual progress each year towards their target.

5.1.3 *Offsetting*

Increasing GHG removals through *activities external to the company (or organisation) involved* to compensate for 'hard-to-eliminate' GHG emissions so that its net contribution to global emissions is reduced or eliminated is known as offsetting. Offsetting can be implemented through a market for carbon credits, tree planting, or other mechanisms.

Offsetting claims are only valid under what is supposed to be a rigorous set of conditions, including that the reductions/removals involved are

additional, not overestimated, and that they are exclusively claimed and not double-counted. Fortescue Future Industries (https://ffi.com.au/) claims to be the only major heavy industry company with a real plan to reach **'real zero'**. It aims to have invested $6.2 billion by 2030 to eliminate fossil fuel risk and reduce annual operating costs.

So, what is to be done? The answer (in brief) is that strategy in banks, issuing houses, asset owners, asset managers, and insurers (the asset side of the balance sheet) and underwriting strategy in insurers (the liability side of the balance sheet) must be aligned with a decarbonisation pathway consistent with a maximum 1.5°C of warming. But what does this statement mean? Such alignment involves several assumptions, including the choice of decarbonisation pathway that each financial firm chooses and the particular way in which it sets targets and achieves the required targets each year. A financial firm could choose a lesser reduction before 2030 and in the decade to 2040 and hope to catch up by 2050. But clearly, such a firm's portfolio would have resulted in greater emissions and temperature rise on Earth by 2030, 2040, and 2050 than one in which a massive effort was made by the asset owner to ensure that its portfolio companies took powerful steps to cut carbon emissions immediately.

5.1.4 *Carbon Budgets, Net Zero, and 'Pathways'*

In December 2020, the UK's Climate Change Committee (CCC) published its 1,000-page study entitled 'Sixth Carbon Budget – the UK's Path to Net Zero', which is a report required under the UK's Climate Change Act. Among other things, this study provides government ministers with advice on the volume of greenhouse gases that the UK can emit during the period 2033–2037.

The study suggests that a 100% cut in GHG emissions would cost 0.5% of 2050s GDP. The same committee in 2008 estimated that to achieve an 80% cut in GHG emissions would cost 1–2% of GDP. The cost of action has thus reduced very substantially (in part due to falling costs of both solar and wind power). The cost of inaction is likely to have risen substantially.

To stabilise global temperature at any level versus the pre-industrial level, there is a finite number of gigatons of emissions that can be released before we achieve net zero. The 'e' after CO_2 means CO^{2+} the impact of other greenhouse gases converted to their *equivalent (e)* damaging effect if they were in fact CO_2. This is referred to as a carbon budget.

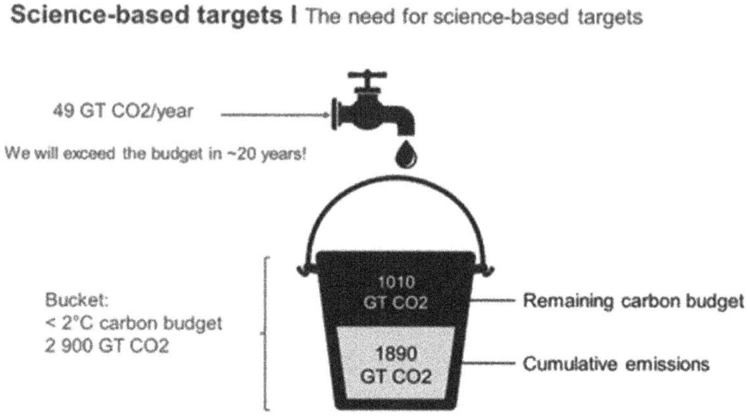

Figure 5.1. The relationship between annual emissions and carbon concentration in the atmosphere.

Bill Gates, when he was interviewed on BBC radio, put it very simply: We have to reduce global GHG emissions today from 53 Gt per year to zero by 2050 if we were to have any chance of achieving a temperature rise of less than 2°C. We also needed to have 2030 targets and 2040 targets rather than having a *transition pathway* that involves little action until 2040. With current concentration of 427 ppm (0.0427%) in 2024, we will exceed the IPCC's estimated 2°C warming threshold of 450 ppm in the next 10 to 20 years though very possibly more quickly.

Figure 5.1[1] from 'Science-Based Targets' shows that to limit global warming, we need to limit the concentration of carbon and other GHGs in the atmosphere to achieve the temperature objective. The relationship between emissions and carbon concentrations is a well-established one, so it is not difficult to work out the remaining 'budget'. To avoid breaching the 2°C target, the maximum amount of CO_2 in the atmosphere (the stock or cumulative emissions) must not exceed 2,900 gigatons (1,890 + 1,010).

Since, as shown in the graphic, there is already 1,890 GT in the atmosphere, the remaining budget for 2°C is 1,010 GT. Since the emission level globally was at 49 GT when this graphic was prepared (or 53 GT by the time Bill Gates wrote his book), this means that if we continue at this rate of emission, in 20 years' time (in practice, probably within 10 years),

---

[1]Corporate Citizenship (SBT, CDP, Diageo) https://vimeo.com/218954934.

we will breach the 2°C target and temperatures will continue to rise beyond 2°C. A 1.5°C target gives a carbon budget of only around 350 GT. At the current rate of emission (and emissions have grown each year since the Paris Agreement), this gives the world less than seven years before breaching the 1.5°C target. Today (2024), we may indeed already be past the 1.5°C point. If there is any budget left, then the questions that to be answered are 'over what time should it be spent?', 'who should be entitled to spend it?', and 'how can countries agree on such budgets?'. This was the subject of political bargaining such as at the COP 26 Glasgow in November 2021 (with little agreement). With the US federal government seeming to have abandoned any attempt to limit emissions, and indeed intent on increasing them, the situation is now worse than it was in 2024. Note that statistics for global carbon emissions vary depending on what is included and whether it is a CO2e (equivalent) amount, which includes other GHG gases as well.

5.1.5 *Externalities*

An externality is a consequence of an industrial or commercial activity which affects other parties (more commonly in a negative way but could also be in a positive way) without this being reflected in market prices. It is a cost, or occasionally a benefit, that affects someone who did not choose to incur that cost or benefit. In the case of climate change, it is the use of the atmosphere as a 'sink' in which to dump carbon dioxide, methane and other GHGs from vehicles, power stations, home heating, agriculture, etc., which imposes costs on everyone in the world which they did not choose to incur. In the case of river pollution, it is the impact on people downriver from the polluter who do not ask for this pollution to be dumped (obviously not), and the externality is that the company is not paying for the cost of the resulting damage caused by its pollution. This externality can be internalised (that is, put into the company's own costs) either through a change in the law to make dumping of waste materials in a river or dumping CO_2 into the atmosphere illegal or through a tax which makes it economic to find alternative means of disposal of toxic materials such as dyestuffs used for dyeing fabrics to supply to 'fast fashion' companies or, in the case of CO_2 emissions, either converting to zero-carbon sources of energy or payment to have emissions 'sequestered' underground, for example, in old coal mines or in empty gas fields in the North Sea or elsewhere.

In the case of CO_2 emissions, internalising the externality would generally involve taxing such emissions, thereby transferring a cost currently imposed on all individuals across the world. This then raises the question of what the cost to society of such emissions is – the social cost of carbon? Most models are very conservative in their assumptions and neglect many likely costs that will be incurred in the future. But without some agreement on what these costs are, it is difficult to develop government and regulatory policy in this area. Effective climate policy must balance the cost of taking action (such as regulation and taxes) against the economic damage from inaction. Thus, we need to *balance costs against benefits*.

5.1.6 *The Need to Scale Up Climate Finance*

A key question that must be asked at this point is how, in practice, the world can achieve the 'real economy' transition that is required and **the maximum contribution from the financial sector in helping to achieve the scale up of efforts to avert the worse impacts of climate change**. Put another way, the question is how to eliminate fossil fuel assets and replace them with 'clean' assets at the lowest possible cost and at the maximum possible speed. The real asset sectors – electricity supply, transport, agriculture, and manufacturing (in particular the energy-intensive industries such as cement, steel, aluminium, and paper) – need to increase the speed of their transition to minimum CO_2 emissions. This is proving difficult technologically.

The chairman and CEO of the world's largest asset manager, Laurence Fink's has, in the past, referred to *long-term prospects* and *a reallocation of capital*. If investors and banks agree with Fink that the financial world is on the edge of a revolution, then they need to show a determination to reallocate the inflows from investors from funds which finance new fossil assets to ones which are financing the necessary carbon-free assets within the needed timespan. However, Fink has greatly reduced his emphasis on the need to reduce carbon emissions rapidly in line with the new 'thinking' at the White House.

Institutional equity investors will need to assess the carbon mitigation plans of companies in which they might invest, and banks will need to move away from refinancing corporate bonds and loans of corporates (and sovereigns) with high exposure to (and hence risk from) climate change. Importantly, financings of all types will increasingly move towards companies taking advantage of climate-related *opportunities* – such as the new and existing technologies which help mitigate climate change. As we noted above, insurers will do the same on the asset side of the balance sheet where they hold the assets that hedge the risks on the liability side (insurance policies). On the liability side, they will avoid covering risks, where it becomes no longer possible to evaluate the climate risk they would be talking on, and hence it cannot be fairly priced. An example of this can be seen in an announcement made on 1 March 2021, by Amanda Blanc, CEO of the large insurer, Aviva, when she said that the firm would no longer underwrite insurance for firms which have more than 5% of their revenue from coal, tar sands, shale gas, deepwater oil, and oil shale – unless they have signed up to SBTi (see Section 6.4).

5.2 Task Force on Climate-Related Financial Disclosure: Disclosure of Strategies and Scenarios[2]

Task Force on Climate-Related Financial Disclosure (TCFD), which is concerned with the disclosure of relevant science-based climate-related information, was created by the Financial Stability Board (FSB). This is an international body that monitors and makes recommendations about the global financial system and, through its efforts, a route to achieving relevant disclosure by corporates and financial firms has been created. The FSB was established in 2009. The Board includes members from all G20 economies and the European Commission. Hosted and funded by the Bank for International Settlements, the board is based in Basel, Switzerland, at the premises of the Bank for International Settlements (BIS). The creation of the FSB itself represented the G20 leaders' first major international institutional innovation. Former US Treasury Secretary Tim Geithner described it as 'in effect, a fourth pillar' of the architecture of global economic governance, alongside the International Monetary Fund, the World Bank, and the World Trade Organization.

As climate change increases climate change risk and hence the potential financial impact on companies and financial firms, **it also has the potential to create global financial instability**. In accordance with its mandate to minimise risks to financial stability, in 2015, the FSB created

[2] https://www.fsb-tcfd.org/.

the TCFD to improve and extend reporting of climate-related financial information. The reason for doing so is that **one of the essential functions of financial markets is to price risk to support informed, efficient capital-allocation decisions**. To carry out this function, financial markets need companies to provide accurate and timely disclosure of climate risks just as they do with other aspects of their performance. Without the right information, investors and others may incorrectly price financial assets, such as equities, bonds, and loans, leading to a misallocation of capital. The FSB created the TCFD to develop recommendations on the types of information and hence disclosures that companies should provide to support investors, lenders, and insurance underwriters in appropriately assessing and pricing a specific set of risks – risks related to climate change. **These disclosures are then intended to guide decision-making in financial firms.**

It was intended by TCFD that its 'Recommendations' **should drive changes in the structure of real assets and also financial asset portfolios**. It may be, however, that taking an approach which, as a result of operating principally through changes in financial asset prices, is simply too slow a mechanism to achieve the Paris (time and temperature rise) objectives. **The TCFD framework is a gradualist approach similar to that favoured by the IFRS Foundation (see later), which focuses on enterprise value.** All frameworks that rely on information disclosure leading to price changes and thus in the cost of capital, with those price changes then intended to lead to real economy changes, raise the question of whether such frameworks can on their own effect the necessary change in the real asset structure of an economy **rapidly enough** to mitigate the climate crisis. In practice, given that the transition is well behind schedule, the answer to date must be no. Nonetheless, TCFD has played a valuable role in 'kicking off' the need for disclosure. It has become the basis for other organisations involved in climate change risk and for governments and the EU. Information flows are expected to lead to appropriate shifts in capital flows. Such disclosures are expected to lead to corporates and the financial sector (banks, asset owners, asset managers, and insurers) making different decisions than they would have in the absence of a consideration of sustainability disclosures.

Changes in decision-making processes and outcomes on (real) capital allocation start with the corporate becoming more aware of the risks and opportunities in their company as the environment changes because of climate change. Companies also must consider **other environmental**

factors as well as changes in social norms and values which are relevant to wider environmental and social risks. Such changes are expected to result in changes in the real sector's investment decisions on capital allocation to physical assets (away from fossil-fuel-dependent assets and towards non-fossil-dependent assets) and thus lead to a 'reorientation of capital flows towards a more sustainable economy'. This is what the Paris Agreement meant by 'making finance flows consistent with a pathway towards low greenhouse gas (GHG) emissions and climate-resilient development'.

Although we remain doubtful that disclosure of information by itself will have the desired impact and that further legislation may be necessary to speed up the transition, nonetheless, all action will start with information and understanding. Understanding is particularly germane to the good governance of climate change risks, as the issues are generally not well understood or else treated cursorily. Understanding the information derived from data is relevant not only for the board of a company but also to all levels in the organisation. Good governance is a central aspect of TCFD. What is likely, however, is that the increased transparency and the introduction of taxonomies (classifications of different types of asset) will lead to litigation against companies that are not conforming with the new requirements. The likely outcome will be that more such cases will succeed – though probably slowly. The European Commission's new Directive on Corporate Sustainability Due Diligence (CSDD) is likely to be a game-changing directive which will also encourage increasingly more successful litigation against corporate laggards (see Appendix 2).

The first key objective of *TCFD* is to avoid a sudden and dramatic policy response by governments and regulators such as the 'Inevitable Policy Response', which could lead to financial instability. This is the almost inevitable response of governments, which will be forced to act much more aggressively to limit physical climate damage as the realities of climate change become ever more obvious to voters. The policy response will lead to a much higher level of transition risk than companies might have expected (as well as physical risk). Corporates, banks, investors, asset owners, and insurers could suffer substantial losses as assets become valueless (for example, oil and gas reserves not currently being exploited but on their balance sheets at a substantial value). Such assets would then be what are known as 'stranded assets'.

The objective of disclosure, therefore, is to reduce risks related to climate change. This is to minimise potential losses to corporates and financial firms from climate and environmental change by making the previously unrecognised risks taken on by insurers, investors, and lenders more obvious both to management of these companies and to companies in the finance sector. In the case of lenders, it should discourage them from lending to or refinancing high-GHG-producing companies, and in the case of equity investors, from increasing their holdings in such companies or participating in additional equity fund raising. It is hoped that, combined with regulatory action such as limits on CO_2 emissions from electric power generation (e.g. less than 100 g of CO_2 per kilowatt-hour) and a rising cost of insurance on fossil assets, companies (and lenders and investors) will reduce or cease to invest in new projects which are high GHG emitters and insurers will cease to insure such projects.

The second key objective of TCFD is to encourage corporates and financial companies to think about the **opportunities** presented to them by contributing to the reduction in global emissions by taking advantage of the transition to a low-carbon economy. Thus, in parallel, they hope that by requiring financial firms to follow TCFD, they will be encouraged to expand their provision of 'green/climate finance'. This is once again in line with the statement in the Paris Agreement; one of its objectives is *making finance flows consistent with a pathway towards low greenhouse gas emissions and climate-resilient development.*

The amount of work required by corporates and financial companies to produce carbon statistics and work out the implications for the profit and loss account and balance sheet of a number of scenarios is seriously high. But while corporates are, of course, already required to provide detailed information on all the material risks they face, until recently there was no requirement, or only a very limited requirement, to specifically consider the **financial impact of climate change as a risk factor**. But as a result of today's new appreciation of these risks, corporates, asset managers, asset owners, bank lenders, non-bank lenders, securities underwriters, and insurers who are all exposed to such risks need to confront the new reality.

Climate change risk (and, importantly, other environmental risks, including natural capital risks) and also social and human capital issues have become differentiating factors in any investor assessment of corporate risk and hence security pricing. The good news is that appropriate responses to climate change by corporates or the lack of such responses should lead to the potential for those investors focused on such issues to generate alpha for their clients, relative to those investors that are not focused on these risks.

5.2.1 *TCFD: Recommendations, Industry Sectors, and the Four 'Themes'*

Financial markets live off information. It is information that enables investors to assess risk and then price shares and bonds. It also enables lenders to underwrite loans appropriately, i.e. assess risk and price that risk in terms of the economic interest rate to be charged and for insurers to charge an appropriate premium for cover. While companies in most countries have been obligated to disclose all risks to which they are subject, and this should include climate-related risks, in practice this has generally been perfunctory or absent.

Financial firms themselves do not have a large environmental impact (bitcoin miners are an exception), but their portfolios of corporate equities, bonds, and loans may include companies that do have such an impact. Only if such risks and opportunities are properly disclosed (and in a consistent way) can financial companies make an assessment of the risks and potential opportunities in their portfolios. In turn, it is only through such disclosures in the published financial reports of banks that the bank's own equity and bond investors can calculate the risk premium that should be attached to a bank's equity and debt.

TCFD focusses in the first instance on those industry sectors that are most subject to climate change and hence to physical and transition risks. It categorises such industries into four groups:

Energy;
Materials & Buildings;
Transportation; and
Agriculture, Food, and Forest Products.

In addition to the corporate sectors above, TCFD covers the financial sector. A key reason for this is that companies in the four sectors above are likely to need to use the services of lenders and investment banks for financing or refinancing – either borrowing from a bank or arranging for an equity or bond issue through a bank (ECM & DCM services) and also insurance services. **Thus, banks and insurers are being asked to include climate risks in their underwriting processes and thereby act as gatekeepers protecting a nation's savers (and the nation's banks and insurers) from being subject to undisclosed risks.** Investors in equities and bonds also need to be able to evaluate the climate-change-related risk to which companies in these sectors are subject before making investment decisions. The reason these types of financial institutions are included in TCFD is that both the expected and unexpected impact of climate change on corporates in which they have a financial investment can have a large effect on the value of their portfolios. Investors and lenders need to be able to consider this risk in their investment decisions, which requires them to have the relevant information available to estimate the appropriate price for the asset relative to the perceived risk.

The intention of the TCFD Recommendations was to establish a global framework for disclosure within the *mainstream financial filings* (such as the annual report and accounts) of corporates but based on the existing standards set out by various bodies such as the Sustainable Accounting Standards Board (SASB), now incorporated into the International Sustainability Standards Board (ISSB). Essentially, central banks are trying to unify voluntary disclosure principles which are gradually becoming mandatory rules or, in the case of the EU, Directives. Thus, while its recommendations are 'voluntary' and also incorporate the principle of materiality, such 'voluntary' approaches generally become mandatory over time.

Types of Risk

The TCFD report divides climate-related risks into two categories:

- transition risks, i.e. risks related to the transition to a lower-carbon economy;
- physical impact risks of climate change.

But it is also concerned that companies (and financial institutions) must take advantage of the opportunities that mitigation and adaptation will offer. In consequence, 'Opportunities' is a key theme of TCFD since if the finance industry does not seek out such opportunities and work with the entrepreneurs, corporates, and others trying to develop the new types of green asset, then the energy transition will not happen. Figure 5.2[3] shows how risks and opportunities feed into the strategic planning and risk management of a business and hence have a financial impact on the balance sheet and profit and loss account. We explain these different risks as follows.

Transition Risks

Transitioning to a lower-carbon economy may entail extensive government policy and legal changes and will also result in technology and

[3] https://assets.bbhub.io/company/sites/60/2021/10/FINAL-2017-TCFD-Report.pdf.

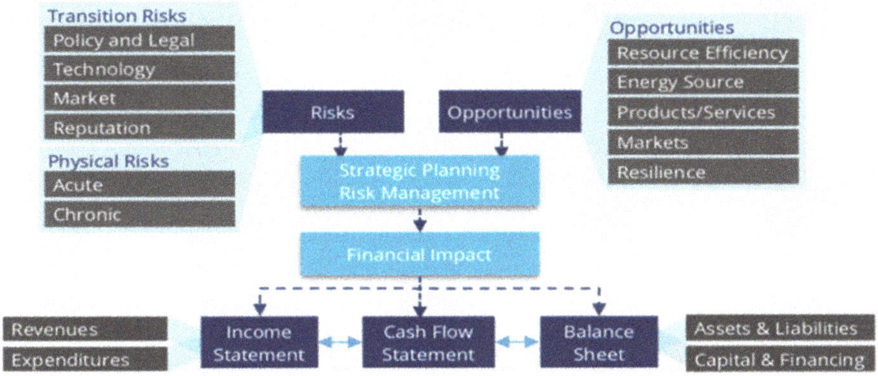

Figure 5.2. TCFD climate related risks, opportunities, and financial impact.

market changes. Policy will be designed to address mitigation and adaptation requirements related to climate change.

Policy Risks

Policy actions around climate change continue to evolve. Their objectives generally fall into two categories: policy actions that attempt to constrain actions that contribute to the adverse effects of climate change or policy actions that seek to promote adaptation to climate change. Some examples include implementing carbon-pricing mechanisms to reduce GHG emissions, shifting energy use towards lower emission sources, adopting energy-efficiency solutions, encouraging greater water efficiency measures, and promoting more sustainable land-use practices. The risk associated with, and the financial impact of, policy changes depends on the nature and timing of the policy change.

Legal (Litigation) Risks

Another important risk is litigation or legal risk. Recent years have seen an increase in climate-related litigation claims being brought before the courts by property owners, municipalities, states, insurers, shareholders, and public interest organisations. Reasons for such litigation include the failure of organisations to mitigate the impacts of climate change, failure to adapt to climate change, and the insufficiency of disclosure around material financial risks. As the value of loss and damage arising from climate change grows, litigation risk is also likely to increase. There are currently some hundreds of potential cases being considered by lawyers and courts.

Market Risk

While the ways in which markets could be affected by climate change are varied and complex, one of the major ways is through shifts in supply and demand for certain commodities, products, and services as climate-related risks and opportunities are increasingly taken into account.

Technology Risk

Some new technological changes will result in some companies thriving while others disappear. For example, many brands of automobile which are unable to match the best new electric vehicles will not survive.

Reputation Risk

Climate change has been identified as a potential source of reputational risk tied to changing customer or community perceptions of an organisation's contribution to or detraction from the transition to a lower-carbon economy.

Physical Risks

Physical risks resulting from climate change can be event-driven (acute) or longer-term shifts (chronic) in climate patterns. Physical risks may have financial implications for organisations, such as direct damage to assets and indirect impacts from supply chain disruption. Organisations' financial performance may also be affected by changes in water availability, food security, and extreme temperature changes affecting their premises, operations, supply chain, transport needs, and employee health and safety.

Acute Risk

Acute physical risks refer to those that are event-driven, including increased severity of extreme weather events, such as cyclones, hurricanes, droughts, or floods.

Chronic Risk

Chronic physical risks refer to longer-term shifts in climate patterns (e.g. sustained higher temperatures) that may cause sea level rise or chronic heat waves.

Opportunities

Opportunities arise during the transition such as the conversion of power plants from coal, oil, and natural gas to the low-carbon energy produced by nuclear. ANT-HTMRs have a very low carbon impact as well as overcoming the 'intermittency' problem of wind and solar and are well able to **'re-power'** existing coal, oil, and gas power stations using the same steam turbines and electricity transmission capacity. Opportunities also arise from hydrogen production, solar, wind, nuclear, storage, and other technologies, as well as the need to reduce the energy intensity of space heating in buildings through insulation, ground source heat pumps, and other means.

5.2.2 *TCFD's Four Themes*

Table 5.1[4] shows the four themes and related disclosures which TCFD has developed. Ultimately, the aim is greater transparency in reporting by the financial industry and non-financial groups to improve the market pricing of securities and loans by reflecting the risks that climate change presents:

> The **governance** theme asks companies to ensure that their governance structure from board level down, enables them to assess climate change risks. The effect may be negative (e.g. reduction in the value of assets) or positive (creating opportunities).
>
> The **strategy** theme asks companies to consider how material climate change is to them and its impact on the business, its strategy, and its financial planning. This should be conducted through forward-looking scenario analysis. We cover 'scenarios' in a following section.
>
> The **risk management** theme stresses the importance of not considering climate risk as something separate from other risks to which a company is subject but to integrate it into their existing risk framework.
>
> The **metrics and targets** theme importantly says that financial reporting should include climate-related metrics and that targets should be set in line with the risks and opportunities identified through the themes of the recommendations.

5.2.3 *TCFD Strategy: Strategy and Scenarios in Corporates and Financial Firms*

We focus on the middle two sets of recommendations above – the TCFD Strategy and Risk Management, rather than the other two themes, as these two are the key ones needed as a foundation.

Before it is possible for a corporate or financial firm to follow the recommendations of TCFD or any statutory framework, i.e. the recommendations in the graphic above, it is necessary to understand what are the so-called 'scenarios' which may come to pass and which will affect different firms in different ways. While normally in finance we use probability distributions when thinking about the future, given the

---

[4] https://assets.bbhub.io/company/sites/60/2021/10/FINAL-2017-TCFD-Report.pdf.

Table 5.1. TCFD recommendations and supporting recommended disclosures.

Governance	Strategy	Risk Management	Metrics and Targets
Disclose the organisation's governance around climate-related risks and opportunities.	Disclose the actual and potential impacts of climate-related risks and opportunities on the organisation's businesses, strategy, and financial planning where such information is material.	Disclose how the organisation identifies, assesses, and manages climate-related risks.	Disclose the metrics and targets used to assess and manage relevant climate-related risks and opportunities where such information is material.
a) Describe the board's oversight of climate-related risks and opportunities.	a) Describe the climate-related risks and opportunities the organisation has identified over the short, medium, and long term.	a) Describe the organisation's processes for identifying and assessing climate-related risks.	a) Disclose the metrics used by the organisation to assess climate-related risks and opportunities in line with its strategy and risk management process.
b) Describe management's role in assessing and managing climate-related risks and opportunities.	b) Describe the impact of climate-related risks and opportunities on the organisation's businesses, strategy, and financial planning.	b) Describe the organisation's processes for managing climate-related risks.	b) Disclose Scope 1, Scope 2, and, if appropriate, Scope 3 greenhouse gas (GHG) emissions, and the related risks.
	c) Describe the resilience of the organisation's strategy, taking into consideration different climate-related scenarios, including a 2°C or lower scenario.	c) Describe how processes for identifying, assessing, and managing climate-related risks are integrated into the organisation's overall risk management.	c) Describe the targets used by the organisation to manage climate-related risks and opportunities and performance against targets.

uncertainties of how GHG emissions will impact different industries and firms, it is more appropriate for each firm to analyse its operations based on what might happen under different circumstances (scenario analysis).

So, rather than thinking in terms of probability distributions, we should be thinking about what are the different possible outcomes that a company faces which then enables its management to consider how they would **respond to different temperature outcomes over future years.** This leads on to developing strategies based on a variety of scenarios.

5.2.4 *What Is Strategy and What Are Scenarios?*

A strategy is a plan to achieve a long-term goal under conditions of uncertainty. It can be applied by corporates (corporate strategy) or by governments planning strategy for climate change mitigation or for future pandemics.

In the case of climate change strategy, we are not talking about strategic planning as conventionally understood in business but rather about how to understand the new uncertainties and, in particular, the changing climate environment that will shape business and society over the next 80 years. This could be likened to making decisions today in view of what might happen beyond the next quarter rather than, as many managements do, focusing on the next three months' revenue and profit figures at the possible expense of the longer term.

The core of TCFD is strategy based on scenario analysis in which multiple plausible future scenarios 'build contrasting stories around key uncertainties'. Following on from this, the issue is how resilient the chosen strategy will be for a particular corporate or financial firm under different climate temperature scenarios.

For investors, TCFD suggests that scenario analysis may be applied in different ways, depending on the nature of the asset(s) being considered:

- Some investors may develop energy transition pathways that they believe to be either optimal and/or likely and use those pathways to measure individual potential investments and drive engagement activities.
- Other investors may consider how climate-related scenarios relate to the future performance of particular sectors, regions, or asset classes.
- The results may show that some portions of a portfolio are set to benefit from a particular scenario, while others face a loss in value.
- Such results, while not conclusive, can be a useful additional factor in determining where to prioritise risk management activities and where to consider making additional allocations.

Financial companies are recommended by TCFD to disclose their strategy in relation to the impact of climate change on their business (normally their financial asset portfolio), in line with the following:

(a) the company has to identify the possible impacts over short-, medium-, and long-term time periods;
(b) it has to describe how the risks and opportunities identified will impact their strategy;
(c) finally, it has to disclose the resilience of its *strategy under a range of different scenarios*, in this case different °C of global warming, including one where future temperature increases by less than 2°C through the period up to either 2050 or 2100.

Disclosure (c) above includes a 'less than 2°C scenario' because that would imply a very substantial **transition risk** for many companies as a result of the 'inevitable policy response'. This is a regulation which forces the pace of climate change mitigation efforts much more strongly than before and would result in a substantial repricing of many securities. Physical risks would be correspondingly reduced if such a response were to materialise.

In summary, TCFD is concerned with climate-change risks and opportunities, strategic planning, risk management, and the potential financial impact of these so-called 'non-financial risks' ('sustainability risks') on the financial aspects of a company, i.e. revenues, expenditures, cash flows, assets and liabilities, and capital and financing. The key to this whole exercise is trying to identify uncertainties and establish the types of risk to which an entity is subject *under a range of scenarios* and then manage these risks. Equally important is a mechanism for taking advantage of opportunities. TCFD aims to minimise the risk of 'the tragedy of the horizon' (Mark Carney's expression), i.e. the use of too short a time horizon in evaluating climate change risk.

Taking such a forward-looking view (based on scenarios) is very different from simply looking at today's 'carbon footprint'. The forward-looking view is critical in assessing climate change risk, as it is *a unique category of risk* where actions taken today have financial implications not just a decade away but two or even three decades away and has many other features that make it different from customary risks. Thus, a forward-looking approach, beyond the normal horizon of corporate boards, investors, and bankers, is necessary.

A scenario is not a 'best guess', i.e. not a forecast of the future, but a tool to encourage critical thinking about the range of possibilities of some variable in the future, in this case, the Earth's temperature. What we need when we are in a world of uncertainty are the possible futures and their impact compared with Business As Usual (BAU). It also notes that the purpose of scenario analysis is to consider, and better understand, how a business might perform under different future states (i.e. its resiliency/robustness). Climate-related scenarios allow an organisation to develop an understanding of how the physical and transition risks and opportunities of climate change might plausibly impact the business over time.

What we are saying, in summary, is that traditional methods of dealing with uncertainty based on probability distributions around a best guess of the future are unlikely to be appropriate.

5.3 TCFD – Risk Management: Black & Green Swans, Radical Uncertainty, New Types of Risk Caused by a Warming Earth

While the shift from human and animal labour to fossil fuels and electric power has brought the world to its present atmospheric GHG concentration level in August 2024 of approximately 427 parts per million (PPM), that level cannot be reduced quickly and indeed will continue to increase for many years and perhaps decades. Even if all emissions stopped today, the concentration level in the atmosphere would take many decades to decline. Thus, the climate would continue to deteriorate over that time. Climate warming changes the frequencies of hot and record-hot weather, the change in the shape of the bell curve, resulting in an increase in the mean and the variance of temperature on Earth (with an increase in the mean and variance along the bottom axis and the probability of occurrence on the vertical axis).

This graphic implies much more hot weather and also much more record-hot weather. Unfortunately, this climate effect (induced by humans) cannot be reversed in the way that damage to the financial system as a result of large credit losses can be overcome over time. The likelihood of a future repeat of the financial crisis of 2007–2008 can also be reduced by more effective regulation, improved supervision, and enhanced equity capital levels.

Climate change risk is different and requires a shift in thinking about this type of risk which at one time was thought of as being similar to credit cycle risk but is not. **Climate change risk actually increases year by year and does not cycle back.**

Climate change risk is different from traditional risks, such as credit or market price risk, which vary through the business cycle (cycle up and down) and **which can be estimated by reference to past lending and investing experience. In the case of climate change risk, historical data are of little relevance**. This is because what we are concerned with is **secular** (ongoing), rather than **cyclical**, changes in the state of the world, which are unprecedented historically. Instead of using the assumption that 'the future will be like the past' or reversion to the mean as most finance risk models assume, we must instead rely on the work of climate scientists and the outputs of their climate models to tell us the most likely climatic impact of secular changes due to greenhouse gas emissions. In this regard, climate change impact is very similar to the COVID-19 pandemic impact in that it is only scientists who can advise on likely impacts under **different scenarios**.

Climate models warn us that warming trends are not only a secular (rather than a cyclical) phenomenon, but that the impacts are also **potentially nonlinear**. They warn that as warming has been seen to accelerate and vary considerably over most of the twentieth century, then abrupt, and possibly irreversible, change in the future is a key risk. We thus have to be not only concerned with predicting the future of climate change based on what has happened to date on this front but also even more concerned about what happens if we have what, in finance, is known as a 'Minsky Moment'.

This is a reference used by the former chairman of the Financial Stability Forum, Mark Carney, when talking about climate risk in relation to the financial sector. A Minsky Moment refers to Dr Hyman Minsky's financial instability hypothesis. This recognises that the nature of the capitalist financial system means that it is always prone to financial crises, such as in 2007–2008 when the financial economy shifted abruptly from a seemingly benign stable system to an unstable one. He was correct in respect of the Global Financial Crisis (GFC), though few believed his hypothesis until financial instability hit the world unexpectedly.

Unfortunately, climate change risks include a Minsky Moment risk, i.e. the risk that we move abruptly from a stable to an unstable system because of, for example, the melting of the polar ice cap or the melting of the Russian tundra, which would result in huge volumes of previously stored GHGs being released into the atmosphere. Such risks require that investors and lenders make decisions based on a range of possible futures (scenario planning). This can only be achieved by **basing decisions on scientific evidence (using climate models) rather than, in the case of credit risk, the use of expected loss models based on historic loss data.** This raises the need to introduce scenario planning for possible future climate change scenarios (such as a range of 1.5–5°C rise in global temperature). Doing this in practice is a difficult transition for many companies.

The 5°C rise in temperature is, hopefully, not what will happen. But it is a warning to everyone in the finance industry that preventing this catastrophe is on their shoulders as well as on the shoulders of governments, regulators, corporates, the farming industry, consumers, and others.

5.3.1 *Types of Risk: White Swans, Black Swans, and Green Swans*

As climate change risk is different in nature from 'traditional' financial risks, we must consider what these differences are. In the field of finance, the usual assumption when modelling risk is that financial risk follows a 'bell curve' and that this curve is stationary over the long term. Models based on the bell curve are used in finance to determine Value at Risk (VaR) in a portfolio. Such (Bayesian) models work well over the space of short periods of time, such as a day, and also when markets are behaving 'normally'. Unfortunately, although such models allow us to predict our maximum loss with, for example, a 99% likelihood, what we should be really worried about is what might happen in the 1% tail of the distribution of possible losses (this tail may not be part of a Gaussian bell curve). Within the 99% likelihood bound, conventional statistical techniques based on historical evidence are reasonably reliable. Beyond this, there is unlikely to be any statistical evidence since it concerns, essentially, with extreme events that have not happened before. In this region of the distribution, probability statements are meaningless. So, *within* the 99% bound, a financial company will be taking and holding risk (which is indeed a key function of financial companies) in order to generate an expected return. Beyond the 99% probability point, the aim of a financial firm is survival.

Nassim Taleb, the author of *The Black Swan*[5] which was published just before the GFC in 2007, was one of the very few people who predicted the collapse in asset prices along with John Paulson (see his Wikipedia entry), who personally made a profit of $4 billion from shorting the US subprime mortgage market in the lead up to the GFC. It is likely that it was no surprise to Taleb that Goldman's risk models were producing 'nonsense' information. His insights are highly relevant to the 'forthcoming' climate crisis. A climate crisis is not unforeseeable (but when, in what form, where, and with what consequences are not foreseeable), while action to forestall the likely damage to the Earth, human society, animals, and plants is urgently necessary yet action on this front to date is well behind what is necessary.

In just over a decade, everyone in the world has been impacted by two global crises – the GFC and the global COVID-19 pandemic, and many people in the world have already been badly affected by global climate change. Global climate change is likely to cause another global crisis that may either overwhelm the world or may be mitigated by actions taken by governments, financiers, businesses, consumers, and others. Central banks can help prevent the crisis leading to financial instability (that is part of their mandate), but they cannot prevent mass migration from land which becomes uninhabitable – the type of migration we consider using the expression 'tipping points'.

Now to more consideration of 'types' of swan. The first type of 'swan' we consider is the 'white swan'. This is the type of event that financial firms and others can protect themselves against through adequate equity capital on their balance sheet (as in the graphic above) and by the employment of standard risk management techniques. 'Adequate' capital has to be calculated by regulators. Value at Risk (VaR) is one approach to determining how much equity capital is necessary. Value at risk (VaR) is a measure of the risk of loss on a portfolio. It estimates how much a portfolio might lose (with a given probability), **given normal market conditions**, in a set time period such as a day.

However, the real existential risks are 'black swan' or 'green swan' risks. Black swans (the actual birds) were once thought to not exist. Thus, there was no need to consider them as a possibility. But just as someone

[5]Taleb, N. N. (2008). *The Black Swan: The Impact of the Highly Improbable*. Penguin Books.

one day spotted a real black swan on a lake, subsequently many others were noticed by other people.

'Black swans' are events that are initially thought to be so unlikely that they will not occur in any living person's lifetime and so can be disregarded. The financial crisis showed that such events in finance did exist and were indeed more common than had been believed. Nassim Taleb was the first person to discuss events such as the GFC in terms of what he called Black Swan risks, even before the GFC had hit the world.

What we call a Black Swan event is one with the following three attributes:

> *First,* it is an outlier outside the realm of regular expectations, as the past is not a guide to such an event.
>
> *Second,* it carries an extreme 'impact'.

He summarises these attributes as: rarity and extreme 'impact', with only retrospective and not prospective predictability. He goes on to say that a small number of Black Swans explains almost everything in our world, from the success of ideas and religions to the dynamics of historical events and elements of our own personal lives. such as choosing a marriage partner.

The term 'Green Swan' is defined as:

> … a climate event that is outside the normal range of expected events. Green swans are different from black swans because there is some certainty that climate change risks will one day materialise.

Decision-Making under Conditions of Risk and Radical Uncertainty

For financial analysts and boards of directors, having to operate today and make decisions in a world where historical information on asset price changes may be quite irrelevant to future asset prices and where, therefore, traditional VaR models have little relevance, is a totally new experience. We don't know with certainty what the mass of future GHG emissions will be (tons of carbon dioxide equivalent (tCO2e), although we can make reasonable forecasts based on the past. We don't know how any particular degree of warming will translate into physical damage (impact assessment). We also don't know what the socioeconomic consequences of policy, regulatory, and social changes will be. COVID-19 impact, regional wars, displaced people, and many other factors will determine actual outcomes for the planet. We also don't know what impact there will be from technological developments, such as the possibility of low-cost power generated from nuclear fusion, or from policy changes (such as a very high carbon tax imposed very broadly across industries in a majority of countries). But we can make informed guesses as to what *could* happen under various scenarios (not forecasts). That is going to be critical for success both in terms of avoiding the worst impact of climate change on corporates and financial firms *and in mitigating global warming.*

The Precautionary Principle is an approach to decision-making under uncertainty and is in fact a core principle of EU environmental law. It is also part of the 1992 Rio Declaration on Environment and Development, which states:

> Where there are threats of serious or irreversible damage, lack of full scientific certainty shall not be used as a reason for postponing cost-effective measures to prevent environmental degradation (UNEP, 1992).

This principle is certainly one that could be sensibly applied in the case of climate change, where there are great uncertainties but also 'tail events' which could wipe out life on Earth or, at best, make life on Earth very difficult for people, animals, sea life, and plants. Insurance is about low-probability, high-impact events. By insurance we don't mean an insurance company policy but what actions 'the world' should be taking now given how severely the worst-case scenario of global warming could be. We have to be most concerned about known unknowns. But there are also unknown unknowns (with credit for this expression to former US Secretary of Defense Donald Rumsfeld). And if one looks throughout history, it is the unknown unknowns that tend to be the difficult ones.

Chapter 6

TCFD Metrics and Targets: Carbon Accounting and the Role of Financial Firms

IFRS S2 (Climate-Related Disclosures) came into force in January 2024. The objective of IFRS S2 is to require an entity to disclose information about its climate-related risks and opportunities that is useful to users of general-purpose financial reports in making decisions relating to providing resources to the entity. It also requires an entity to disclose information about climate-related risks and opportunities that could reasonably be expected to affect its cash flows, access to finance, or cost of capital over the short, medium, or long term.

The objective of TCFD is for a company to tell its investors whether and how it has addressed climate-related risks and opportunities in its strategic planning and decision-making, how it is resourcing its plans, and how it is progressing against the plans that it has in place. An example of the sort of disclosure we mean here is providing information about any transition (to net-zero) plans that a company has in place, such as its greenhouse gas emissions targets – what those targets are, how it plans to achieve them, and what progress it is making over time. To achieve this, it is necessary to be able to measure those emissions.

One of the four 'themes' of TCFD, therefore, is 'metrics and targets', as shown in an earlier graphic. However, the TCFD does not specify how companies should measure GHG emissions or how they should set a target. In the following chapter, we cover International Financial Reporting

Standards (IFRS) in relation to climate change risk, as this is likely to be the framework used by most corporates worldwide (with US corporates being among the exceptions) for meeting the requirements for disclosure laid out by TCFD.

As IFRS has stated that its 'standard', S2, fully incorporates the TCFD concepts, we have the same issue with IFRS that it does not specify 'how' to achieve the objectives it sets. Like TCFD, IFRS asks for information to be provided about metrics and targets, as well as governance, strategy, and risk management. IFRS also states that if a corporate follows the TCFD recommendations, it will be very familiar with those concepts because it took them and incorporated them into S2.

While neither TCFD nor S2 specify what metrics should be used or how targets should be set, they do state that Scopes 1, 2, and 3 GHG emissions, as defined by the **Greenhouse Gas Protocol** (GHGP; see later), should all be disclosed, along with the risks to the company associated with such emissions.

TCFD does, however, in its Guidance for All Sectors in relation to Metrics and Targets, note the following:

> Organizations should provide the key metrics used to measure and manage climate-related risks and opportunities.
>
> Organizations should consider including metrics on climate-related risks associated with water, energy, land use, and waste management where relevant and applicable.
>
> Where relevant, organizations should provide their internal carbon prices as well as climate-related opportunity metrics such as revenue from products and services designed for a low-carbon economy.
>
> Organizations should provide their Scope 1 and Scope 2 GHG emissions independent of a materiality assessment, and, if appropriate, Scope 3 GHG emissions and the related risks.
>
> GHG emissions should be calculated in line with the **GHG Protocol** methodology. Organizations should describe their key climate-related **targets** such as those related to GHG emissions, water usage, energy usage, etc.
>
> Other goals may include efficiency or financial goals, financial loss tolerances, avoided GHG emissions through the entire product life cycle, or net revenue goals for products and services designed for a low-carbon economy.

In describing their targets, TCFD says that organisations should consider including the following:

> Whether the target is absolute or intensity-based, the time frames over which the target applies, the base year from which progress is measured, and key performance indicators used to assess progress against targets.

6.1 Greenhouse Gas Protocol

The starting point for all measurements of GHG emissions and their impact is the work that has been undertaken by the GHGP. This body has set the globally agreed standards to measure and manage emissions. This is a critical starting point since it has been relatively easy for suppliers to claim they offer carbon-free power when in fact such a claim may be dubious.

As an example, some energy suppliers offer what they call a clean energy carbon-free electricity (CFE) power purchase agreement (PPA), even though it is not a **24/7** CFE PPA. The issue here is knowing what this actually means. However one looks at it, it is very different from true 24/7 CFE. The reason that they can make the claim they do is that the rules that were set by the GHGP many years ago have allowed it. The GHGP is a framework to enable standardised accounting for, and reporting on, GHG emissions. It was, somewhat surprisingly, launched as long ago as 1998 by the World Resources Institute (WRI) and the World Business Council for Sustainable Development (WBCSD) as a voluntary system of accounting and reporting.

The view taken by the joint promoters of this key initiative was that by obtaining the agreement of companies that they would account for their emissions and report them, thus making them widely available, this would lead to **impact** by incentivising more companies to reduce their GHG emissions and utilities to close gas and coal power stations and replace them with solar and wind. At that time, it was probably not particularly relevant to set the rules in such a way that they prohibited the contradiction of a 100% CFE **supply** through a pay-as-produced (PAP) PPA. The key to this issue is that **purchasing** 100% CFE power may not mean, and often does not mean, that the purchaser was actually **consuming** 100% CFE **24/7**. The purchaser may have been selling surplus CFE power from the PAP PPA in the summer but using those sales as a credit to enable it to claim to be using CFE power in winter when in fact it was using power produced using a high percentage of fossil fuel.

Today, this issue has become critical. In July 2024, GHGP issued a study entitled 'Market-based Accounting Approaches Survey Final Summary Report'.[1] 'Market-based' means using 'offset'-type instruments to reduce reported emissions. Accounting for such market-based approaches is a complex business, but the GHGP hopes to issue new rules based on the results of their survey. The GHGP also offers internships worldwide for those with relevant master's degrees with a view to future employment.

While the GHGP started as a voluntary programme, it has become the worldwide basis for GHG reporting. But more than this, it has become the basis for mandatory reporting in the EU commencing in 2024 and, at some point, also in the US through the expected requirements from the financial market regulator in the US, the Securities and Exchange Commission.

The protocol is a framework for the accounting and reporting of ***full lifecycle emissions*** of a company throughout its value chain. It divides emissions into those that can be attributed to the company's direct emissions and indirect emissions which are those from its suppliers (including its power supplier), on the one hand, (upstream in the value chain) and customers (downstream in the value chain), which would include the burning of an oil and gas company's petrol in a motor car. Figure 6.1[2] shows the complete product cycle in more detail, starting with upstream activities (inputs to the production process) coming into the company and then downstream activities, which is everything that happens after production through to 'end of life', such as the scrapping of a motor car, the demolition of a building, or the storage of nuclear waste from a reactor. Also be noted at the top of this graphic are all the GHGs which this standard considers, such as methane and nitrous oxide.

Scope 1

Direct GHG emissions that occur from sources owned or controlled by the reporting company, i.e. emissions from combustion in owned or controlled boilers, furnaces, vehicles, etc.

---

[1] https://ghgprotocol.org/blog/now-available-market-based-accounting-approaches-survey-final-summary-report.

[2] https://ghgprotocol.org/blog/you-too-can-master-value-chain-emissions.

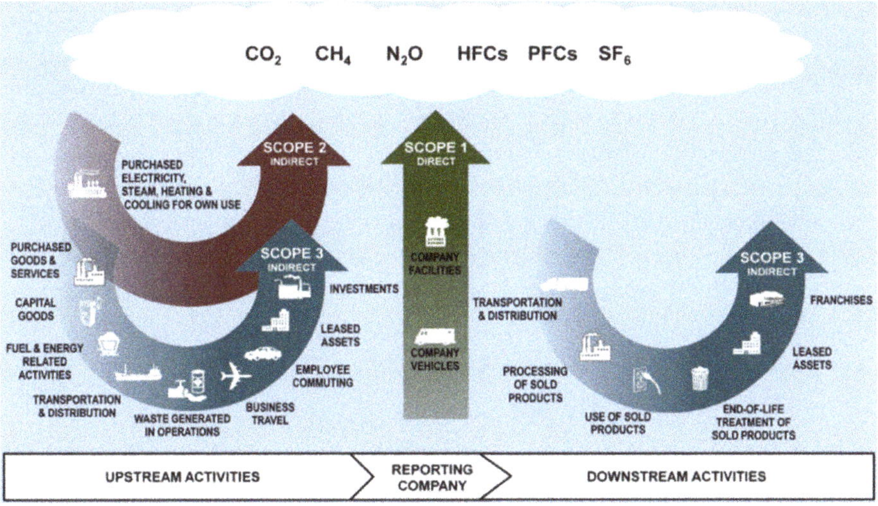

Figure 6.1. The complete product cycle, starting with upstream activities (inputs to the production process) coming into the company and then downstream activities and related Scopes 1, 2, and 3.

Scope 2

Indirect GHG emissions from the generation of purchased or acquired electricity, steam, heating, or cooling consumed by the reporting company.

Scope 3

All other indirect GHG emissions (not included in Scope 2) that occur in the **value chain** of the reporting company.

Scope 3 can be broken down into upstream and downstream emissions. Upstream emissions include all emissions that occur in the life cycle of a material/product/service up to the point of sale by the producer, such as from the production or extraction of purchased materials. Downstream emissions include all emissions that occur as a consequence of the distribution, storage, use, and end-of-life treatment of the organisation's products or services. In particular, in the case of an oil company producing petrol (gasoline), diesel, and kerosene (for planes), its reporting of Scope 3 emissions would cover emissions by the customers for its products, i.e. the CO_2 produced when people drive their petrol of diesel cars or travel in an aircraft using kerosene, etc. Frequently, however, oil and gas companies, at least in the past, have reported only their own emissions, which are much smaller than those of their customers burning fossil fuels produced by them.

One consideration when trying to calculate the relevant numbers for reporting on a company's emissions is the risk of double counting. All of a company's Scope 3 emissions are someone else's Scope 1 or Scope 2 emissions. Not only is there the risk of double counting, there are also considerable difficulties in obtaining data on many of the emissions which are required to be attributed, in particular, inputs to a company's value chain which are produced overseas. However, our main concern at this point is with 'purchased electricity for own use' in Scope 2.

6.2 The Role of Energy Attribute Certificates

The difficulty with knowing the attributes (in particular the carbon content) of power coming to a customer through the transmission and local distribution systems arises because an electron is an electron is an electron and does not carry any indication of how it was generated. It is not possible to 'tag' electrons as they enter a customer's premises. As a result, without some mechanism which can link the source of supply to the point of use, it is impossible to know what type of energy, fossil or carbon-free, produced the electricity being used. The solution that was arrived at some 20 years ago was the development of what are known as Energy Attribute Certificates (EACs), which are a means of attributing a supply of electricity from a particular source to a consumer purchase. However, until recently, this could only be achieved on an annual basis, which meant that surplus solar energy produced during the day in summer could be claimed for use as clean power even during the darkest winter evenings.

There are three different certificates according to the region in which they were issued:

Guarantee of Origin (GO): Instruments of the European Energy Certificate System used in 30 European countries.

Renewable Energy Certificate (REC): In the United States and Canada.

International Renewable Energy Certificate (I-REC): Used in countries outside Europe and North America.

On the demand side of the market for EACs, this primarily comes from corporations wishing to show that their power comes from 100% renewable sources and can then be accounted for using the GHGP Scope 2 emissions methodology to enable them to provide disclosures, normally through the Carbon Disclosure Project (now known simply as CDP) for climate disclosures or RE100 for renewable energy usage.

What is required in the **next transition to 24/7 CFE**, which is currently underway, is a shift to what is known as **temporal granularity**, meaning the scale or level of detail in a data set. This means being able to 'timestamp accurately and regularly' (down to 15 minute intervals).

In summary, the complete value chain in the power industry – regulatory authorities, 'green' certificate-issuing bodies, generators, TSOs/DSOs, wholesale and retail consumers, and others – is currently working towards standards for the future in order to enable the transition to 24/7 CFE. This requires new processes and tools to manage ***Time-Based EACs***.

6.3 EnergyTag[3] Time-Based Certificates

We could not, until recently, know if an offer of power by a supplier of power to provide 24/7 clean power was a genuine offer of clean power at any time of day and in any season or simply 'greenwashing'. To help overcome such problems, *EnergyTag* has been developing an hourly (granular) matching system. Their goal has been to establish the Granular Certificate (GC) as a common, tradable instrument that provides traceability across markets for power, flexibility, and carbon. The GC Scheme Standard provides the first version of the criteria that those implementing GC schemes need to meet to prove, through a third-party audit, that they are *Compliant with the EnergyTag Standard.*

EnergyTag is an independent non-profit organisation which is defining and promoting net-zero-aligned electricity *carbon accounting*, enabled by real-time tracking with *Granular Certificates.* Its initiative is

[3]The information in this section is based on EnergyTag's publication Carbon accounting & tracking for 24/7 clean grids https://energytag.org.

seeking to define a framework for adding a timestamp to energy attribute certificates (EACs) for both clean power and clean hydrogen, which will make them more reflective of the physical availability of clean power or clean hydrogen. Participants believe this may improve public perceptions of clean energy claims and could incentivise energy storage and support new carbon accounting methodologies.

The problem as they see it, and the reason behind EnergyTag's creation, is that the more successful the world is in deploying renewables, the harder they are to integrate into the grid. Current methods for procuring clean energy, as considered in the sections above, **only match supply and demand over a year and over large geographical boundaries. This results in a disconnect between carbon accounting and grid realities and fails to address the hardest hours of decarbonisation, i.e. at night or in daytime when the wind speed is low or the sun is obscured more than usual by cloud cover. This is clearly a critical issue.**

Even when claiming to offer '100% renewable', an organisation may still be reliant on fossil fuels for a significant portion of the day, particularly in winter, enabling substantial underreporting of real-world emissions. The lack of data about the *time of electricity production* on EACs is a key root cause of this issue. Given that corporates purchase over 50% of the world's electricity, fixing these carbon accounting issues is crucial for achieving deep grid decarbonisation. **The failure of companies to contribute towards additionality despite their REC purchases and yet seemingly achieving their SBTs** is **a key issue.**

EnergyTag's granular accounting (accounting in ever smaller time steps) is designed to ensure consumption is matched with clean energy in the same hour (or half-hour or quarter-hour) from a location where electricity can be delivered and *encourages new clean energy supply*. This is the next frontier of clean energy sourcing and will be enabled by time-stamped GCs, also known as Time-based Energy Attribute Certificates (T-EACs). Its system is designed to enable consumers drive deep decarbonisation of the grid – every hour of every day. This is a way of:

- enabling the production of clean products, not only power but also hydrogen and its derivatives such as liquid hydrocarbon fuels and ammonia, without any fossil reliance;
- providing more transparency by linking production to consumption in 'real-time';
- supporting storage and flexibility by providing a new price signal by ensuring the market price for power in times of shortage, even over

short periods, reflects this scarcity and thus makes the provision of energy storage on the grid or the provision of generation capacity at such times provide an adequate reward to investors;
- facilitating better risk management by avoiding volatile fossil-based electricity pricing.

6.4 Science-Based Targets Initiative

TCFD (and ISSB), on their own, do not provide protocols to measure carbon emissions or achieve targets for emissions reduction. The initial work of the Science-Based Targets Initiative (SBTi) was to enable corporates (the main GHG emitters) and other emitters to have a science-based target (SBT) which provided a clearly defined pathway that would be sustainable. It would achieve this by specifying how much and how quickly a corporate needed to reduce its GHG emissions to be 'Paris-compliant'.

Nearly 1,000 companies in 50 sectors, from coal and gas to pharmaceuticals, with a cumulative market cap of $15.4 trillion, have pledged to align their decarbonisation plans with the Paris Agreement by adopting SBTs. But no such target framework was available for banks or other financial institutions – until 2020. We need to reduce emissions, well beyond even what was agreed by all the world's governments in Paris in 2015 to achieve a <2°C target. There is a limited 'carbon budget' (see Section 5.1) left for the world as a whole, and even if one company uses more of it than calculated as allowable under its SBT, there is less available for other companies, which have few or no options for carbon reduction. The consequence then would be that the 1.5°C target would not be achieved, nor even the 2°C target, resulting in substantial costs to everyone due to physical damage.

Targets adopted by companies to reduce GHG emissions are considered 'science-based' if they are in line with what climate science says is necessary to meet the goals of the Paris Agreement – to limit global warming to well below 2°C above pre-industrial levels and pursue efforts to limit warming to 1.5°C. SBTi calculates this target, or 'mini budget'. SBTi makes clear the benefits from financial institutions being actively involved with corporate decarbonisation through the influence they can exert on corporates to which they lend or in which they invest. They see financial institutions as being the vital link in enabling the kind of system-wide change we need and enabling the redirection of capital to support the carbon-free energy technology solutions, which we cover in later chapters. The financial sector SBTs framework *is designed to enable financial institutions to accelerate the transformation required* by aligning lending and investment portfolios with the level of ambition required by science. However, note that the focus of this initiative is less on the risk to financial firms' portfolios but more on the extent to which financial firms can drive climate change mitigation through their lending and investing decisions.

An obvious question is whether corporates are, in practice, actually cutting their emissions by the amount required to hit Paris targets, particularly given many of the necessary tools, such as SBTs, are now available. It would seem that in practice many have not set targets, and others which have do set targets have nonetheless found it impossible to meet them. This is part of the explanation why GHG emissions continue to rise almost every year.

In June 2021, SBTi published a report entitled 'Taking the Temperature: Assessing and Scaling-Up Climate Ambition in the G7 Business Sector'. SBTi notes that a dramatic gap in ambition remains when it comes to 2030 targets set by corporations and countries. Recent analysis shows a significant difference – a gap of 20–23 gigatons of carbon dioxide (GtCO2e) – between the level of GHG emissions expected in 2030, under current policies and country commitments, and where they should be to limit global warming to 1.5°C.

SBTi has now advanced one of its objectives, which is to provide a **net zero standard for financial companies**. The framework was introduced in 2020 for lenders, investors, and insurers. These will now be able to verify that their climate targets are aligned with the Paris Agreement through the SBTi. One of these approaches is the **temperature rating approach**, which financial institutions can use to determine the current temperature rating of their portfolios and take actions to align their portfolios with long-term temperature goals by engaging with portfolio companies to set targets.

Approved targets should result in financial companies committing to reduce their Scope 1 (direct) and Scope 2 (power-related) emissions by 4.2% annually under the 1.5°C pathway and 2.5% annually under the 'well below 2°C' pathway. Scope 3 target requirements will vary based on the kinds of assets which the organisation has holdings in. It should be noted that a revised standard was introduced in May 2024. To help speed up the transition, one change in the revised framework is that validated targets will have a reduced timeframe window for Scope 1 and 2 emissions, down from 5–15 years to 5–10 years.

6.5 Partnership for Carbon Accounting Financials: Financed Emissions Standard and Carbon Disclosure Project

There is a considerable difference between the issues in accounting for the GHG emissions of a corporate and accounting for the carbon intensity of a lender or investor's financial asset portfolio. Another organisation, *Partnership for Carbon Accounting Financials* (PCAF), has developed a standard for **financed emissions**[4] (emissions resulting from financing provided by a financial institution).

The PCAF is an industry-led initiative. Created in 2015 by Dutch financial institutions (FIs), PCAF extended to North America in 2018 and scaled up globally in 2019. More than 300 banks and investors participate in PCAF. As Table 6.1 from the PICAF publication *Global GHG Accounting and Reporting Standard for the Financial Industry* shows, without agreed

Table 6.1. A list of the most common financed emission metrics applied in the market.

Metric	Purpose	Description
Absolute Emissions	To understand the climate impact of loans and investments, and set a baseline for climate action	The total GHG emissions of an asset class or portfolio
Economic Emissions Intensity	To understand how the emissions intensity of different portfolios (or parts of portfolios) compare to each other per monetary unit	Absolute emissions divided by the loan and investment volume, expressed as e.g. tCO_2e/€M invested
Physical Emissions Intensity	To understand the efficiency of a portfolio (or parts of a portfolio) in terms of total carbon emissions per unit of a common output	Absolute emissions divided by an output value, expressed as e.g. tCO_2e/MWh, tCO_2e/ton product produced
Weighted Average Carbon Intensity (WACI)	To understand exposure to carbon-intensive companies	Portfolio's exposure to carbon-intensive companies, expressed as tCO_2e/€M company revenue

[4]PCAF (2022). The Global GHG Accounting and Reporting Standard Part A: Financed Emissions. Second Edition. https://carbonaccountingfinancials.com/files/downloads/PCAF-Global-GHG-Standard.pdf.

standards there are many possible 'metrics' that corporates might choose to use in their company's GHG emissions reporting. In the case of financial firms, they would be reporting on the carbon intensity of their portfolio (lender asset book or equity and bond funds). In the case of financial firms, the input into their own disclosure is, of course, based on the reporting of the corporates that they hold in their portfolios.

Using PCAF's Financed Emissions Standard equips financial institutions with standardised methods to measure financed emissions and enables them to:

- assess climate-related risks in line with the recommendation of the TCFD;
- set SBTs using methods developed by the SBTi and other science-based methodologies;
- report to stakeholders, such as the CDP.

6.5.1 *CDP*

Originally called 'Carbon Disclosure Project' but now using as its strapline 'Disclosure, Insight, Action', CDP runs the global environmental disclosure system. Each year CDP supports thousands of companies, cities, states, and regions to measure and manage their risks and opportunities with regard to climate change, water security, and deforestation. It does so at the request of their investors, purchasers, and city stakeholders.

CDP believes that improving corporate awareness through measurement and disclosure is essential to the effective management of carbon and climate change risk. It requests information on climate risks and low-carbon opportunities from the world's largest companies and others. Over 23,000 companies representing two-thirds of global market capitalisation report through CDP on climate change, forests, and water security.

CDP has also produced a position paper on carbon credits – a highly contentious issue – in which they provide clear global guidance and accounting on what constitutes a 'credible' carbon credit and a standardised approach to determining the quality and use of such credits. Without this, it is difficult to evaluate efforts towards deep decarbonisation and whether companies are falling short of where they need to be. If the use and reporting of carbon credits are not credible, comparable, and consistent, it risks damaging the integrity of emissions accounting more widely.

PCAF and CDP collaborate closely to enable transparent disclosures, allowing stakeholders to better understand the climate impact of financial institutions' portfolios. CDP's disclosure framework integrates *The Global GHG Accounting and Reporting Standard for the Financial Industry* into its Climate Change Questionnaire for financial services, which asks for the disclosure of Scope 3 portfolio emissions.

While it might be expected that, through the various target-setting, disclosure, and reporting organisations that we have covered above, all corporates and financial firms would have clear targets for emissions reduction, this has not so far been the case for banks.

6.6 'Big Tech' Case Studies – Google and Microsoft: 24/7 Clean Baseload PPAs

Many case studies of 24/7 CFE use a 'tech' company as an example. In particular, the tech companies that provide 'cloud' services, cryptocurrency companies, and, more recently, ChatGPT-type services using large language models (LLMs), which consume **multiple amounts of power** compared with more traditional services. This would be the case, for example, for the increase in power required for LLM-type searches compared with the traditional Google search model. In the case of Ireland, the International Energy Agency has forecast that data centres will account for 32% of Irish electricity demand by 2026. Such data centres also need cooling using water. Thus, shortages of power and water around the world remain critical for such companies' growth and for the green and energy transitions. As we noted in the previous chapter, however, **while the tech companies have tried hard to achieve their targets, much of their reporting seems to have been based on the use of 'offsets' rather than reductions in real GHG emissions. (Note: We cover the sudden increase in interest in 2024 by the 'tech' companies in advanced nuclear technology for generating carbon-free power in Section 15.2.)**

As a result of competition for power supplies between households in a region and data companies, local authorities have in some cases refused to allow new data centres unless the owner of the intended data centre can source all its power from **additional clean generating capacity and, ideally, 24/7 CFE**. As a result, the 'big five' US tech companies – Alphabet (Google), Amazon, Apple, Meta, and Microsoft – and Alibaba, Tencent, and Xiaomi in China have all been involved in 24/7 CFE initiatives. It has been reported that in London, a Datacentre Task Force has been set up to try to limit the issues that can arise through competition for power supplies. This can be exacerbated if a potential location has good fibre-optic cable provision, which is critical for any data centre.

6.6.1 *Google*

Google is an example of a company that is supporting EnergyTag's work. As more organisations shift to hourly clean energy purchasing, Google emphasised that there needs to be a way to track and certify their progress. To contribute to this effort, Google teamed up with M-RETS, an energy certificate registry in the US, to pilot T-EACs. These instruments track how, where, and when electricity is produced. T-EACs are designed to bring transparency to granular clean electricity production data and consumer claims. Google is also supporting efforts on a global industry scale, including a €530,000 grant from Google.org to EnergyTag, which is the leading organisation developing an international standard for T-EACs.

Figure 6.2[5] shows how the energy needs of one of Google's data centres in one day in March were met over each of the day's 24 hours. The straight line shows the steady demand from their data centre. The coloured bars show what type of energy met this demand. Their objective here, as

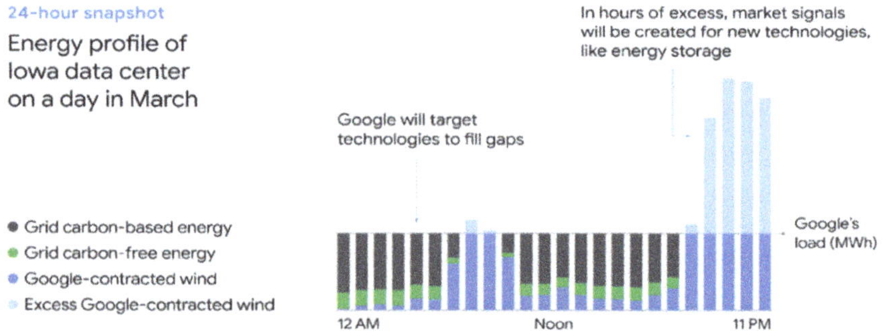

Figure 6.2. Hourly snapshot of the energy profile on one Google data centre on one day in March.

[5]https://cloud.google.com/blog/topics/sustainability/t-eacs-help-drive-around-the-clock-carbon-free-energy.

elsewhere, is to eliminate the black bars, which they are suggesting could be achieved by storing the surplus contracted wind power and using it to power their data centre at times when the grid is using fossil fuels.

6.6.2 *ENGIE and Google Enabling 24/7 Carbon-Free Energy Solutions*

The energy company, ENGIE, signed a contract in 2021 to provide Google's data centres in Germany with a **guarantee to provide 80% CFE supply on an hourly basis**. This requires ENGIE to develop a CFE portfolio on Google's behalf.

Unfortunately, the advent of LLMs, such as Google's DeepMind model, Gemini (or OpenAI's ChatGPT), has resulted in a surge in electric power requirements. As a result, it has been reported by Google that their GHG emissions have increased by nearly 50% over the five years, from 2019 to 2024. The reason is that a generative AI system (LLM) can use 30 times as much power as computers running task-specific software. The IEA expects that by 2026, data centres could be using as much electricity as Japan.

6.6.3 *Microsoft Cloud for Sustainability*

In 2020, Microsoft made a commitment to shift to 100% renewable energy supply in its buildings and data centres by 2025, although sadly this no longer seems achievable even in the medium term. On this journey, the company has recognised that how it tracks its progress is just as important as how it gets there. It is now the first hyperscale cloud provider to track hourly energy consumption and renewable energy matching in a commercial product using the Vattenfall 24/7 Matching solution for its new data centre regions in Sweden. Microsoft itself has committed that, by 2030, it will have 100% of the energy consumption of its data centres, 100% of the time, globally matched by zero-carbon energy purchases – its '100/100/0 by 2030' goal.

The critical issue as the world aims to move towards 90–100% clean power every hour of the day is 'how to fill the gaps'. Thus, for Google, Microsoft, and others attempting to achieve this aim, it remains an aim except in a small number of regions of the world which have dispatchable power from geothermal, run-of-the-river, or nuclear power.

6.6.4 *Location-Based Versus Market-Based Carbon Footprint*

It is clear that tech companies have committed to making progress on reducing emissions related to their activities (and to reducing water consumption for cooling). But as always, 'the devil is in the detail'. Although some of the big tech companies **do have targets** to eliminate CO_2 emissions by a specific date, the evidence suggests that while reported numbers may have been declining, actual physical emissions have been increasing dramatically. There are two approaches to measuring carbon footprint, which can give entirely different results. The GHGP has the following definitions of the two approaches:

> A **location-based** method reflects the average emissions intensity of grids on which energy consumption occurs (using mostly grid-average emission factor data).
>
> A **market-based** method reflects emissions from electricity that companies have purposefully chosen (or their lack of choice). It derives emission factors from contractual instruments, which include any type of contract between two parties for the sale and purchase of energy bundled with attributes about the energy generation, or for unbundled attribute claims.

While the evidence suggests that Big Tech is not doing nearly enough to decarbonise, in November 2024, a number of the Big Tech companies announced nuclear power initiatives, which are likely to outdate the current discrepancies between the two approaches over the course of the next decade. We cover nuclear power financing, including initiatives by Big Tech, in Section 15.2.

A useful article with an excellent set of graphics demonstrating the difference between location-based and market-based methods in relation to large tech firms can be found at: 'Big Tech's bid to rewrite the rules on net zero FT August 2024' (https://www.ft.com/content/2d6fc319-2165-42fb-8de1-0edf1d765be3).

Chapter 7

Accounting and Disclosure Frameworks: International Accounting Standards Board, IFRS, and ISSB, and the European Commission Double Materiality, CSRD, and ESR[1]

In the previous chapters, we covered some of the risks to which any investment may be subject, with a particular focus on climate change risk. The TCFD Recommendations covered the approach that needs to be taken by both corporate and financial firms in trying to assess the impact of such risks on their businesses. Once the risks have been assessed, it is necessary to report them in an appropriate way in order that investors can make decisions based on a full disclosure of risks, including climate risk, that the company foresees. Of course, this is all easier said than done. We now consider the reporting side of such risks.

7.1 IASB, IFRS & the Creation of the ISSB

The International Accounting Standards Board (IASB) is an independent international group of experts responsible for the development and publication of International Financial Reporting Standards (IFRS). The Board

[1]The material on IFRS and ISSB is all available at https://www.ifrs.org/sustainability/knowledge-hub/introduction-to-issb-and-ifrs-sustainability-disclosure-standards/.

is also responsible for approving interpretations of IFRS as developed by the IFRS Interpretations Committee. IFRS are currently required in more than 140 jurisdictions and permitted in many more.

In November 2021, IFRS issue a press release in which they announced:

> The formation of a new International Sustainability Standards Board (ISSB) to develop – in the public interest – a comprehensive global baseline of high-quality sustainability disclosure standards to meet investors' information needs;
>
> The publication of prototype climate and general disclosure requirements developed by the Technical Readiness Working Group (TRWG), a group formed by the IFRS Foundation Trustees to undertake preparatory work for the ISSB. These prototypes are the result of six months of joint work by representatives of the CDSB, the International Accounting Standards Board (IASB), the Financial Stability Board's Task Force on Climate-related Financial Disclosures (TCFD), the VRF and the World Economic Forum (Forum), supported by the International Organization of Securities Commissions (IOSCO) and its Technical Expert Group of securities regulators. The TRWG has consolidated key aspects of these organisations' content into an enhanced, unified set of recommendations for consideration by the ISSB.

These announcements mark a major step forward in achieving a unified standard for sustainability disclosure globally (though the US is not a party to IFRS). They also closely follow the TCFD approach to disclosure.

In March 2022, two exposure drafts were issued by IFRS.

Objective

The objective of [draft] IFRS S2 *Climate-Related Disclosures* is to require an entity to disclose information about its exposure to significant *climate-related risks and opportunities*, enabling *users* of an entity's *general purpose financial reporting*:

(a) to assess the effects of significant climate-related risks and opportunities on the entity's *enterprise value*;
(b) to understand how the entity's use of resources, and corresponding inputs, activities, outputs, and outcomes support the entity's response to and strategy for managing its significant climate-related risks and opportunities; and

(c) to evaluate the entity's ability to adapt its planning, *business model*, and operations to significant climate-related risks and opportunities.

Strategy

The objective of climate-related financial disclosures regarding strategy is to enable users of general-purpose financial reporting to understand an entity's strategy for addressing significant climate-related risks and opportunities.

To achieve this objective, an entity shall disclose information about:

(a) the significant climate-related risks and opportunities that it reasonably expects could affect its business model, strategy, and cash flows, its access to finance, and its cost of capital over the short, medium, or long term;
(b) the effects of significant climate-related risks and opportunities on its business model and *value chain*;
(c) the effects of significant climate-related risks and opportunities on its strategy and decision-making, including its *transition plans*;
(d) the effects of significant climate-related risks and opportunities on its financial position, financial performance, and cash flows for the reporting period, and the anticipated effects over the short, medium, and long term – including how climate-related risks and opportunities are included in the entity's financial planning;
(e) the *climate resilience* of its strategy (including its business model) to significant physical risks and significant transition risks.

It is likely that the new climate-related disclosure requirements will be implemented in many countries by around 2024. This will lead to the need for auditors, compliance staff, and others to understand the rationale behind the introduction of these rules as well as the rules themselves.

In April 2022, the ISSB held a live webinar on its proposed standards for general sustainability-related and climate-related disclosures It should be noted that the focus is on the information needs of investors, **not** the 'double materiality' of the EU's approach but purely on 'financial materiality'. In due course, this approach may change due to public and investor pressure. It is also noticeable that it refers to the TCFD structure, which is now the overarching framework for disclosure.

In May 2022, the chairs of the IFRS Foundation's IASB and ISSB announced plans for the two boards to work together to agree on how to build and integrate the Integrated Reporting Framework into their standard-setting projects and requirements. The chairs of the IASB and ISSB actively encourage the continued adoption of the Integrated Reporting Framework, given the importance of its principles and concepts to both high-quality corporate reporting and the connectivity between financial statements and sustainability-related financial disclosures.

The IFRS Foundation has said that the standards being drawn up by the ISSB **do take account** of the impacts of a company's activities (on society as a whole) if they are material to its value. They have also said that while enterprise value is the key issue in their standards, it is also the case that this can be affected by the reverse impact of a company on society. This, however, is quite different from reporting on 'Environmental & Social Materiality', as required by the EU.

7.1.1 *Graphical Representation of the IFRS Approach*

Figure 7.1[2] shows an interesting graphical representation of the amalgamations into the IFRS framework of five bodies originally involved independently in developing sustainability standards. The diagram is called 'the house'.

On the roof of the house is the Integrated Reporting (IIRC), whose aim is to be an 'umbrella' organisation acting to integrate all information relevant to economic decision-making within one report. Thus, it acts as an umbrella covering both financial accounting standards (IASB and FASB) and sustainability frameworks and standards organisations (SASB and CDSB). In turn, these latter two organisations build on data provided by the Global Reporting Initiative (GRI). The GRI is an independent, international organisation that helps businesses and other organisations take responsibility for their impacts by providing them with a global common language to communicate those impacts. It provides the world's most widely used standards for sustainability reporting: the GRI Standards.

The five bodies referred to above have since been integrated. They acknowledge that there are two types of users with different *materiality* needs. For a wide range of potential users, what is material to them (what matters to them) is how any particular company impacts the environment, social systems (people), and the overall long-term health of the economy (sustainable development) in a material way. Thus, for this large group of people, anything which materially impacts these factors is important.

---

[2] https://sasb.ifrs.org/blog/ifrs-foundation-aims-for-coherence-not-complexity/.

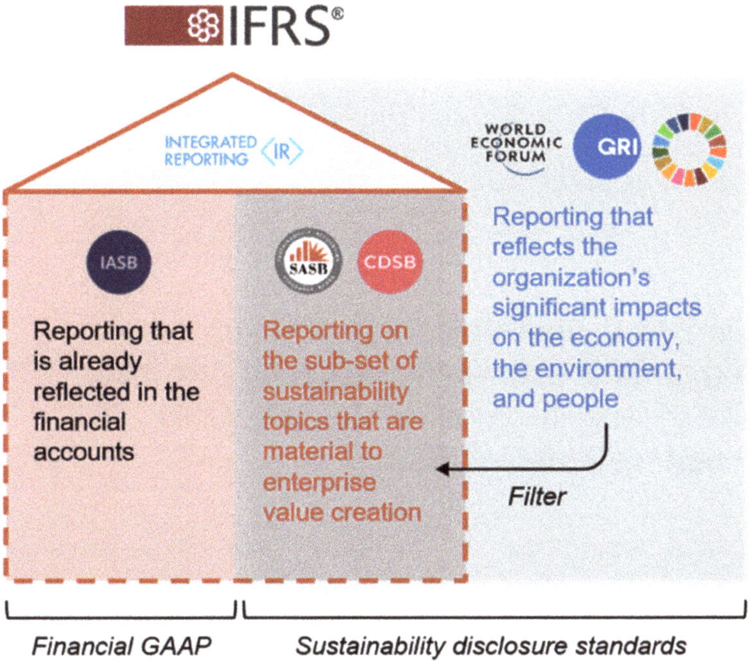

Figure 7.1. A coherent, comprehensive corporate reporting system.

This is the European approach we covered above. While acknowledging this type of materiality, this does not mean that the ISSB will adopt the double materiality approach, and currently, it has not.

For the second group (mainly investors), what is relevant are factors which *materially* impact enterprise value creation or destruction. What is new here is that it is now widely accepted that 'non-financial factors' impact corporate value just as financial factors do. So, just as financial accounts recognise financial factors which materially impact corporate value, material sustainability topics are also relevant to enterprise value creation or destruction.

One of the five bodies (IIRC) has been focused on the integration of financial and non-financial reporting into a single report using the International Integrated Reporting <IR> Framework. Indeed, the International <IR> Framework has now become part of the IFRS Foundation. Its purpose is to bring both sets of material factors affecting long- (or short-) run enterprise value into the annual report and accounts of a company. While this might seem like an acknowledgement of the EU double materiality concept, it is not, in fact. It is simply an acceptance that environmental factors impact corporate value. While they acknowledge that corporates also impact the environment, they do not agree at this point that the ISSB should require that this impact should also be reported.

7.2 The EU's Double Materiality Concept, Its Corporate Sustainability Reporting Directive (CSRD), and the European Sustainability Reporting Standard (ESRS)

There is a key area of controversy over the approach taken by the TCFD and the IASB/IFRS/ISSB, on the one hand, and other approaches, particularly the one taken by the EU in the European Commission's *Corporate Sustainability Reporting Directive (CSRD)* and its European Sustainability Reporting Standards (ESRS), which we cover in the following. The controversy centres on whether sustainability-related disclosures should focus **only** on the possible impact of climate change on the profit and loss account or balance sheet of an entity or whether the reverse impact should also be considered, i.e. the impact of the entity (normally a corporate) on the environment and, hence, on society. In more detail:

- what is the impact of sustainability issues, such as climate change risk or water supply risk, on the entity and its financial position? or;
- should it also disclose the opposite side of the equation, namely the impact of the entity itself (the corporate) on civil society, employees, consumers, and presumably also investors?

This divide regarding this issue between these two perspectives is considerable and even fundamental. The EU has been very clear in its view of the duality of disclosure required.

The concept of 'materiality' is important in this debate in order that both corporates and society should be able to avoid wasting resources on activities that may not contribute much to mitigation or, alternatively, failing to focus resources on aspects that are widely believed to matter. Thus, the meaning of the word 'material' in the context of finance is something we need to consider. One definition would be any data or information which if not provided, might prevent investors or others from making appropriate decisions.

The **financial impact of climate change on a corporate** arises primarily from physical risks (material physical damage to its real assets) and transition risks (from loss in value of its real assets and from material changes in regulations and technology). In addition, risks arise from impacts on populations (including employees and consumers) in the environment where they work and live, which may be a result of productivity losses due to climate change, such as heat, insufficient food or drinking water, and the possibility of new diseases. **Thus, if such potential impacts are 'material' to the financial well-being of the company (however that is defined), information on such potential impacts should be reported in a standard IFRS annual report and accounts.**

In the view of many, appropriate action on climate change risk means addressing concerns about not only the impact of climate change on a company's value **but also the impact of a company's behaviour on society**. The European Commission (EC) argues that corporates need to be concerned with not just one perspective on materiality, the traditional one, i.e. the impact of climate change on the company, but *two perspectives (double materiality as they call it)*, meaning also covering the impact of the company on society. In fact, it is being increasingly recognised that both impacts affect sustainability and hence *enterprise value* and are therefore material. So, to the extent that both of these impacts are material (under the IFRS definition), they must be reported.

Figure 7.2[3] is from Nordea, a large investment management company. Nordea itself must follow the reporting and other requirements of the European Union's CSRD and ESRS, even though Norway, its home country, is not part of the EU. The two perspectives on materiality in the figure are:

1. on the right-hand side, the; external impacts of the company's activities' (labelled as 'environmental & social materiality', whose audience consists of consumers, civil society, employees, and investors);
2. on the left-hand side, the potential or actual impacts of climate-related risks and opportunities on the 'performance, development, and

[3]Nordea *How Sustainability Risks Affect Investments* 2023 https://www.nordea.com/en/doc/final-sustainability-risks-and-how-they-affect-investments-q2-2023.pdf. Adapted from Ksapa.org: Understanding the Concept of Double Materiality.

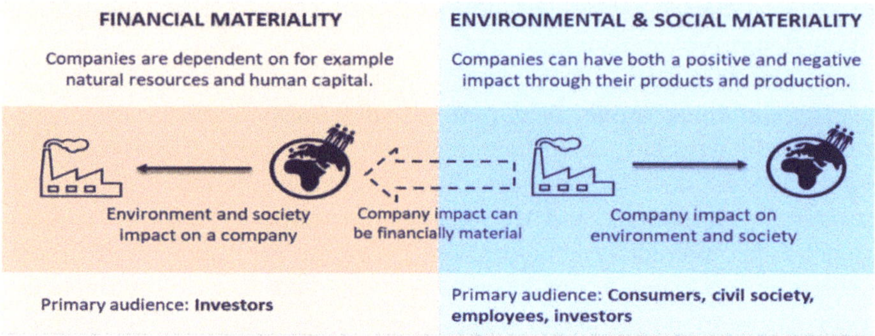

Figure 7.2. Understanding the concept of double materiality.

position' of the company indicated by 'financial materiality', with a focus on the interests of investors (TCFD's main focus).

When reporting on environmental and social information, an entity must use the European Sustainability Reporting Standards (ESRS) as described in the following.

7.2.1 *Corporate Sustainability Reporting Directive and European Sustainability Reporting Standards*

In April 2021, the EC published its proposals for a new directive on corporate disclosure (CSRD), which amended and greatly strengthened the previous sustainability reporting requirements in the EU under the Non-Financial Reporting Directive (NFRD). It represents a major step change in corporate reporting which will affect businesses not only in Europe but also worldwide. The additional reporting requirements focus on providing more detailed reporting on double materiality (inward and outward impacts), alignment with the Paris Agreement, and forward-looking sustainability information. At the implementation date, large businesses will need to disclose their strategy and business model in relation to sustainability-related risks and climate scenarios, their sustainability targets and indicators, and the extent of progress towards achieving them, environmental-related matters affecting or affected by their business, social considerations such as equal opportunities, their approach to business ethics, their governance processes and control mechanisms, and their risk management approach.

The ESRS (Reporting Standards) will be used to meet the requirements of the CSRD. The CSRD provides the reporting requirements and obligations, while the ESRS provides a framework and methodology for undertaking the reporting.

Chapter 8

SDGs, DEI, and ESG: Financial Rating Agency Approach to ESG and Other Risks

8.1 The United Nations Sustainable Development Goals

There are many societal issues which are relevant to achieving the best possible quality of life for all people on Earth and might, therefore, be relevant to those who run corporates and manage equity or bond portfolios. Achieving, or not achieving, these wider goals is likely to impact all countries and also corporates and financial companies, as goals such as minimising hunger and poverty, reducing inequality, and having strong and diverse institutions governing countries are all likely to reduce the likelihood of wars and mass migrations, all of which can improve corporate profitability. Even without wars or migrations they affect corporates and, in particular, large multinational corporates based mostly in Western economies and in China.

Just as the United Nations (UN) was central to the achievement of the Paris Agreement, it has been very active in other initiatives designed to make the world a better place for humanity. In the following, we consider areas which are highly relevant to lenders and investors.

An asset owner, an asset manager, an insurer, or a bank needs to have an underlying set of principles or goals to operate under when it comes to defining its 'purpose' and its long-term strategy for success. Trying to develop a set of principles or goals or define a company's

'purpose' is difficult in the absence of some generally agreed framework to aid decision-making. Having a framework has been made easier by two things:

> First, the development of 17 global goals by the UN, agreed by 193 countries at the UN General Assembly in September 2015;
>
> Second, the Paris Agreement in December 2015, which we covered earlier.

The ideas underpinning the *Sustainable Development Goals (SDGs)* were developed by Paula Caballero, director of economic, social, and environmental affairs at the Ministry of Foreign Affairs in Colombia. The fact that the most participatory intergovernmental process in human history ended up with 17 agreed goals is nothing short of miraculous (Figure 8.1).

We now have, as one of the expressions of 'society's goals', the framework (the SDGs), which **was used as an input by the 23 banks that, in 2019, developed *Principles for Responsible Banking*** (Principles for Responsible Investment came into existence four years before the SDGs were developed).

The UN notes[1] that:

- 'Sustainable development has been defined as development that meets the needs of the present without compromising the ability of future generations to meet their needs.
- Sustainable development calls for concerted efforts towards building an inclusive, sustainable and resilient future for people and planet.
- For sustainable development to be achieved, it is crucial to harmonize three core elements: economic growth, social inclusion and environmental protection. These are inter-connected and all are crucial for the well-being of individuals and societies.
- Eradicating poverty in all its forms is an indispensable requirement for sustainable development. To this end, there must be sustainable, inclusive and equitable economic growth, creating greater opportunities for all, reducing inequalities, raising basic standards of living, fostering equitable social development and inclusion, and promoting integrated and sustainable management of natural resources and ecosystems'.

A number of investment managers offer SDG-focused funds, which employ strategies that focus on investing for positive outcomes. This is likely to be companies that are providing solutions to address the world's

[1] https://www.un.org/sustainabledevelopment/.

Figure 8.1. The UN sustainable development goals.

challenges and are helping to meet the environmental and social targets set out in the UN SDGs. Water and food security are two key SDGs.

8.2 Diversity, Equity, and Inclusion

Diversity, Equity, and Inclusion (DEI) is a concept that dates back to the mid-1980s, particularly in the US. The aim of the DEI movement is to promote fair treatment and full participation of all people, especially those who have historically suffered from discrimination on the basis of their skin colour, disability, sexual orientation, gender, etc. particularly within corporate business but more widely, including in universities. In the corporate environment, the argument is not only about fairness but also the belief, supported by research, that a diverse workforce should be seen as a competitive advantage rather than simply as a legal constraint.

Today, the term is increasingly seen as a demographic and economic imperative, particularly in the US, where traditionally white people (mostly of European descent) have constituted the majority of the population. It is now the case, given existing demographic trends in the US, that by 2027 there will no longer be any ethnic or racial majority group in the under-30 age range. Over time, that trend will extend across all age groups and increasingly to other countries as well. What the DEI movement is, in effect, saying is that corporate America's workforce should reflect the diversity of America. This would mean that every level within an organisation, from new hires to the C-suite (including those with 'chief' in their titles, such as CEO), should be representative of the country as a whole and, hence, of the potential customers for the company's products or services. Looking at DEI as 'Demographic and Economic Imperatives' essentially means that a business must be aware of the characteristics of its customers, suppliers, and other stakeholders when considering hiring and training policies. If not, it may have a reduced financial performance compared to when following DEI considerations. Many Republican politicians disagree with the principles underlying DEI, while in Europe it is much more widely accepted as quite normal to apply such principles.

8.3 What Are ESG Principles and How Are They Applied by Financial Rating Agencies?

The abbreviation ESG stands for 'Environmental, Social, and Governance' factors. ESG is now a widely used term in the finance industry. Our particular interest in ESG is in the creation of new assets which meet genuine criteria for sustainability and provide ESG funds with suitable assets. Figure 8.2[2] summarises some of the factors considered in ESG assessments, many of which will be relevant to any company.

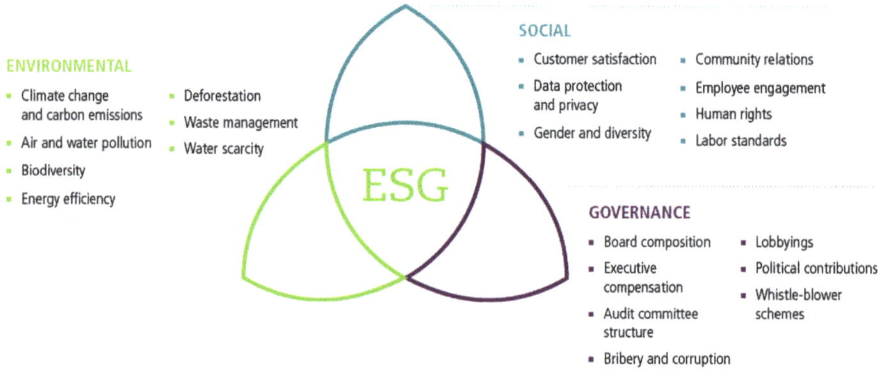

Figure 8.2. ESG principles.

[2]Trade Financial Global https://www.tradefinanceglobal.com/posts/sustainable-finance/.

Figure 8.3. The three P's: people, profit, and planet.

The word *Sustainable* at the centre of Figure 8.3,[3] could be defined as integrating E&S criteria as well as normal economic or financial criteria (achieving the investor's required rate of return for an asset in the particular risk class), into **financial investments, lending, insurance, and corporate (real) investment decisions for the long-term benefit of investors, beneficiaries such as pension plan holders, and society at large**.

The implication is that those investors (and corporates) which do not consider E & S principles and the G (governance) principle of the company, **including governance of climate-change risks**, may not generate as high returns **in the long term,** or even in the short term, as those who do. Indeed, a long-term versus short-term perspective is critical in this new approach to investing/lending and to the viewpoint of bank prudential regulators – hence one aspect of a required **transition in thinking**.

Sustainability (as in the centre of Figure 8.3) **is most often defined as *meeting the needs of the present without compromising the ability of***

[3]https://www.yelp.com/biz_photos/microgreensla-hermosa-beach?select=WNSb0ZJmVFmJD9aT3zUK-g.

future generations to meet theirs. Its three main pillars are social, environmental, and economic (a company must make a profit to survive). These three pillars are informally referred to as 'people, planet, and profits', or **3Ps**, as shown in the figure.

An alternative way of looking at **interdependencies between people, planet, and profit** is to consider the dependence of economic systems (manufacture, electric power generation, food production, etc.) on the maintenance of life support ecosystem services, such as climate regulation, land quality, water resources, and biodiversity. The Earth needs to operate within its 'sustainability limits'.

The first use of ESG ratings is to enable **security selection decisions** for investors and to provide criteria for IPOs and secondary issues of new equity and appropriate loan and insurance **risk underwriting**. The availability of an ESG rating on a company, i.e. the company as a whole (unlike a credit rating, which is focused on a particular security), enables an investor to make use of both financial and sustainability information that is relevant to the valuation of that company's securities. But just as the *availability* of data remains a serious issue, so too is the issue of the particular *metrics* used and their *reliability* and *comparability* across companies. The production of indices based on this data and ESG funds which track these indices is also an area full of 'minefields'. We look at some aspects of rating agencies' approach to constructing their ESG indices in Chapter 8.

Most companies, banks, insurers, and investment managers have a sustainability department devoted to matters relating to environmental [E] and social [S] issues, along with those related to how the company is governed and to what **purpose** [G], collectively constituting ESG. While climate change may be the **existential issue for mankind**, in terms of analysing the potential return on an investment, social and governance issues, as well as other environmental issues, are also critical. In the EU, the European Commission has made it clear in its Directives and regulations that it believes social issues are almost or as important for *society* as environmental issues.

While this book focuses on climate change, since this is the existential issue for mankind, it is only one of several issues that have attracted investor attention in recent years. **Other environmental [E]** issues are biodiversity, drinking water contamination, river pollution, water shortages for domestic and agricultural use, air pollution, loss of rainforest, ecosystem destruction, land degradation, and damage from mining and waste disposal, many of which reflect a loss of **natural capital**. All these issues are environmental [E] issues.

Social [S] issues in particular have come to the fore more recently in financial and corporate circles. The 'S' in E&S includes matters such as diversity (ethnic origin, gender, sexual orientation, disability, etc.) in boardrooms, senior management, and other areas of business; equal pay (males and females doing the same job to be paid at the same rate); human rights; health and safety; consumer protection; income and wealth distribution; the terms and conditions of 'gig' workers, such as those working for Uber; and minimum pay rates.

Some objectors to including social issues such as workforce diversity consider that such issues are irrelevant to financial performance and that the job of directors of a company is simply to maximise profit with no mention of the risks involved in focusing solely on absolute profit and not also on risk or timescales of impacts. Recent research has, however, demonstrated fairly conclusively that in the case of gender balance in companies (the proportion of females in a company), diversity pays off handsomely.

Whether companies with more gender-balanced workforces outperform their least-balanced peers remains a contentious subject. But some evidence suggests that they have outperformed by as much as two percentage points. However, this could also depend on which year or years were chosen for research.

However, ESG is not yet a science and may never be, and an investor should not make any assumptions about high-scoring ESG companies or ESG funds providing higher returns over any specified period than low-scoring ones. Indeed, some recent research suggests that returns may be comparable with other types of assets rather than exceeding them. But the past may be no guide to the future as the world's climate changes.

Investors are increasingly demanding E&S information from companies before deciding whether to invest or not, and companies are becoming more aware of the impact of their behaviour on both of those fronts if they lack concern for, and do not take appropriate action, on such issues. Investors are generally seeking alpha from the companies they invest in (unless they are index investors) by using 'non-financial information' as well as financial data. Conversely, corporates are concerned about the impact, whether positive or negative, on their share price if they neglect such issues. As we explain later, the EU now uses the term '**sustainability-related financial disclosure**', as it is more meaningful than 'non-financial disclosure'.

The increasing focus on diversity in boardrooms in many countries, in management, and on the 'shop floor' is a result of the belief that companies **should** reflect the diversity of the population, not only in routine jobs but also right up to CEO level. The importance of treating all people equally, coming under DEI policies, arises also from the belief that different ways of thinking from diverse populations are likely to lead to better decision-making and potentially a higher level of profitability. This is partly because the customers of most firms are households, and households are diverse. Thus, having a corresponding diversity in the workforce through all levels within an organisation may result in relevant factors that were not previously considered becoming a normal part of the process of designing, marketing, and selling new products.

Morgan Stanley research, as well as other analyses, reports: 'Our research suggests that using racial and ethnic board diversity factors can improve US large-cap equity stock selection. There may be additional benefit in tilting toward more diverse companies across all four developed markets'. As a result of such evidence of the impact of social issues on securities' prices, understanding and analysing non-financial factors (sustainability factors) as well as financial factors is likely to be a means of achieving improved financial performance in investment portfolios and, for banks, a higher return on capital from lending. For insurers, it should lead to a lower loss ratio through better pricing of risks. High-scoring companies in an industry are more likely to be protecting natural capital and considering social factors, which in many cases could lead to outperformance relative to their peers.

Corporate governance [G]: The third letter [G] in ESG is always considered along with E&S in the analysis of a company to create a set of ESG principles. Governance can be thought of as how the company strives to ensure that its *purpose* is achieved. One key purpose will be achieving the appropriate risk-adjusted return for shareholders. But it also includes ensuring that the company survives profitably **in the long term in the interests of its shareholders and other stakeholders**. This highlights the distinction between profitability over relatively short periods (quarterly reporting) and longer-term risks and returns for shareholders. As many of the large investors are pension funds with long-term liabilities, such funds need to be concerned with risks perhaps one or two decades away, which implies that they are concerned about knowing that returns will be sustainable long into the future.

A specific governance focus on environmental and social risks is included in TCFD's Recommendations, where issues of governance are one of the four areas it focuses on. It requires companies to develop a framework specifically for ensuring appropriate **climate risk governance**.

The objective of all four of TCFD's sets of disclosures is to 'nudge' directors and officers of a company to think harder about climate change risk rather than fudging the issue by issuing bland statements which might amount to no more than 'greenwashing', i.e. pretending that something was being done about these issues.

8.3.1 *Ratings, Rating Agencies and Users of ESG Ratings*

S&P describe how they undertake an ESG assessment, detailed as follows, and also a joint venture by S&P with Boston Consulting Group, called Climate Credit Analytics[4]:

> S&P Global Ratings' environmental, social, and governance (ESG) evaluation is a cross-sector, relative analysis of an entity's capacity to operate successfully in the future and is grounded in how ESG factors could affect stakeholders, potentially leading to a material direct or indirect financial impact on the entity. ESG factors typically incorporate the **entity's impact on the natural and social environment and the quality of its governance**. **Our definition of stakeholders goes beyond shareholders to include employees, the local community, government, regulators, customers, lenders, borrowers, policyholders, voters, members, and suppliers, among others**.
>
> First, we establish an ESG Profile for a given entity, which assesses the exposure of an entity's operations to observable ESG risks and opportunities, accounting for the governance structure in mitigating risks and capitalizing on opportunities. Second, we assess the entity's long-term *Preparedness* namely its capacity to anticipate and adapt to a variety of long-term plausible disruptions.

Clearly, S&P has considered the sustainability issues that arise if a company is or is not prepared for eventualities, and they consider the degree of *preparedness* of a company as an important contributor to the ESG rating, which may raise or lower the final evaluation. It is important to note that the ESG evaluation is **not a credit rating, a measure of**

---

[4]https://www.spglobal.com/_assets/documents/ratings/ratingsdirect_environmentalsocialandgovernanceevaluationanalyticalapproach_1176513_apr-10-2019.pdf, and for Climate Credit Analytics, https://www.spglobal.com/marketintelligence/en/solutions/climate-credit-analytics.

credit risk, or a component of its credit rating methodology. However, the information it gathers for an ESG evaluation *can inform its credit analysis of rated entities.*

ESG factors may impact industry risk, competitive position, cash flow and leverage, and management and governance. They identify these as the areas where ESG factors can influence the capacity and willingness of issuers to meet financial commitments (interest and principal payments on debt securities or loans).

The next step after the production of an ESG rating for a company is the aggregation of companies in an index based on their ESG performances. For example, FTSE Russell Indices (part of the London Stock Exchange Group, as is Refinitiv) overcome the difficulty many investors would have in trying to select appropriate stocks for a large, diverse portfolio based on ESG characteristics. An example of the use of such indices was demonstrated recently by BlackRock when one of its new funds – *BlackRock US Transition Readiness* exchange-traded fund (ETF), launched in April 2021 – soon became the largest ETF launch ever as it quickly reached $1.2 billion. The fund was based on the Russell 1000 index. The fund was created by taking this index and then assigning weights to its components to reflect their 'carbon transition readiness score' rather than the usual E, S, and G components.

Chapter 9

Financial Portfolios and Climate, Temperature, and Paris Alignment: Case Study of InfraRed Climate Aligned Fund

9.1 What Is the Meaning of Climate Alignment?

The data, metrics and targets and the corresponding climate-related disclosures recommended by TCFD that we considered earlier are all designed to ensure 'alignment'. Alignment is a much used but somewhat ill-defined expression. The *Center for Climate Aligned Finance*[1] (part of the Rocky Mountain Institute) based in New York provides the following useful definition: 'Climate-aligned finance is the process of bringing a financial institution's portfolio of lending and investment activities into alignment with a 1.5°C consistent emissions pathway'. It is a short and simple definition, but achieving it requires an arduous journey and, in fact, a transition in thinking by those working in the finance industry.

The risk for lenders and investors is that many assets that they hold may not be valued correctly, as lenders and investors may not be pricing them in anticipation of this real economy transition. 'Market efficiency' may be low in this regard because of the myopic view that many investors seem to have about the longer term.

[1] https://climatealignment.org/#:~:text=Climate%2Daligned%20finance%20is%20the,%C2%B0C%2Dconsistent%20emissions%20pathway.

9.1.1 *A Way Ahead? Portfolio Temperature Metrics/Portfolio Climate Alignment*

The Investment Leaders Group (ILG), a group which includes large financial companies such as Pimco, Robeco, Zurich, State Street and HSBC, has, like many others, concluded that reporting a temperature score for a fund, based on standardised metrics, is the best way for stakeholders to know if the fund manager is 'aligned with the Paris Agreement', i.e. 2°C or less. But why use this particular metric? One good reason is that non-specialists can understand it in an intuitive way, and it is a relatively simple measure.

As an increasing number of countries do legislate for net zero in all sectors of the economy, corporates, financial companies, households and central, regional and local governments will all need to adapt. For investors and lenders, this will mean identifying risks and opportunities in the transition to a net-zero economy. This will require a framework for measuring the alignment of companies and banks and investor portfolios with net zero targets or with temperature targets, as well as targets for households and sub-national authorities.

With such a framework:

- Investors and lenders will be able to assess the position of companies and portfolios on the path to net zero – to compare them with peers and track progress over time.
- This will, in turn, increase the ability of financial institutions to effectively allocate capital in a way that supports climate goals, assess transition risks and engage with companies to facilitate real economy emission reductions.
- This will also increase the ability of financial institutions to track their own contribution to limiting global temperature rises and reaching net zero.
- While a number of methods exist today, their fragmentation and methodological differences prevent the different methods from fulfilling their fundamental purpose, as outputs across companies and portfolios lack consistency and comparability.
- Methodological differences could also increase greenwashing risks, as the metric could be adjusted to fit different purposes.

The text above is from 'The Portfolio Alignment Team' publication 'Measuring Portfolio Alignment',[2] led by David Blood and Iren Levina.

---

2 https://www.tcfdhub.org/wp-content/uploads/2020/10/PAT-Report-20201109-Final.pdf.

Portfolio alignment is, indeed, the objective we have focused on, as it should enable more effective action to be taken in respect of the real economy. We have, however, changed the text from 'without such a framework' to 'with such a framework', and thus we are listing the positive benefits of having such a framework rather than the negatives of having no framework. It is important to understand however that, as the fourth point above notes, at present there is no settled way of measuring alignment. We are dealing with a relatively new field of endeavour – minimising the risks to humanity posed by climate change. Everyone is learning as they go but based on an understanding of different approaches that have been developed.

There are three possible approaches to measuring alignment with net zero, as the Portfolio Alignment Team (PAT) explains:

- Measuring the percentage of the total value of the portfolio which reflects companies in the portfolio with net zero targets. This does not, however, measure the alignment of the part of the portfolio without targets nor different de-carbonisation transition pathways of individual companies. It also reflects only a 'commitment' and does not consider the 'transition', i.e. it does not tell us where the company is on its path towards net zero.
- Measuring the deviation of the portfolio from a target or sector benchmark, which then involves measuring the deviation of each company within the portfolio. For example, the EU Climate Benchmarks for portfolios, from which index companies can then create indices for investors to replicate, require them to decarbonise at an average rate of 7% per year, as well as being aligned with the IPCC 1.5°C scenario.
- Using degree warming metrics to which we referred above. This can use the output from the second approach above – 'deviation from a sector benchmark', which can be converted to a temperature score to provide *degree warming metrics.* SBTi provides guidance on how to set targets and also on how to convert to a temperature score.

Degree warming metrics seem to be more 'intuitive' to most people than deviations from a benchmark. Those investors who have already started experimenting with the degree warming approach, such as the insurer and asset manager Axa, find that the approach is useful not only to assess climate-related risks but also to facilitate capital allocation in line with climate objectives and, importantly, to provide a good starting point for engagement with a company on its decarbonisation strategy, plans, and actions.

When we think about the impact of a single company's contribution to global warming, we are faced with the awkward fact that if only one company in a sector decarbonises, it will make little difference to global emissions. As a result, when it comes to translating a carbon budget into a benchmark, what we are in fact doing in assessing a company is comparing it with its peers around the world. What we need is for all companies in the sector to decarbonise at the scientifically calculated rate that will enable the target to be met. So, the benchmark sets what the industry needs to do, and the company's position relative to this index tells us if it is acting at the rate at which, if all companies in the industry acted, then the industry would keep within its carbon budget.

9.2 Portfolio Temperature, System Transformation, and Paris Alignment

The Paris Agreement alignment is not merely about asset portfolios. It is really about a system transformation and alignment. It is also an approach to ensuring that governments of developed and developing countries institute policies that should ensure 'better development'. The slogan 'build back better' has been used by politicians in several countries, and while the use of these three words as a slogan may simply be 'politics', it is nonetheless important to ensure that what the finance industry achieves serves people and the planet in a wider sense. It is not simply about a narrow concept of project-based finance flow alignment. It is about making countries and the planet less susceptible to crises and more resilient in the face of crises which do arise, such as global pandemics, tsunamis, water shortages, famine, sudden rises in energy costs, and other events that disrupt human, animal, and plant life. A good summary of this approach would be the United Nations Sustainable Development Goals (SDGs).

Specifically on climate finance, we emphasise once more that there are two key desired outcomes: one is for investors and bankers to reduce the 'temperature' of their portfolios, and the other is to shift their portfolios in a transition over time by **increasing their allocation to new issues of shares and bonds or refinancings of and lending to companies creating climate mitigation and adaptation assets for the real economy**. Without that second step, while we would see falling emissions, we would also see a fall in living standards, which would arise if we simply shut down GHG-emitting companies without 'building back better'.

The development of ESG ratings and the challenges involved is a good starting point for considering the issue of how to transition the world to achieve the Paris objectives: 1.5°C/2050 Net Zero Paris Alignment, by ensuring finance flows are consistent with low GHG emissions and a climate-resilient pathway. Unfortunately, there are many issues in the creation, construction, and use of indices.

9.2.1 *Paris-Aligned Funds*

The EU Technical Expert Group (TEG) Final Report observes:

> Benchmarks have an indirect but important impact on investments. Many asset managers and investors rely on them as investment monitoring solutions to track their return (passive and smart beta investment strategies) or to define the investment universe and to measure the performance of an investment fund/portfolio (active investment strategy). A more recent use case for benchmarks that has emerged in the industry is their use as an incentive tool to encourage companies at scale to improve their ESG performance. Benchmarks play a significant role and can be a key lever in aligning the investment and analyst community with long-term sustainability considerations and the transition towards a low-carbon economy. Since conventional benchmarks do not reflect low-carbon considerations in their methodologies and are not appropriate to measure the performance of sustainable investment strategies, over the past decade benchmark administrators have designed hundreds of ESG and 'low-carbon' benchmarks.
>
> While the supply of low-carbon benchmarks is there, lack of harmonisation of the methodologies (especially lack of consensus on how comprehensive the assessment of a carbon footprint should be) and lack of clarity on the objectives pursued with regard to the impact on global warming have affected comparability, reliability and adoption of low-carbon indices. Furthermore, the varying °C of reporting hinders market players' ability to compare indices and choose the adequate benchmarks for their or climate-related investment strategy.
>
> Finally, the benchmarks currently offered in the market do not necessarily align with the financing needs implied by the limitation of global

> warming to well below +2°C pursuant to the Paris Climate Agreement, as current benchmarks are likely to be more aligned with a 'business-as-usual' scenario, where temperature rises range from 4°C to 6°C, leading to catastrophic damage to the Earth.

The TEG arrived at a need for two different but related benchmarks which asset owners or asset managers could use to develop indices and/or funds, based on the requirements of the two indices. The two types of climate benchmarks pursue a similar objective but have different levels of restrictiveness and ambition. The so-called Paris Aligned Benchmarks (PABs) are designed for highly ambitious climate-related investment strategies and are characterised by stricter minimum requirements, while EU Climate Transition Benchmarks (CTBs) allow for greater diversification and serve the needs of institutional investors in their core asset allocation. A summary of the minimum standards is shown in a Standard & Poor's (S&P) graphic in Table 9.1.[3]

The main users of EU CTBs are likely to be institutional investors such as pension funds, with the objective of protecting assets against various investment risks related to climate change and the transition to a low-carbon economy, i.e. transition risks. The main users of EU PABs are likely to be institutional investors at the forefront of the immediate transition towards a +1.5°C scenario. The differentiating features are as follows:

- EU PABs allow for a higher decarbonisation of the investment relative to the investable universe;
- EU PABs have additional activity exclusions;
- EU PABs have a stronger focus on opportunities with a significantly enhanced green share / brown share ratio.

A benchmark is not an index or a specified list of securities which match requirements. It is simply a 'set of rules' according to which an index provider can undertake security selection and portfolio allocation to create a benchmark-compatible fund. As an example of indices based on the EU benchmarks i.e. following the requirements of the CTB and PAB benchmarks, S&P is one of several index providers that have produced such indexes based on these benchmarks. S&P call these Paris Aligned Climate Transition (PACT) indices:

> To meet the EU's proposed minimum standards for CTBs and PABs, we have launched two new index series: the S&P Paris-Aligned Climate

[3] S&P Dow Jones Indices LLC, The EU Technical Expert Group on Sustainable Finance. Table is provided for illustrative purposes.

Table 9.1. EU Paris-aligned and climate transition benchmarks.

Minimum Standards	CTB	PAB
Minimum Scopes 1, 2, and 3 GHG Emission Intensity Reduction Compared with the Underlying Benchmark	30%	50%
Scope 3 Phase-in Period	Up to four years	
Baseline Exclusions	Controversial weapons and societal norm violators	
Activity Exclusions	None	• Coal (≥1% revenues) • Oil (≥10% revenues) • Natural gas (≥50% revenues) • Electricity producers with lifecycle GHG emissions higher than 100 gCO_2e/kWh (≥50% revenues)
Year-on-Year Self-Decarbonisation of the Benchmark	At least 7% on average per year: in line with or beyond the decarbonisation trajectory from the IPCC's 1.5°C scenario (with no or limited overshoot)	
Minimum Green Share/Brown Share Ratio Relative Compared with Investable Universe (Voluntary)	At least equivalent	Significantly larger (factor 4)
Exposure Constraints	Minimum exposure to sectors highly exposed to climate issues is at least equal to equity market benchmark value	
Corporate Target Setting	Weight increases considered for companies that set evidence-based targets under strict conditions to avoid greenwashing	
Immediate Disqualification	If misaligned with trajectory for two consecutive years	

(PA) Indices and S&P Climate Transition (CT) Indices, respectively, collectively referred to as S&P PACT: the S&P Paris-Aligned & Climate Transition Indices. Both methodologies meet the respective EU requirements and other climate objectives to become compatible with a 1.5°C scenario and recommendations from the TCFD.

9.3 The Impact Investment Spectrum

A portfolio aligned with the Paris Agreement is critically important for the Earth and its inhabitants – humans, animals, and plants. But there are other important environmental and social issues that should be considered when making investment decisions in what we might call, in shorthand, an ESG world. These considerations apply as much to corporates making real asset investment decisions as to financial firms.

While climate change may be the existential issue for companies and humankind, in terms of analysing the potential return on an investment and also the impact it may have on the lives of those impacted by it (impact), social and governance issues, as well as other environmental issues, are also critical.

There is a wide range of approaches to creating sustainable portfolios, but all of them should be designed to have some 'impact' on the wider environmental and social issues that society faces. Figure 9.1[4] shows that there is a wide range of investor objectives. If there is no impact of investment policy on the wider world outside of a company, project, or fund management house, then it might as well simply apply a classic approach. This can be thought of as an approach where impact creation is not considered at all, and the only concern is financial return. On the other

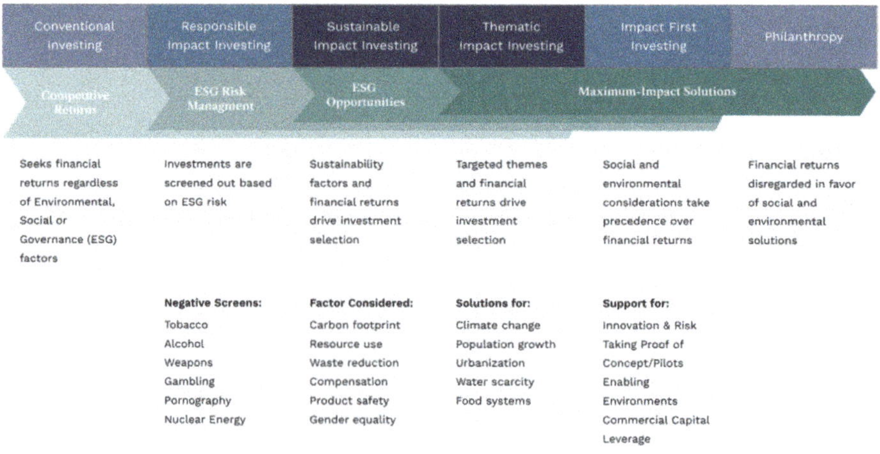

Figure 9.1. A range of investor objectives, from conventional to philanthropy.

[4] https://sonencapital.com/impact/our-approach-to-impact-investing.

hand, subsequent approaches are all designed to have an impact in varying degrees.

Sonen Capital defines 'impact investing' as the full range of investment opportunities which address an investor's desire to realise financial return while achieving meaningful and measurable positive social and/or environmental impact. The **Impact Investing Spectrum** in Figure 9.1, after conventional investing, shows different categories of impact investing:

> **Responsible** impact investing avoids negative impacts. It was once the main policy of fund managers who sought to avoid investing in 'bad' things such as tobacco, oil & gas, and weapons manufacture. This approach was often driven by the underlying asset owner (a pension fund, an endowment, etc.), stating that this was a requirement of the mandate. Note that 'nuclear energy' is a 'negative screen' in this diagram. However, many managers are now reversing that exclusion and instead including nuclear. Weapons manufacture may also start to be included rather than excluded.
>
> **Sustainable** impact investing involves the integration of ESG considerations into investment decisions. In the past, the data available for doing this were very limited, but they are increasingly becoming available. Factors that would be considered include the impact of a company on climate change risk and the well-being of the employees of the company.
>
> **Thematic** impact investing considers what main activity the company is involved in rather than just how they ensure that their investments have a sustainable impact. Here, we are concerned only with themes with a (positive) social or environmental impact.
>
> **Impact First** investing considers the impact of an investment to be more important than the ROI of the investment. As with thematic investments, there will be a clear connection to a particular E or S impact theme.
>
> **Philanthropy** generally involves providing a grant to a project. There is no expectation of any financial return. The only consideration is its impact. But many private philanthropists are willing to give money to projects they support.

Once environmental and social issues are considered as factors in investment decisions, all approaches apart from the 'classic' ones involve *avoiding harm*. When we move on to 'sustainable' and 'thematic', then we have an approach that *considers all stakeholders* (investors, suppliers, customers, employees, the local area, and residents) and tries to benefit them or, at worst, not disadvantage them in some way. With 'impact first' investment, the focus is on contributing to 'solutions' to particular issues. Two issues on which an *impact first* investor or fund might focus would include the energy transition and ensuring freshwater supplies.

9.4 Case Studies: InfraRed Capital Partners, Gore Street Energy Storage Fund, Gresham House, Schroders Greencoat UK Wind, and NextEnergy Capital

9.4.1 *Infrastructure Equity*

Most clean energy projects start life as unlisted assets (project finance company). Institutional investors can access such unlisted projects either directly or subsequently through funds, which can also be listed. We focus on unlisted investments. Private market assets may have different characteristics from both listed infrastructure assets (such as utilities) and the companies which underlie indices such as the S&P 500 or the FTSE 100. Particularly relevant is that their liquidity is lower than that of listed assets, and thus there is no 'market price', as there will normally be with listed equity. In theory, they should offer a higher expected return for a given level of risk to compensate for their illiquidity, but in practice, they may not.

9.4.2 *InfraRed Capital Partners*

We have covered many aspects of financing clean energy infrastructure in the previous chapters. We have also covered project finance (infrastructure bonds) and financing companies at a development stage prior to TRL9. We now have a case study of a large infrastructure manager which, among other private (unlisted) investment funds for institutional investors, manages a large London Stock Exchange (LSE)-listed clean energy infrastructure fund (TRIG) as well as a general infrastructure fund (HICL).

The company was established in 1990 when there was little clean energy infrastructure available and continues to manage such assets. But today, it is an international infrastructure investment manager, with over

190 professionals operating worldwide from offices in London, New York, Sydney, and Seoul. InfraRed manages over $14 billion of **equity capital** for investors around the globe in **listed and private funds** across both income and capital gain strategies. It has invested in over 300 infrastructure assets across 18 countries and realised more than 90 investments.

The Renewable Infrastructure Group (TRIG)

This fund, managed by InfraRed Capital Partners, is a FTSE 250, London-listed UK investment company. It was one of the first renewable energy investment companies to be listed on the LSE.

It is important to understand the difference between this listed fund and 'private market funds', as both types of fund are managed by InfraRed. Private market funds are for investment on a wholesale scale, and secondary market sales of participations involve bilateral agreements between buyer and seller, as there is no market for pricing such share participations. On the other hand, TRIG's shares are priced daily on the LSE. Private market funds should generate an 'illiquidity premium' over listed shares, which is compensation for the illiquidity. However, depending on the demand and supply balance, there may or may not be such a premium, and indeed a quick sale might require a discounted price and hence an illiquidity discount.

TRIG's portfolio comprises established renewable technologies: 48% of the portfolio is onshore wind, 34% is offshore wind, 13% is solar, and 5% is flexible capacity. The factsheet above gives more information on TRIG. The company's purpose is to generate sustainable returns for its shareholders from a diversified portfolio of renewables infrastructure that contributes towards a net zero carbon future. TRIG's highly diversified portfolio consists of onshore and offshore wind farms, solar parks and flexible capacity in the UK and across Europe. These assets generate revenue from the sale of electricity and from government-backed green subsidies and incentives. TRIG employs a differentiated approach to portfolio construction, assembling a mix of assets that contribute to decarbonisation and security of energy in Europe whilst aiming to provide investors with long-term, income-based returns with a positive correlation to inflation.

We noted above that TRIG had 5% of its assets in 'flexible generation'. This type of generation is an absolute requirement in systems transitioning to renewables, as it can 'fill in the gaps' in renewable generation power output. It can be provided by energy storage or by natural gas turbines or, as in the following case, storage as an instant source of electricity and natural gas generation afterwards. An example of this type of company that was in its portfolio (but has now been sold to realise a capital gain) is Statera Energy. For more information on flexible production, see Section 10.2.

9.4.3 *Gore Street Capital and Its Energy Storage Fund*

Just as Infrared Partners is the investment manager for TRIG, so Gore Street Capital (GSC) is the manager for the Gore Street Energy Storage Fund. Gore Street Capital was one of the first asset managers in energy storage and is the only UK-listed energy storage fund with an internationally diversified portfolio. Since the inception of Gore Street in 2015, it has increased its assets under management at the time of writing to 1.7 GW spread across five grid systems, including in Ireland, where the company manages assets for the Ireland Strategic Investment Fund (ISIF).

The company's internationalisation seems to have helped it weather falling UK revenues in 2023, when storage capacity on the UK grid almost tripled and prices in a number of market sectors fell. In a similar way to diversification policy in equity or debt securities as a means of improving the risk/return profile of a portfolio, Gore Street has taken a proactive approach to diversifying revenue sources across different revenue streams within a country's market structure and through revenue streams from a range of countries.

In terms of geographical diversification, it has storage assets in the UK, Ireland, Germany, Texas, and California. It is this approach to asset management (physical assets) that has enabled it to differentiate its performance from many competitors. Its target portfolio includes decarbonisation, digital inclusion, regeneration, health and education, resource efficiency, and waste solutions.

9.4.4 *Gresham House: Specialist Investment*

Gresham House is a specialist alternative asset manager providing funds, direct investments and tailored investment solutions, including co-investment across a range of differentiated alternative investment strategies. They note that significant investment is needed in new sustainable infrastructure that can support and accelerate the changes and innovative solutions required to address key environmental and societal challenges. They identified six subsectors where real asset-based solutions and their 'modern' approach to infrastructure have the best opportunity to create the biggest impact – and therefore the biggest financial reward.

Investing in sustainable infrastructure with Gresham House offers appealing characteristics to buy-and-hold investors and provides the potential for stable, long-term returns. They note that sustainable infrastructure investments can (meaning may or may not) benefit from:

- downside protection through substantial asset-backing,
- the potential for capital growth,
- attractive long-term cash flow yields,
- material positive social and environmental impact,
- returns with a low correlation to the broader economy or financial markets.

As well as its sustainable infrastructure funds, it also offers investors new energy funds. These funds can be accessed through listed vehicles, unregulated collective investment schemes (UCISs), or co-investment.

9.4.5 *Schroders Greencoat UK Wind*

Schroders Greencoat LLP is a specialist manager dedicated to the renewable energy infrastructure sector. It has offices in London, Dublin, Germany, the Netherlands, and Chicago, with approximately £10 billion under management, making it one of the largest dedicated managers in Europe. It was founded in 2009 and currently has fund mandates with strategies investing in bioenergy, renewable heat, and solar and wind energy infrastructure across the UK, Europe, and the United States.

Greencoat UK Wind invests in operating UK wind farms. Given the nature of the company's income streams, the board has increased and intends to continue increasing the dividend in line with the retail price index (RPI) inflation. The company also aims to preserve capital on a real basis by reinvesting excess cash flow in additional operating UK wind farms and through prudent use of portfolio leverage.

The company offers exposure to UK wind generation through a 'premium' listed vehicle. Greencoat UK Wind is invested solely in operating onshore and offshore UK wind farms, which are currently producing income. Greencoat UK Wind was the first renewable infrastructure fund to list on the LSE main market and is the only infrastructure fund or renewable infrastructure fund domiciled in the UK. The company is a constituent of the FTSE 250 index. Greencoat UK Wind is incorporated in England and Wales and is a UK **investment trust**. The company has a broad investor base that includes numerous blue-chip institutional investors and also retail investors.

9.4.6 *NextEnergy Capital*

This fund manager in total has $4 billion in assets under management (AUM). It has one listed fund and five institutional funds (unlisted). In total, it holds 400 assets with a total capacity of over 3 GW distributed over nine countries.

NextEnergy Solar Fund Limited (NESF) Listed on the London Stock Exchange

This is a specialist solar and energy storage climate impact fund, which is listed on the premium segment of the LSE. It currently has an installed capacity of over 983 MW, spread among 102 individual operating assets in the UK and Italy, with an asset value of over £1 billion. NESF is one of the largest listed solar and energy storage investment companies in the world.

NextPower Private ESG Funds

NextPower V ESG is NextEnergy Capital's fifth investment vehicle focused on the solar sector. It is targeting mid- to double-digit returns. But beyond the financial returns it generates, it expects to contribute to the decarbonisation of the power generation sector, reducing electricity prices and increasing energy security. It restricts itself to OECD country solar assets and energy storage expertise. It is a 10-year closed-ended vehicle that qualifies as an Article 9 Fund under the EU SFDR.

NextPower UK ESG is a private, unlevered fund investing in greenfield subsidy-free solar projects with PPAs in the UK. It is a 10-year closed-ended private fund launched in December 2021 and has raised total commitments of £733 million. Interestingly, the UK Infrastructure Bank provided cornerstone match-funding of £250 million. The fund is classified as Article 9 under the EU SFDR.

Part III

The Energy Transition Challenge: Converting Variable Renewable Energy into 24/7 Dispatchable Power, Decentralised Systems

Chapter 10

What Is the Energy Transition and What Are Its Technologies?

In this and subsequent chapters, we move on to consider the actual issues in the energy system, particularly in the electricity system, that require the very large volume of funding we referred to in an earlier chapter. Without some understanding of the existing system and its characteristics and the structure of the New Energy Economy and its quite different characteristics, it will be difficult for debt and equity financiers to undertake due diligence on projects which are offered to them. Equally, for entrepreneurs, intrapreneurs, corporates, utilities, and others, without some understanding of the financing side and how due diligence is undertaken, which we covered earlier, it is less likely that they will convince financiers to provide the necessary capital for their project at the lowest possible cost of capital. Following this chapter, we cover the individual technologies and their techno-economic features.

10.1 A Key Issue: The Loss of 'Firm Power': Integrating Variable Renewable Energy (VRE) into the Electricity System, 'Nameplate Power', Capacity Factors, LCoE, and the Energy Trilemma

The current energy transition (a key component of the sixth wave of innovation in Figure 1.3) has been ongoing for some decades, in fact, since 1990 and will hopefully have achieved society's objectives for mitigating further climate change by 2050. It must be remembered that in 1990 (some 35 years ago), the world did not have solar power or wind turbines that could be used at scale to generate power. The technologies had to be developed. The speed of development has been impressive relative to earlier transitions. Nonetheless, the current transition is going so much more slowly than is required to even come 'within the last mile' of the Paris Agreement objectives. It is for this reason that a massive step-up in investment in electricity systems and other clean energy technologies requires a massive increase in the funds allocated by investors to this task. However, as we have noted before, this will only be achieved either through a massive increase in regulation which forces companies to replace their fossil fuel assets or through much higher and broader carbon taxes, or by governments being willing to take on many of the risks to which investors would be subject if they were to singlehandedly speed up their investment in carbon-free assets.

In the absence of fossil fuels to provide energy for heating, cooking, and lighting, transport and industry (as fossil fuels are planned to be phased out by 2050), there is a need to find alternative means of generating electric power and producing heat, enabling transportation and continuing to have an industrial base but without using coal, natural gas, or oil. Electrification will have a major role, for example electric transport (cars, trucks, and rail), electric heat using heat pumps for space heating, and electric industrial processes. In addition, it is likely that hydrogen will be used across all sectors, though currently its price is very high relative to the price of fossil fuels. Hydrogen will not only be used directly but will, to a limited extent, also be used to manufacture synthetic liquid hydrocarbon fuels for use in transport.

Hydro, solar, wind, nuclear, geothermal, and biomass as sources of primary energy have to be converted to energy carriers, as shown in Figure 1.2. The first, and main, such carrier is electric power. Clean hydrogen can then be manufactured from a proportion of that power along with water and a proportion of that hydrogen can then be converted to hydrogen derivatives, including synthetic methane, methanol, ammonia, and synthetic liquid hydrocarbon fuels.

In the past, the whole power system was based on using heat energy generated from fossil fuels to produce steam to drive a steam turbine, which then drove an electricity generator. Such systems provided continuous 24/7 power generation (dispatchable power), but solar and wind, which will eventually make up a high proportion of power generation, are not continuous but intermittent, i.e. variable renewable energy (VRE). Today, the only large sources of non-fossil continuous power generation are geothermal, hydro, and nuclear.

In more detail, generating companies provide electric power with one of two characteristics:

> **Firm power (dispatchable, round-the-clock, 24/7 supply):** Fossil fuel heat sources (coal, oil, and natural gas) connected to steam turbine generator sets, flow-of-the-river hydropower, nuclear, concentrating solar power (CSP), and geothermal heat sources;
>
> **Variable renewable energy:** Wind turbine, solar photovoltaic (PV), CSP which are non-dispatchable without energy storage and/or demand-side response (DSR) – these sources of power are not constant and indeed, in the case of solar, clearly produce no power after sundown.

The key question in today's power system design is how to convert VRE into firm power. This is necessary if power systems are to be able to 'keep the lights on' at all times. As all countries and regions need to take steps to mitigate climate change risk which must involve changes to their energy mix and sources of power, they must, at the same time, **ensure security of the energy supply**.

The term 'ensure energy security' is a key one but can refer to two different causes of energy 'insecurity':

> *First,* it means **minimising the probability of 'the lights going out' (brownouts or blackouts) by ensuring that countries have adequate 'firm power sources', i.e. 'dispatchable' power with 24/7 availability**.
>
> The energy transition requires the elimination of the use of petrol, natural gas, diesel internal combustion engines (ICEs), and, of course, coal, all of which have provided dispatchable power for many decades (with the old energy system now being phased out). In addition to providing power, the turbo-generator sets associated with generation using steam to drive a steam turbine, which then drives a power generator (turbogenerator), have also provided a key service to the grid by having 'inertia', i.e. the weight of the turbogenerator enables it to continue to turn and provide what are known as 'frequency response services' to the grid. This is required to avoid damage to the power network when power generation on the network fluctuates over short time periods. In the UK,

in one period when solar and wind (which do not provide inertia) were generating too much power, the National Grid still had to request gas turbine generators to start up to provide inertia. National Grid then had to pay a high price to encourage European countries to absorb our surplus power.

Second, in Europe for many years, it has meant avoiding the risk of Russia (or other countries) turning off gas supplies to Europe or the Middle East cutting oil supplies (although many politicians chose to neglect these risks). One of these risks was realised after Russia invaded Ukraine and in March 2022. The European Commission then announced that countries in Europe would cease buying oil and gas from Russia as soon as possible. We are not covering the issues involved in coping with the second type of energy security but only with the first.

VRE sources can, however, provide dispatchable power if linked to an energy-storage facility co-located with generation or located separately on the grid. At present, storage makes economic sense only for relatively short periods – up to 4 or sometimes 8 hours. But long-duration energy storage (LDES), often described as making power available for 8 hours or more but also defined as being available for at least 100 hours, is likely also to be required.

Even in sunny locations, the power generated by solar may also be much lower than the stated maximum power output of the solar panels for quite long periods. As a result, without longer duration storage, power users may have 'brownouts' or 'blackouts'. Indeed, solar produces only around 25% of its stated maximum power over a 12-month period. The fact that solar panels do not generate any power when it is dark is well known. But they also generate less than their stated power for much of the time. Their stated (nameplate) power is also called peak power (MWp). This is the theoretical maximum production of the panel on a clear sunny day with the panel at a right angle to the sun. In addition, the stated nameplate power in the northern hemisphere is the power that will be generated at latitude 35° north (N) in summer. No part of Europe falls into this category, with Morocco being the closest to Europe that falls within this definition. In addition, the temperature of the panels which heat up under sunshine has a considerable effect on output. The peak power will only be generated at a temperature of 25°C or below. Thus, even global warming will affect output. It is for these reasons that the actual output of solar panels known as their capacity factor is much lower than MWp and is around 25%, as shown in Figure 10.1.

10.1.1 *System Costs – The True Cost of Different Power Sources: The Energy Trilemma*

What the above makes clear is that there are substantial issues with intermittent power as a result of costs which grow disproportionately in a nonlinear way as the proportion of VRE sources of power increases. This could be partly ameliorated by reducing the historical standards of electricity reliability and on-demand availability by accepting the economic, socio-industrial, and environmental consequences of more frequent optional interruptions and non-optional brownouts or blackouts. The high capital cost for solar and wind arises from the fact that their low capacity factors, which means that the average fraction of time for which they are generating at their rated full output stated on their 'nameplate' can be as low as ~25%. This is known as the **capacity factor**, and the low capacity factor of VRE (wind and solar) means that for a stated maximum MW power output, much more VRE plant investment is required to achieve the required megawatt hours during a year. Thus, if the capacity factor of solar is ~25%, as shown in Figure 10.1,[1] then four times the amount of plant must be built to achieve that stated amount of output *and, even then, only when there is adequate sunshine.* In addition, that output is often not produced at a time when demand for power is at its highest.

A low capacity factor also means more land must be used to achieve a given true capacity which in Europe is an issue. In the case of offshore wind, the capacity factor is likely to be higher than shown here and may, indeed, reach 50%. However, on those days or weeks when there is little wind even offshore right across the European mainland, any electricity system has a severe problem if it is too reliant on wind. Indeed, Mads Nipper, a previous CEO of Ørsted which is one of the largest North Sea wind generators, said that April, May, and June 2021 'had been the third worst quarter in the last 88 quarters. It was an extraordinarily poor wind quarter. It's very serious. It is like being a farmer and it doesn't rain'. Lost earnings were said to have amounted to $220 million.

---

[1] https://www.energy.gov/ne/articles/infographic-capacity-factor-energy-source-2019.

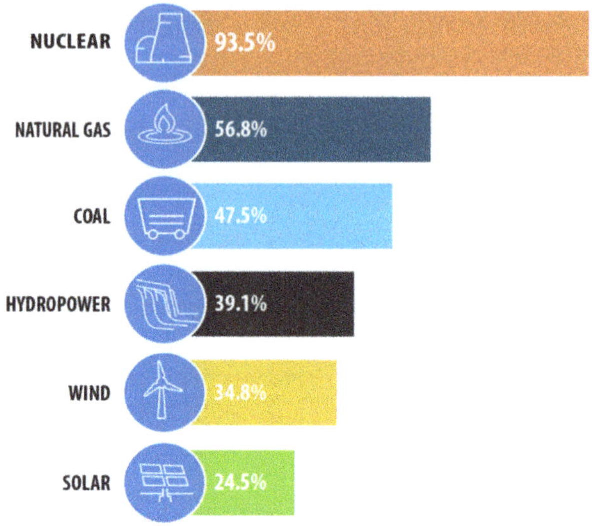

Figure 10.1. Capacity factor by energy source.

An example of the difference that location can make is shown by statistics on the location of an e-fuel plant in Chile. It has been reported that in Chile, a wind turbine runs at full load for an average of 270 days a year, whereas in Germany it runs at full load for 66 days a year due to geographical and meteorological conditions. At 74%, the utilisation rate of the wind turbine in Chile is therefore four times higher than in Germany at 18%.

The Energy Trilemma shown in Figure 10.2[2] is a graphical representation of the need for decision-makers to balance the different features of a power generation system. As with most things in life, there is a need to optimise in determining an energy and electric power policy. The interconnectedness of the three goals shown in the graphic underscores the relationship between economic growth, social equity, and environmental stewardship, including tackling energy poverty.

System cost is the additional amount that must be added to the 'farm gate price' of any form of power generation to find the full cost of that

[2] World Energy Council Trilemma https://www.worldenergy.org/assets/downloads/Trilemma-what-is-the-energy-trilemma.jpg.

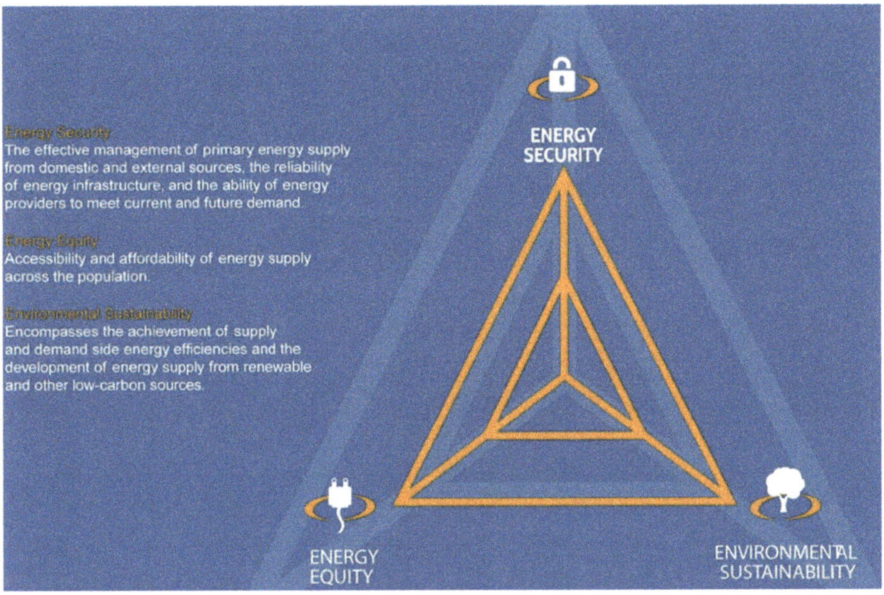

Figure 10.2. Balancing the 'Energy Trilemma'.

source. Such additional costs include long-distance transmission costs, local distribution costs, and the cost of 'balancing the grid' (as covered later). In the case of a high proportion of wind and solar on a grid system, balancing costs can be high. Transmission costs may be high since the source of generation is likely to be distant from industrial demand locations. Large amounts of battery energy storage will be required, the costs of which are substantial. There is also a need for the grid operator to pay considerable sums of money to wind and solar sources, known as 'constraint payments', when there is potentially too much power flowing into the grid and wind and solar sources are paid to switch off (when it is sunny and windy). The national grid was only ever intended for fossil fuel power stations that are situated near urban centres, not remote offshore wind farms.

As an example, in the UK, on a windy day in June 2025, one wind farm (Ocean) in Scotland received £72,000 to switch off for 30 minutes, yet because of the lack of transmission capacity (a common problem in many countries now), a natural gas plant in the south of England was paid £43,000 to produce more power. £65 million was paid out by the grid operator last year to Seagreen, which is Scotland's largest wind farm, for being asked to restrict output for more than half of the time. Total payments to renewable generators to stop generating when there is too much wind and solar and to natural gas generators to 'switch on' when there is too little wind and solar amounted to over £1 billion in 2024, and this additional cost is expected to be considerably higher in 2025.

While the cost of electricity from each generating source is frequently evaluated using a measure known as 'Levelised Cost of Energy' or Levelised Cost of Electricity' (LCoE), which we call the farm gate price, a much more appropriate measure is 'Levelised Full System Costs of Electricity' (LFSCOE). This would show that, for example, nuclear as a source of carbon-free power, which can be located near customers (in particular, the so-called Generation IV small modular reactors covered in Chapter 14) and, therefore, requires less transmission capacity and no battery energy storage, may in fact have a full system cost slightly lower or slightly higher than natural gas. In addition, Generation IV nuclear can be 'embedded' in industrial facilities or within cities, thereby cutting down both transmission and distribution costs substantially and eliminating government taxes on electricity.

10.2 The 'Holy Grail' of 24/7 Fossil-Free Electric Power

In the previous chapter, we covered the convertibility of one type of energy to another – mechanical, electrical, and thermal. In addition, we have potential energy sources, including chemical, nuclear, and gravitational, which are also relevant to power systems. There are two main problems with achieving the 'holy grail', i.e. the need to overcome insufficient power generation at some times of the day, month, or season and excess power generation at other times of day. This is the **'balancing problem'** – how to balance supply and demand of a commodity (power) which has traditionally been considered incapable of being 'stored' on a power grid system except in small amounts using traditional lead-acid batteries as used in fossil-fuelled road vehicles. The balancing problem has always been with us; it's just that if we have a high proportion of renewable energy, the problem becomes much more severe.

Solar energy is the most abundant energy resource on Earth – nearly 200,000 TW of solar energy strikes the Earth continuously. That's more than 10,000 times the world's total energy use. Offshore wind has a resource potential of some thousands of gigawatts. Thus, there is a huge amount of energy available on Earth without the need to use fossil fuels. However, neither the generation profile of wind nor of solar matches the profile of demand. One partial solution to this problem can be achieved by combining wind energy (which often has an evening peak) and solar energy (which usually has a daytime peak) to create a 'hybrid' wind–solar complex, which can combine to better match daily demand. It might also require battery storage; indeed, many projects today are now combining solar, wind, and storage as the cost of storage has declined.

Many people would argue that even solar, wind, and storage together does not provide the **optimal trade-off** between energy security (a guaranteed continuous supply regardless of the weather), cost (the cost of a purely renewable solution has rising marginal costs), and environmental sustainability. They would argue that an electricity **system** needs to have a proportion of dispatchable power from nuclear and geothermal to achieve the best trade-offs.

10.2.1 *Balancing the Grid: Dispatchable Power and Flexibility*

Managing the power transmission system in a country or region is necessary because electric power cannot be stored in the same way as a water or gas supply. Both water and gas can be easily stored in 'containers' such as reservoirs for water or caverns for gas. Without some means of 'storing' power, if demand rises unexpectedly at short notice, a country or region will suffer a 'brownout' or even a 'blackout', i.e. we will not achieve the energy security objective in the trilemma. Sudden absence of power cannot be prevented unless an additional supply of power from dispatchable generation or so-called balancing technologies such as energy storage is made available immediately or if some users agree to have their power switched off (demand response).

Equally, if too much power is generated relative to demand, this can overload and damage the grid, and consequently, the transmission system operator (TSO) will need to request that some producers curtail their production in return for a pre-agreed **curtailment** payment. As a result of this need to balance supply and demand, a country or region must manage the system on a second-by-second, minute-by-minute, hour-by-hour, day-by-day, and month-by-month basis. This management process involves forecasting both demand and supply at each future time period, starting with 'day ahead'. **With increased wind and solar power feeding into the network, there is also an ever-greater need for accurate forecasts of the expected available supply of energy from sunlight and wind.** Equally, the TSO will be trying to estimate demand at each future moment in time based on weather conditions (including the need for home heating or air conditioning), whether an important sports fixture is going to be on television at a certain time, and the likely average time that people will go to bed, which may be later at weekends. The need to continually **balance the grid** existed long before renewable energy became an important source of power – it is simply, as we noted above, a much larger problem today.

The issue with the ever-increasing proportions of VRE is how to provide the same ability as fossil to balance the grid when wind and solar are the main source of power but are not dispatchable sources. This is becoming an ever more critical factor as the proportion of VRE (solar and wind) rises.

An FT report on 13 September 2021 noted that the UK had seen the worst (low) wind conditions in the North Sea for more than 20 years. Most of Europe suffered from the same phenomenon. Unfortunately, the stated capacity factor for wind (which is much higher than for solar) is an average over a whole year and thus tells only part of the story and not the most important part. More important is the issue of wind droughts, which can last for a week or a month, even if the average capacity factor of a wind farm over a period of 12 months is high. Wind speeds can also fall dramatically for longish periods. For example, while wind accounted for 18% of the UK's power in 2020, by September 2021, it was only 2%.

To date, gas generators have been 'fired-up' during wind droughts to enable the grid to be balanced, but in future, the other solutions, which we cover in the following, will hopefully ensure that fossil fuels do not have to be used at such times. In addition to the need to balance supply and demand, there has always been a need for so-called ancillary services that are frequently needed by the grid operator to enable the smooth transmission of power. These services include frequency control and voltage stabilisation.

The first solution to the variability of renewables is to have an adequate supply of clean, **dispatchable power**, perhaps of the order of magnitude of 20% or more of total demand. In many countries most storage is **pumped hydro**, which is stored potential energy in a reservoir at a height above sea level. For reasons of technological diversification, having a substantial proportion of nuclear in a country may be an economic means of ensuring that 'the lights remain on'. However, even nuclear can go offline, as happened in France on a large scale in 2022 due to the need for maintenance of, and repairs to, a high proportion of the nuclear fleet.

There are also limits on both wind and solar quite apart from the high capital costs. The first reason for this, in the case of wind, is a shortage of suitable land and seabed areas for wind and also planning issues. In the case of solar, not only are planning issues a cause for concern but also low solar irradiance in Europe, measured through both summer and winter. But a further reason for the limits to solar and wind as sources of electricity supply is that as the proportion of VRE exceeds about 70% of total power generation, the cost of achieving 100% carbon-free energy (CFE) using renewables rises rapidly. It is widely believed that the cost of using nuclear, which is a dispatchable power source with a **high capacity factor**, is likely to be on a par with, or lower, than the alternative VRE route as the proportion of VRE on a grid increases towards 100%.

The high energy density of nuclear means that it can take up less land per unit of energy output than existing fossil plants or wind or solar. For example a 1,000 MWe (1 GW) nuclear plant could be built on a factory-sized plot, whereas the equivalent solar plant would require 50–100 sq. miles and a wind plant 200–300 sq. miles. Such large acreages of land (or seabed) in Western Europe, as would be required for high proportions of solar and wind, would be unlikely to be available. In addition, due to its exceptional energy density, nuclear has an extremely low ecological footprint to build, run, fuel, and decommission. This applies, in particular, to Generation IV reactors (which we cover shortly) and to the plant itself, its fuel/material mining, and final waste streams. In contrast, the manufacture, transport, and combustion of methane gas used by a gas turbine all create considerable emissions of methane from accidental leaks and other releases, in addition to CO_2. In addition, methane is 20–30 times more potent as a greenhouse gas than CO_2, so 'fugitive' emissions (leaks) do matter. At present, gas turbines powered by methane gas (natural gas) are the main alternative to nuclear for providing the necessary dispatchable (firm) power, with Li-ion and other storage mechanisms being an alternative for relatively short periods.

While nuclear clearly raises issues of disposal of waste fuel, even windfarms raise environmental issues when the turbine blades start to fail, perhaps within 20 years, and have to be buried and left to decay while still ensuring that the plasticisers and other chemicals used in their construction do not contaminate the water table.

Imports of power can change the percentage of power generated by natural gas. This is enabled by interconnectors, which are power cables connecting countries or regions. However, if the whole of Europe is suffering a wind drought, it is unlikely that any one country can supply power to any other. The conclusion we can draw from the above is that much more so than a power system using mostly firm power sources (as was the case when most sources were based on fossil fuels), power systems based on VRE require 'flexibility', which is the capability of a power system to cope with the variability and uncertainty of solar and wind and thereby avoiding curtailment of power and reliably supplying all customer energy demand.

Figure 10.3[3] shows that some of the **'system flexibility solutions'** are applicable to fossil sources of power but all are relevant to VRE. On the left-hand side, it shows some of the factors that can result in the need to balance the grid. The first, **variation in demand**, is an obvious one. On the left-hand side, we have a list of causes of imbalance in the system. On the right-hand side are some ways of ameliorating such shortages, namely system flexibility solutions.

10.2.1.1 *System Flexibility Solutions*

Energy Efficiency Services

There is no doubt that if total demand can be reduced, for example through replacing incandescent filament lighting in offices and homes with LEDs, a lot of power will be saved. But all that does is to reduce the total need for generation capacity, but does little for balancing a grid which has a high proportion of VRE.

[3] European Parliament Energy Storage: Which Market Designs and Regulatory Incentives Are Needed? https://www.europarl.europa.eu/RegData/etudes/STUD/2015/563469/IPOL_STU(2015)563469_EN.pdf.

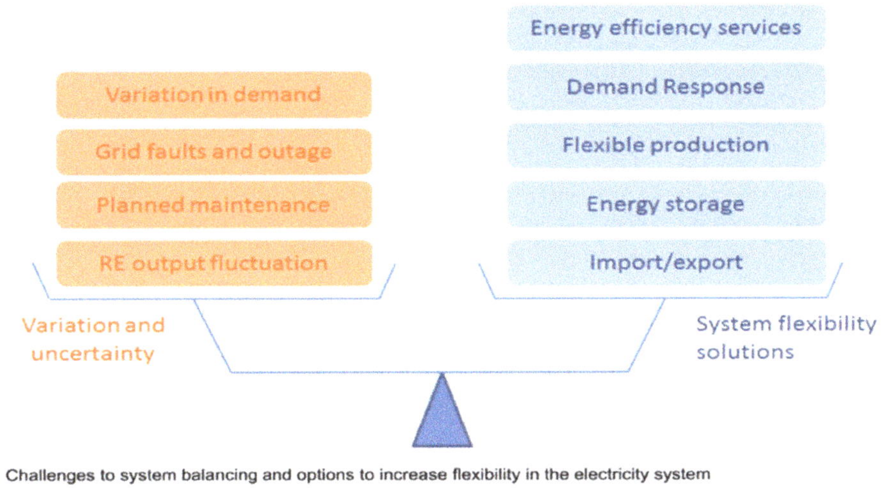

Figure 10.3. Mechanisms to balance supply and demand in an electricity system.

On the left-hand side of Figure 10.4 is a generic graph[4] of power demand against time of day. On the right-hand side is an explanation[5] of DSR from the UK National Grid. Traditionally, the peak load is the most expensive to provide since 'peakers' which use open-cycle gas turbine generators only operate for a short time each day and are paid a higher price per MW than baseload generators. In this (US) graphic, peak demand arises in summer from air conditioning. To enable the power distribution system operator (DSO) to supply the air-conditioning demand *without additional generating capacity*, it is necessary either to reduce other demand during peak hours or to use energy storage or ideally both. The objective is to 'shave the peak'. Increasingly, peak demand can be met by battery storage rather than from natural gas.

Time-of-use (TOU) pricing can make the profile of demand flatter. However, some dispatchable peaking power may still be required. The benefit to the network of such time shifting of demand is to reduce the total

---

[4] https://www.nrel.gov/state-local-tribal/blog/posts/when-does-energy-storage-make-sense-it-depends.html.

[5] https://www.nationalgrideso.com/industry-information/balancing-services/power-responsive/demand-side-response-dsr#:~:text=Demand%20side%20response%20(DSR)%20is,shift%20demand%20in%20real%2Dtime.

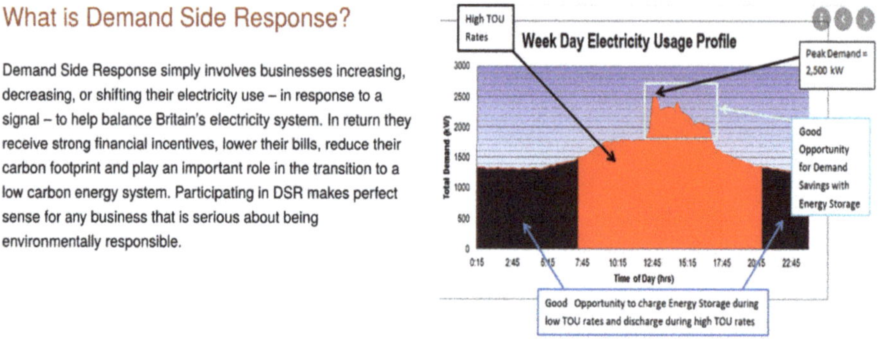

Figure 10.4. Demand-side response and weekday demand profile over 24 hours, also known as Time of Day/Time of use (TOU) pricing.

power generation assets required on the system and thus to cut the capital costs of providing round the clock (RTC) power, including the high demand for the peak times of day. It can be seen in the graphic that there is an arbitrage available for owners of storage systems. They can buy power before 7.45 am cheaply, store it (see storage in the following), and sell it at peak times at a higher price as if they were a dispatchable generating source In this example, power is being stored during low-demand periods (power prices are low) and supplied at peak times (prices are high).

Demand response is designed to encourage customers to shift demand from periods of potential excess demand on the grid to times when there is potentially a surplus of supply. This can be achieved either through TOU pricing or through monetary incentives to allow the DSO or TSO to disconnect certain power demands automatically. For example, some heating might be cut or car charging delayed. Some large technology companies with high demand for power may avoid undertaking non-urgent tasks during peak demand periods. However, TOU pricing does not increase the supply of power and may cause inconvenience and even raise production costs in industry.

When coupled into a power grid, the new technologies which use power or store power can also become components of the power system. For example, electric vehicles (EVs), electric boilers, heat pumps, and electrolysers for hydrogen production can provide flexibility to the power system either by adjusting their demand profile based on price signals (using market forces) or by making any integrated storage (in particular, electric vehicles) a source of energy storage for the power system. By using these resources on the grid, it might be possible to decouple the timing of demand for a proportion of final energy from the total electricity demand.

The approach of the International Renewable Energy Agency (IRENA) is to incorporate demand-side flexibility, grid reinforcements, storage, and sector coupling as additional flexibility sources and potential gamechangers in the action needed to increase the proportion of non-fossil energy in the power system. They note that making a system flexible is likely to require two sources of flexibility: one is coming from the generation side (supply) and the other is coming from the usage side (demand). In addition, the two sides need to be digitally connected (sector coupling) to enable control of power users' demand by the supply side.

Demand side response (DSR), also known as load shifting, is essentially the process of moving electricity consumption **from one time period to another**. For example, postponing an industrial process to another time. The idea is that by shifting the load to another time, the returns generated through energy cost savings or DSR participation are greater than the loss of production. Unlike many energy cost-saving strategies, load shifting tackles the 'when' rather than the 'how much' conundrum.

10.2.1.2 *Flexible Production (Supply-Side Response)*

This means the ability to call on a resource to increase or reduce its supply at short notice – milliseconds, seconds, or a few minutes and normally for a relatively short period of time. Currently, that often means calling on open cycle gas turbine generator sets to start providing power. They have an advantage over the more efficient combined cycle gas turbine (CCGT) of being simple and flexible with quick start-up times and lower capital costs. But in times of excess supply, flexibility can mean requiring wind or solar generators to cease injecting power into the grid to avoid overloading it.

As the proportion of VRE on the grid increases, the issue becomes much greater. Thus, the need for **flexible energy generation and storage technologies**, both of which can help balance the grid, is likely to continue to increase. Storage includes pumped storage in a reservoir, electrochemical battery, and hydrogen. Flexible energy, at present, cannot be easily achieved without resorting to natural gas. As an example of this,

Statera Energy, as part of its portfolio of assets, includes pumped hydro storage but also flexible natural gas generation.

We note here that the UK transmission system operator (NESO) has reported that, while the UK grid will have reduced the share of unabated gas generation to below 5% on average over a year by 2030 (under one of its pathways), the country will need to retain the same capacity in natural gas plants as a strategic reserve for windless days.

10.2.1.3 *Energy Storage*

Storage in a form which can almost instantaneously be converted into electricity is a key supply-side flexibility technology. **Storage does not generate energy** as the grid has to supply the power in the first instance, unless the storage facility is located contiguously to the generating source rather than charging from the grid. A storage system can recharge at times when there is excess power coming onto the grid, which is available at a much lower price than when there is an imbalance is in the opposite direction. Most storage mediums and, in particular, battery storage have the great advantage of being able to provide power almost instantaneously, much like a torch which lights up immediately it is switched on. What it does is to **'bridge time gaps'** by separating the production of power from its use. It may soon be the case that large-scale battery storage rather than natural gas use is the first choice for peaking services, based on flexibility, cost, services to the network, and emissions reduction.

10.2.1.4 *Import/Export*

This is about extending grids to connect with grid systems in other states (as in the US) or areas such as neighbouring countries in the EU or, more recently, distant countries. *Interconnector power lines* between regions of a country ideally in different time zones or with different wind conditions reduce variability in power available relative to power demand. **The portfolio effect of combining sources of generation which are not correlated can enable reductions of investment in power generation required to ensure a continuous supply of power.**

An example of a new interconnector is ElecLink which was opened in 2022 and goes through the Channel Tunnel. Unlike other interconnector projects, which are fully underwritten by electricity consumers, ElecLink bears 100% of the investment risk. The project has been financed entirely through private funds. ElecLink is a wholly owned subsidiary of Getlink, the company that owns the Channel Tunnel concessionaire, Eurotunnel.

It can happen that in Europe, for example, all regions are suffering from a lack of wind or sunshine, or both, and sometimes for many days. All those connected to interconnectors may therefore be seeking power at the same time. Thus traditional interconnectors, **not being storage facilities**, may not be available just at a time when they are needed. However, interconnectors attached to **additional** generation capacity in areas with high levels of incident solar radiation, hydropower, or geothermal power, including countries outside Europe, provide the possibility of what is effectively 24/7 dispatchable power from VRE. In the case of, for example, solar and wind projects in Morocco, it is likely that if they were also manufacturing hydrogen, they could cease such production for periods when power prices in Europe were high, and they could sell the power very profitably compared with manufacturing hydrogen. They would then resume the manufacture of hydrogen when power prices in Europe fell, and thus such manufacture would be more profitable than selling power. The other means of providing power when wind speed drops for long periods would be long-duration energy storage (LDES). This could be either from electrochemical storage or from hydrogen storage linked to electrolysers.

An example of an interconnector with a country distant from Europe and which was at an advanced planning stage at the beginning of 2025 but has now been told by the UK government that they will not support, is Xlinks.

Xlinks[6] – Long-Distance High-Voltage DC Transmission of VRE Power from Morocco to England

Europe is not the best location for solar or wind power generation due to relatively low levels of sunlight and wind strength compared with other locations. Morocco, as well as generating power for local use, is also seen as a source of power generation for use in Europe, with up to three times the intensity of sunlight and the benefit of the strong trade winds from the Atlantic. The Xlinks proposal is one such attempt to bring North African power directly to the UK rather than using the power to produce hydrogen locally, converting the hydrogen to ammonia, shipping it to the UK in

[6] https://xlinks.co/.

tankers then converting it back to hydrogen, either for direct use or for converting back to power. The project's estimated will cost is $22 billion. It would comprise 7 GW of solar and 3.5 GW of wind, along with on-site 5GW/ 20 GWh battery storage, in Morocco. The cable would start in Morocco's Guelmim-Oued Noun region with a UK landing point in Devon. The cable would be around 2,400 miles long (3,800 km). This would make it the longest power cable in the world (currently the longest is a 2,539 km overhead cable in Brazil). It should be noted, however, that during the night in Morocco, no power will be generated, and thus the power supply at such times, if required, would come from energy storage.

10.3 How Clean Are National Power Systems and Is 100% Clean Actually 100% Clean? Time-/Location-Matched 24/7 Carbon-Free Energy/Electricity

National grids across Europe and elsewhere in the world are certainly nowhere near being clean. Not only that, but power sold under a 100% renewable PPA will often not, in practice, be 100% CFE. The need for much speedier action to decarbonise power systems will be clear after looking at Figure 10.5, which reflects the average CO_2 emissions over the 12 months of 2023.[7] This company provides users with actionable data quantifying the carbon intensity and origin of electricity. These data are available on an hourly basis across 50+ countries and more than 150 regions.

The map shows that only two geographical regions – the Nordic region and France – are somewhat close to having a 100% 'clean grid supply', while Poland, which relies mainly on coal, is at the 'dirtiest' end of the graphic, with a carbon intensity of 800 g of CO_2 per kilowatt-hour (kWh).

If we look at the overall carbon emissions intensity of power in different parts of the transmission grid across Europe, we find that they differ substantially by country and by region. In some cases, regions are connected to each other to give an average carbon intensity, which is the same across the

[7] Courtesy of Electricity Maps. https://app.electricitymaps.com/map.

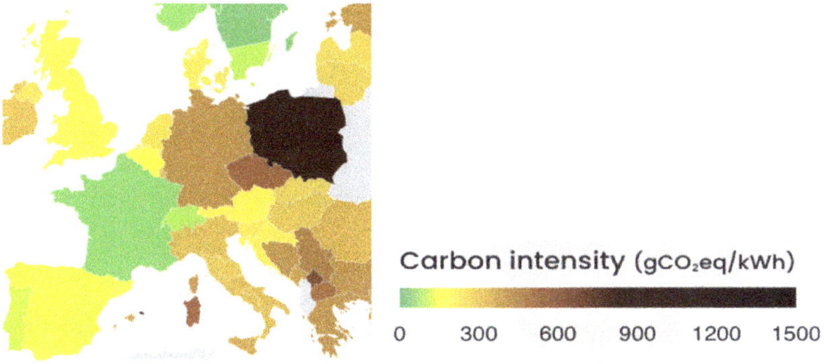

Figure 10.5. Average emissions intensity across European countries.

country, but in others, these regions are not connected. In addition, as we noted earlier, these intensities depend not only on the local generation in that country but also on the carbon intensities of the countries from which they import power through interconnectors. It can be seen that the UK is very far from being decarbonised (sometimes defined as 'less than 100g CO_2/kWh'), despite the fact that it imports from France (mainly nuclear) and Norway (mainly hydro), which are two of the cleanest countries. France has a very low-carbon power system, with around 75% of total French electricity generation being from nuclear. As a result, 93% of electricity generated is carbon-free. In France, in 2019, the emission of CO_2 was only 35 g CO_2/kWh compared to an average of around 350–500 across Europe. Poland is around the 800 g intensity. From having intended to reduce the proportion of nuclear in its fleet, they now plan to maintain or increase it.

Some retail suppliers in a country claim that their power is 100% clean, i.e. CFE. In practice, this is not likely to be the case. Customers must take power from the grid unless they have a direct connection to a renewable source combined with storage or their supplier is using storage to cover gaps in the availability of solar or wind. Otherwise, only a few national or regional grids (in France and Scandinavia) are close to being 100% CFE.

Power is 'fungible', i.e. any one electron coming down a wire looks like any other electron. When charging an electric vehicle, in many countries, the power has been produced by generators using, on average, around 40% fossil fuel or in some cases a much higher fossil percentage. Thus, at present, electric cars are not '100% clean', though many purchasers think they are. They are clean 'at the tailpipe' and do not emit health-endangering substances, such as particles less than 2.5 micrometres in diameter, also known as fine particles or $PM_{2.5}$, which are emitted by diesel engines. The only ways of making electric cars 100% clean is if the **grid** is 100% clean at every hour of the day or if the companies providing commercial charging facilities have what we examine in the following, namely a granular 24/7 CFE contract with a power company.

10.3.1 *Time-Matched/Location-Matched 24/7 Carbon-Free Energy/Electricity*

Increasingly, customers want 100% or near 100% clean power either because they are willing to make an effort to limit climate change or because carbon taxes are high or because regulations require them to report the carbon intensity of the power they consume. If the latter is the case, while it may not be illegal for a corporate to consume high carbon power, it may be that investors and customers of the company will take action against it for doing so.

We now consider what this means in practice. We need to know how PPAs in particular can be adapted to make them equally relevant to the new world that is within reach if we can develop appropriate frameworks and contractual arrangements which will also encourage ***additionality***, i.e. creation of new assets.

10.3.1.1 *According to LDES*[8]

> A new form of PPA — the **24/7** CFE PPA — promises dramatic increases in the effectiveness of emission reduction efforts in the power generation sector and for corporate off-takers. 24/7 clean PPAs have the potential to address key problems in balancing supply and demand for power, as renewables contribute an increasing proportion of generation capacity. ***In the process, they can help providers and off-takers accelerate the overall decarbonisation of the power sector.***
>
> 24/7 clean power PPAs measure electricity consumption and greenhouse gas emissions on a **granular** basis — *e.g.* by the hour (or half or quarter

[8]LDES Council: A path towards full grid decarbonisation with 24/7 clean Power Purchase Agreements https://ldescouncil.com/wp-content/uploads/2025/05/2205_ldes-report_247-ppas.pdf.

hour) — and provide *time-matched* clean power. They represent an improvement from today's Scope 2 decarbonisation with pay-as-produced renewable PPAs, which balance supply and demand on an annual basis but only achieve 40–70% decarbonisation of the off-taker's *actual* electricity consumption.

Time-matched clean power provided through 24/7 clean PPAs addresses both problems but requires a combination of renewables, dispatchable generation and flexible capacity such as storage and DSR to create a more granular match between renewables supply and demand. **Current barriers to wider adoption of 24/7 clean PPAs include cost and a lack of agreed standards**, but these can be overcome.

Traditional clean baseload contracts oblige the seller to enter the power market (including the fossil-generated power market) to purchase power to meet the contract's requirement of a continuous power supply. This may still be called a 100% clean energy contract since all the CFE power purchased by the buyer over 12 months is credited to the PPA contract counterparty. Even when some of the power was produced in excess of the buyer's need in summer, it gives the buyer a **credit** in winter when it is actually using 'high carbon' power much of the time.

The **credit** is the Energy Attribution Certificates (EACs) from this generation (covered in the following) which can be used in winter when as an 'offset' to the use of fossil fuels. This enables it to claim the summer certificates against its total power consumption for the year. In consequence, much of the power may be generated by fossil sources, as shown in Figure 10.6, from LDES.

In contrast, under a **24/7 clean baseload contract (24/7 CFE) contract, clean power must be supplied almost every hour of every day rather than just on average over a one-year period**. However, at present almost no energy provider will guarantee 100% CFE at every hour. Thus, the PPA will state perhaps 70% or 80%. EACs generated in the summer from CFE under this type of hourly contract **cannot be used in the winter 'to make up the numbers to 100% CFE'**. A PPA from a CFE generator to supply CFE on a 24/7 basis will almost certainly require energy storage, including long-duration energy storage to cover wind 'droughts' and other interruptions to the supply of electricity.

The United Nations Energy Compact defines 24/7 CFE as every kilowatt hour being provided by CFE every hour of every day. At present, the

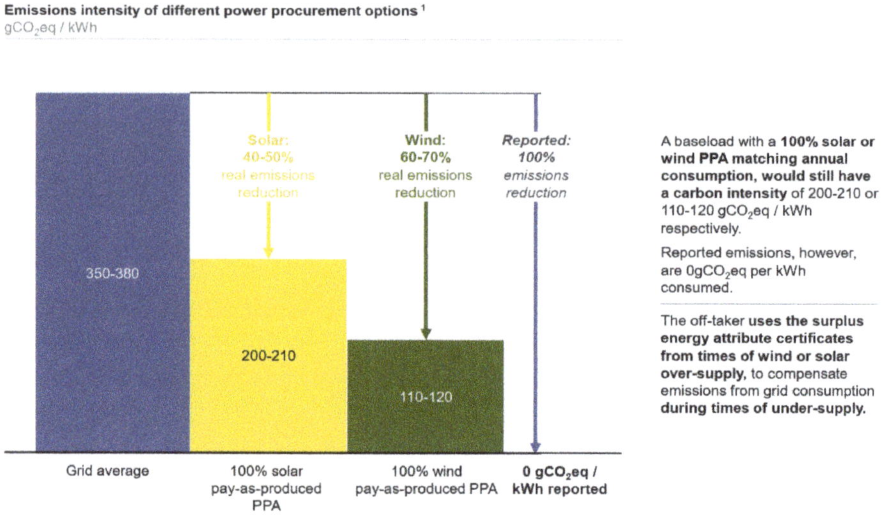

Figure 10.6. Emissions intensity of different electricity generation sources.

Note: [1]Based on 2021 average grid emissions and RES generation data for Germany and California. Emissions intensity of the grid and wind PPA: lower range applies to Germany (offshore wind), and upper range to California (onshore wind). Emissions intensity of solar PPA: lower range applies to California, upper range to Germany.

world does not have 24/7 granular time-matched 100% CFE PPAs except on an experimental basis. What can be provided by a supplier is a 24/7 granular time-matched PPA but not necessarily, or likely yet to be providing CFE during every hour of the contract, i.e. not CFE 24/7. Google was one of the first to set a target of 100% CFE, but in 2020, it was still only able, from a techno/economic point of view, to achieve 67% matching on an hourly basis.

Google first said it was carbon neutral in 2007 through carbon offsets. In 2017, it started purchasing CFE through PPAs in 2017 and is now trying to transition from 100% **annual** renewable energy matching to **24/7 matching** by 2030. This involves achieving **hourly** load matching instead of annual, volume-based matching. By 2020, it had achieved only 67% matching. Since then, the demands of AI large language models (LLMs) have dramatically increased its need for power, making the achievement of improving matching much harder to achieve.

The question today is how to transition towards the point where buyers of power can be guaranteed that they are buying power with an agreed high percentage of CFE. *Eurelectric* has shown how this organisation thinks we could move towards 'co-establishing commitment to a minimum % threshold of hourly CFE matching'. Eurelectric describes itself as 'the federation of the European electricity industry'. Its membership comprises more than 3,500 European utilities active in power generation, distribution, and supply. It recently produced a publication entitled 'Contracting Principles for Moving Towards Location and Hourly Clean Energy Matching', which can be found on their website.

Achieving a system for 100% CFE on an hourly basis is part of a process in which electricity systems move from purely fossil energy to close to 100% CFE. Such a system would only allow for clean energy consumption claims if proof of **generation** origin was matched with electricity **consumption** on an hourly basis and sourced from the local grid. This would be achieving both CFE for consumers of power and additionality of CFE assets on the grid.

10.4 The United Nations 24/7 Carbon-Free Energy Compact[9]

According to the Carbon-Free Energy compact, 24/7 CFE means that every kilowatt-hour of electricity consumption is met with carbon-free electricity sources, every hour of every day, everywhere. It is based on the following principles:

1. **Time-matched procurement:** Matching hourly electricity consumption with carbon-free electricity generation. Hourly matching helps connect clean energy purchasing to underlying electricity consumption.
2. **Local procurement:** Purchasing clean energy on the local/regional electricity grids where electricity consumption occurs. This is the only way to drive the electricity-related emissions that a consumer is directly responsible for to zero.
3. **Technology-inclusive:** Recognising the need to create zero-carbon electricity systems as quickly as possible. All CFE technologies can play a role in creating this future.

[9]This material is sourced from https://www.un.org/en/energy-compacts/page/compact-247-carbon-free-energy#:~:text=24%2F7%20Carbon%2Dfree%20Energy%20(CFE)%20means%20that,hour%20of%20every%20day%2C%20everywhere.

4. **Enabling new generation:** Focusing on delivering additional CFE to drive the rapid decarbonisation of electricity systems.
5. **Maximising system impact:** Addressing the dirtiest hours of electricity consumption where the most fossil fuel is used in generation.

Note in the list above, the third principle is an approach that includes any CFE source, including nuclear, not just renewables. The fourth principle refers to the fact that this system would provide better price signals to the system, as the price of hourly certificates would vary in time, depending on the availability of clean electricity in the system. It should therefore help generate confidence in those considering 24/7 CFE power projects as they move towards the financial investment decision (FID) on the assumption that more such power supplies will be required. But granularity can, with benefit, go further. The European Network of Transmission System Operators for Electricity **(ENTSO-E) has recommended that the granularity target should be set to 15 minutes**, in line with the target imbalance settlement period in Europe. This would enable the certificate market to be more closely aligned with the physical grid reality.

For the following chapters, the important matter to keep in mind is that, with a few exceptions, it is not yet possible in fact to have 24-hour-a-day 100% CFE across a country or region. We have the technology to achieve it, but scaling up that technology **at a cost acceptable to voters** is what is difficult. Thus, 24/7 CFE across the grid, and hence also for the power consumed by electrolysers to ensure that it is 100% CFE hydrogen, is the objective but remains some years away yet. Like so much of the technology we cover, it is an ongoing project requiring the collaboration of many parties involved in the power industry value chain. However, confusion can be caused by the different meanings attributed to the expression '24/7' by different parties. This also makes it more difficult for investors to commit to projects that might move the world closer to being carbon-free.

10.5 Decentralised Electricity and Heat Systems and Their Financing

Back in 2020, Bill Gates was agonising over the inability of countries and districts with incomes lower than average to take advantage of low-cost renewable energy and the lack of improved technologies being developed and made available to such countries by higher-income countries to help drive down costs. In one of his early 'Gates Notes',[10] he explained what he called the 'Green Premium' as:

> the difference in cost between a product that involves emitting carbon and an alternative that doesn't. This difference in cost is what he calls the Green Premium, and understanding it is key to making progress on climate change.

He continues by noting that:

> …they help us measure our progress toward eliminating carbon emissions. The bigger a Green Premium is—especially for lower-income countries like India and Nigeria whose energy needs are growing—the further we are from a zero-carbon future. They also serve as a guide to action. In cases where the Green Premiums are big, we know we need innovations that will close the price gap. In cases where they're small—or where clean products are actually cheaper than the polluting version—it suggests that something other than the cost is keeping zero-carbon products from being deployed, and we need to understand why.

Up to this point in the text, we have been considering the production, distribution, and financing of power in the context of high-income countries and regions which have traditionally employed large, centralised coal or gas power stations, with electricity transmitted through a grid system and then through local distribution networks. This chapter covers some of the technological possibilities offered by decentralised electricity provision, which can be used in countries or regions that do not have connectivity to a national network or where such connectivity would be economically unattractive. Such electricity-generating systems have fallen dramatically in cost since Gates wrote the above words, but today the barriers are still as much in the financing arena as in the required technologies.

Decentralised energy systems, which describe the generation of electricity and heat off-grid closer to where it is consumed, are crucial in progressing the energy transition to enable a zero-carbon global economy and, potentially, achieving energy equity. Such systems can be deployed

[10] https://www.gatesnotes.com/meet-bill/accelerate-climate-innovation/reader/introducing-the-green-premiums?ref=carbonrisk.world.

more rapidly and with less upfront cost than a conventional centralised energy model. They enable renewable generation, distribution, and consumption at a community level, and when using low-carbon generating systems, particularly solar, wind, and battery storage, they reduce reliance on fossil fuels, enable shorter distribution connections and minimise energy losses and costs without the need for high-voltage transmission networks. As well as introducing electricity for the first time to new neighbourhoods, such projects may also involve replacing generating systems currently using diesel generators. They can also enable the elimination of home cooking devices using gas or wood-burning stoves, both of which are harmful to health and also emit CO_2. Their replacement with electric stoves would then also require a capital investment in such home hardware, which would need financing.

Despite the advantages it could bring, both to households and to global decarbonisation, the adoption of decentralised energy remains underfinanced. International climate finance still favours investments in large-scale fossil fuel utility companies or large-scale centralised wind and solar projects. Such capital flows tend to concentrate on high-income countries and high-ticket centralised systems, thereby excluding decentralised renewables which could speed up energy access and the global energy transition. Innovative financing could facilitate the acceleration of the energy transition by extending electrification to parts of the world where energy utilities would find it difficult to justify investment. But progress in this regard is slow and is only likely to proceed with more investment by national development funds and international organisations such as the World Bank.

While aspects of green bonds, yieldcos, and securitisation in renewable energy financing have been studied (though largely in the context of developed markets), most existing research focuses on centralised renewable energy projects and mechanisms for the integration of large-scale renewables into centralised transmission systems and the operational and planning challenges associated with such integration. On the finance side, this has mostly focused on the potential of green bonds as an instrument to promote renewable energy.

10.5.1 *Decentralised Energy Systems*

Decentralised energy systems include any system where energy production occurs at or near the point of use, including micro and mini grids, stand-alone projects (e.g. PV rooftop solar panels), battery storage, mini-hydro, biogas plants, and micro nuclear reactors. They imply not only distributed and dispersed generation and spatial proximity to their consumers but also new ownership, financial, and governance models. Decentralised systems can expedite energy access, particularly in remote and sparsely populated areas, enable climate resilience, and contribute to energy equity, affordability, and environmental protection.

Decentralised energy systems have the potential to enable energy equity (as they can provide clean energy to underserved rural and urban communities), energy resilience (as decentralised grids can be more resilient to climate change than large power plants and transmission lines), and empowerment (enabling communities to become energy producers as well as being consumers, i.e. prosumers.

Traditionally, centralised utilities are often known for being cost-inefficient, conservative, carbon-intensive, lacking transparency, and operating as natural monopolies. They are vulnerable not only to climate change but also to geopolitical risks. Decentralised energy systems, on the contrary, imply a shift towards off-grid generation facilities, from large utility companies to medium-sized and individual enterprises and to private residents as both producers and consumers of energy.

Such structures are highly relevant in developed markets. But in addition, they should be increasingly relevant in emerging markets in countries or districts where GDP per capita is low. In such areas, at present, they may require official financial support to be viable. But in such cases, the increase in small-scale manufacturing and other industries that may become possible as a consequence of electrification may enable such local systems to become economically self-supporting in the longer run and also to contribute to the economic growth of a region or country.

Decentralised models require new financial, ownership, governance, and business models. Rather than being state-owned companies or large utilities quoted on a stock exchange, they would have a widely dispersed financial ownership structure. These types of ownership include:

- Private Ownership (plug-and-play model), where the microgeneration unit is owned and financed by the end user, who may also export energy into the grid (e.g. BeON, the system available from EDF, the French utility).
- An energy services company (ESCO), where energy providers use microgenerators to substitute for central power generation. Consumers provide a site for the microgenerator, which is owned and operated by the energy service company. ESCOs can reduce upfront capital

investment, as they finance the project, and thus customers do not need to pay upfront. Payments to the ESCOs may also be contingent on achieving energy savings if the ESCO is replacing an existing (possibly diesel-fuelled) generator system. By taking on financial risk, ESCOs can make energy efficiency projects more accessible for customers, both domestic and larger organisations such as schools or hospitals.

- Community or corporate ownership (micro-grid model), where all microgeneration units are connected within the microgrid and can be partially owned by consumers. This ownership model combines the benefits of corporate control with consumer involvement and partial ownership, which incentivises the system to operate smoothly.

The current features of international climate funds discourage investment in small-scale, decentralised projects. These funds incentivise large-scale investments in high- and middle-income countries, particularly grid-based and fossil fuel energy, over decentralised renewable energy projects. International climate finance does not prioritise decentralised energy projects. Its institutions, such as Multilateral Development Banks (MDBs), are not designed to, nor do they have the instruments to, invest in small-scale decentralised renewable energy projects.

The largest international climate funds, Clean Technology Funds (CTFs), employ concessional loans to finance utility-scale projects in middle-income countries, which offer better returns and lower transaction costs compared to funding many individual decentralised energy projects.

10.5.2 *Barriers to the Financing of Decentralised Energy*

Decentralised energy projects are perceived as less bankable and require a different range of financing and risk-mitigating instruments; they are perceived as having higher rates of failure and frequently operate in the context of an uncertain regulatory framework. Because of their small size, they may face high transaction costs and, more problematically, high-credit-risk customers. Even established companies such as Mobisol, a Berlin-based company that provides solar home systems, became insolvent in April 2019 despite achieving 500,000 users. It struggled with low consumer repayment rates and an inability to scale a pay-as-you-go (PAYG) off-grid model without ongoing subsidies.

Barriers to decentralised energy also include policy frameworks that favour investment in centralised energy systems, financial instruments that enable only larger investments, a lack of reach by special-purpose agencies capable of channelling funds to small-scale projects, and a lack of capacity building. These risk characteristics are barriers to increasing the appetite of private capital for such projects. Despite all the barriers referred to above, decentralised projects involve less bureaucracy, and their smaller fund sizes can be appealing to some types of impact investor.

10.5.3 *Financing Mechanisms for Decentralised Energy*

10.5.3.1 *Green Bonds*

We covered green bonds earlier in the context of large-scale projects. They were created to fund projects which have positive environmental or climate benefits. Proceeds are earmarked for green projects but can be backed by the issuer's entire balance sheet when the issuer is a corporate, as most have been to date, or may be project financings through a special purpose vehicle (SPV). Green bond terms are based on ICMA Principles and must offer investors transparency, impact assurance, and alignment with global ESG standards. They can provide a powerful bridge between the demand for sustainable investments by particular categories of investors and the need to scale up community-level carbon-free infrastructure, which is a necessity if the world is to come within reach of the Paris Agreement's objectives.

However, for decentralised energy projects, their utility is limited given their high issuance and transaction costs, including legal, verification, and certification requirements. These fixed costs are burdensome for small-scale projects. Costs are increased further with the post-issuance reporting and monitoring requirements for green bonds.

Green bonds are most suited to large-scale infrastructure projects, as institutional investors seek minimum ticket-size investments. Decentralised energy projects, driven by local enterprises, community-based organisations, and local developers, thus have limited access to the bond markets, given their lack of a credit rating and limited market visibility and, consequently, capacity to issue bonds. Aggregation challenges limit their use in small projects, as they do not achieve economies of scale and require complex aggregation structures such as securitisation and green asset-backed securities (ABSs).

As decentralised energy projects are diverse, fragmented, and lacking in standardisation, they are less attractive in bond markets. Finally, there is a lack of uniform standards on legal terms, and some issuers label green bonds without offering credible or verifiable environmental benefits. This clearly undermines investors' trust, leading to greenwashing risks. The limitations of green bonds as financing instruments for decentralised energy systems always come back to the small scale of projects, the lack of creditworthiness, and the lack of liquidity, or even simply a secondary market. Without a credit risk guarantee, they would be unlikely to attract investors. In future, however, it might be possible to develop a 'green'-type bond which would be suitable for smaller projects and attractive to impact investors wishing to improve lives in a community.

10.5.3.2 *Yieldcos*

Yieldcos, short for 'yield companies', are publicly traded investment vehicles that own and operate renewable energy assets (e.g. solar farms, wind, and hydro plants) that have entered into long-term energy delivery contracts with customers and are therefore, in theory, able to generate predictable cash flows. Yieldcos are structured to deliver regular dividends to investors in a similar way to REITs and master limited partnerships (MLPs). Yieldcos are structured with the specific purpose of acquiring operating energy projects from sponsors and third parties. They generate stable revenues through power purchase agreements (PPAs), mitigation of development risks, and price certainty for renewable investment assets arising from long-term contracts to deliver energy to consumers at fixed prices.

To be attractive, decentralised energy projects must deliver large, diversified, stable cash flows via long-term PPAs in order to provide dividend-like returns to attract investors. Yieldcos can aggregate portfolios of small-scale distributed assets, providing scale, standardisation, liquidity and bundling across geographies, types of consumer, and technologies. They allow for capital recycling, as developers can sell operating assets to Yieldcos and use the proceeds to fund new projects, providing an exit route for early-stage investors and acting as platforms for refinancing early-stage projects.

Yieldcos in decentralised energy have, however, high aggregation and transaction costs, as collecting customer payments, achieving standardisation, and undertaking due diligence are all more administratively intensive and costly than for a single, large, centralised plant. In practice, many decentralised energy projects have revenue volatility, as they often lack long-term PPAs and depend on small customers' payment reliability.

10.5.3.3 *Securitisation*

Securitisation bundles renewable energy loans into tradeable securities. It aggregates them to achieve scale, diversification of risk and lower capital costs. In theory, securitisation should provide a creditworthy structure attractive to long-term investors. They are a mechanism to aggregate cash flows from multiple energy assets to create a large-scale investable asset, making these projects accessible to institutional investors. Securitisation allows early-stage financing and capital recycling, as it converts illiquid, long-term project returns into tradable securities, attracting private capital into emerging markets and off-grid projects.

However, securitisation still involves considerable risks for investors, as observed during the 2008 global financial crisis (GFC). First, many potential customers of decentralised energy systems are low income with no formal credit histories, which makes it difficult to make cash flow predictions. To overcome the lack of creditworthiness of many end users, guarantees are needed (e.g. PPAs with corporates or some official backing). On the other hand, if supplying retail customers, guarantees are difficult, if not impossible, to obtain. Second, securitisation involves high transaction and structuring costs, represented in legal, rating, and structuring fees. Further, there is still a limited track record of securitisation in the decentralised energy space, which is still perceived as a high-risk investment.

In summary, these financial instruments should be aiming at increasing the ability of decentralised clean energy projects to raise capital, facilitating interest cost reduction and speeding up the process of raising capital. At present, however, particularly when interest rates in most countries are higher than they were some years ago, even the most innovative instruments may not be able to overcome the essential problem in those countries where consumers have very limited income and hence creditworthiness. On the other hand, these instruments, or modified versions of them, can still work for decentralised projects in high-income countries, and over time as incomes rise in other countries, it should increasingly be possible to bring energy equity to many more people.

This chapter has covered decentralised renewable electricity systems. There is another type of decentralised electricity provision which is currently being developed and based on advanced nuclear technology (ANT) small and micro reactors. Their inherent safety, even compared with legacy reactors, means that they can be located close to demand in an SMR campus or an industrial zone or even be embedded within a data centre itself. This might also require a connection to the grid, though normally its energy storage system would take it through most interruptions to supply. Such a system is also known as the 'private wire' approach and overcomes many issues, particularly the shortage of transmission line capacity in developed markets, which are required to take distributed solar and wind from where it is generated to where it is needed. The private wire approach should also enable power to be provided at well below the price that users would need to pay if they relied on grid power. We cover the private wire approach in much greater detail in Section 15.2.

Chapter 11

Energy Storage as a Contributor in the Conversion of VRE into Carbon-Free Dispatchable Electric Power

To be of greatest value to consumers, carbon-free electricity (CFE) needs to be available continuously, even when not being produced at night or when wind force is low. The 'conversion process' from variable renewable energy (VRE) to 24/7 CFE power becomes ever harder as the proportion of CFE on the grid increases. The storage of energy in, for example, chemical form in Li-ion batteries, which can be converted to electric power virtually instantaneously, has been a key focus for technologists, developers and financiers. In addition, as the proportion of VRE rises, there is an ever greater need for long-duration storage when wind force is low for periods of weeks. In the UK, for example, winter electricity demand is twice that of summer (as the UK does not yet have much domestic air conditioning). However, without a proportion of dispatchable power on the network, the cost of power is likely to be much higher than it would otherwise be if dispatchable power were a significant part of the power mix. In practice, transitioning from 95% clean power to 100% is likely to involve the use of backup natural gas storage and gas turbine generation (hopefully with carbon capture and storage) and would almost certainly be much less costly than attempting to achieve 100% carbon-free electricity using primarily VRE.

11.1 Storage Technologies for 24/7 Dispatchable Electricity from Variable Renewable Energy

There is still much debate about how best to 'fill in the gaps', as we have already observed. The debate concerns those who believe that the lowest-cost means of achieving 100% clean energy is to provide a certain percentage of firm (dispatchable) clean, or at least low-carbon, power sources such as nuclear, flow-of-the-river hydro, geothermal, and fossil plants capturing and storing CO2, as we have noted above. The reason for this is that trying to achieve 100% renewable solely by using storage is unlikely to optimise the energy trilemma. The outcome may be a much more costly power system, yet one which is not as reliable as should be expected by consumers. **Batteries and demand flexibility do not substitute for firm resources.** Without a certain percentage of dispatchable power, electricity costs rise rapidly as a system approaches 100% CFE. Trying to achieve 100% CFE without firm power is not a least-cost strategy.

In the following, we cover a wide range of technologies for energy storage. Some use the force of *gravity,* for example hydro, to release energy from water which has been raised to a height through the energy derived from solar evaporation of water, which then returns to Earth as rain. Some use **energy stored as *heat*, for example Rondo**, which can then be used when required to generate power or provide industrial heat. Indeed, the **Terrapower Natrium** reactor, as well as being able to generate dispatchable power all day, also has a 'boost' function through which heat is stored for release at times when power prices are lower. The heat is then used to generate and sell power, in addition to its normal output, when power prices are high. Some storage mechanisms use the energy stored in *hydrogen*, which can then be released into the grid by passing it through an electrolyser or even an internal combustion engine (ICE). Others use *compressed gas* such as air, which, when released, can then power a turbine.

Some of the storage mechanisms noted above differ from the type of storage that is growing fastest at present – electrochemical storage, in some form of 'battery' or battery energy storage (BES). Much of the electrochemical battery storage capacity 'earns a living' by means of 'financial arbitrage' in power markets by purchasing power at a lower price than it is subsequently sold at (time of day arbitrage).

It is important to be clear that electrochemical storage does not generate any power, in contrast to, for example, 'flow-of-the-river' hydro generation, which is a source of electrical power. In contrast, 'pumped' hydro,

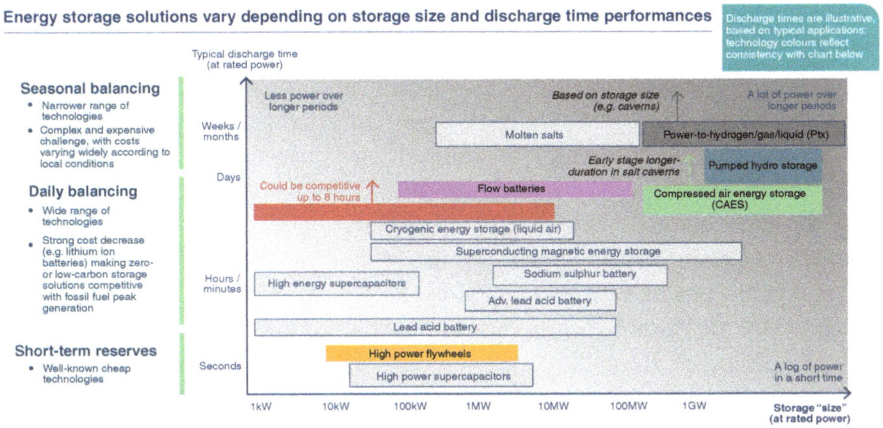

Figure 11.1. Duration of energy storage mechanisms against storage capacity.

such as battery storage, also involves a type of financial arbitrage and does not generate power. Instead, it buys power when it is cheap in order to pump water uphill and then releases it to generate power when the price of power is high. It operates essentially like an electrochemical battery.

In addition, there is a 'balancing market', which is power supplied to help balance the grid at certain times. Traditionally, this has been provided by gas turbine generation. The ETC graphic shown in Figure 11.1[1] summarises some of the main technologies in existence at present, many of which are still being developed and are not yet completely proven in terms of either their technology or their cost-effectiveness.

On the vertical axis, it shows the maximum *period* of storage at present, which is from seconds to months. On the horizontal axis is the maximum *amount* of storage that can be provided by the technology at present at an economic cost. It can be seen that hydrogen has the potential to be at the multi-gigawatt scale because large amounts of hydrogen can be stored and also, therefore, used to cover seasonal needs and not only intra- and inter-day gaps in supply and demand, as is currently the case for battery storage. The only problem with hydrogen at present is that green hydrogen

[1]Energy Transitions Commission (ETC) *Making Clean Electrification Possible: 30 Years to Electrify the Global Economy* April 2021 https://www.energy-transitions.org/wp-content/uploads/2022/07/ETC-Global-Power-Report-Final.pdf. Adapted from Climate Policy Initiative for the Energy Transitions Commission (2017), *Low-Cost, Low-Carbon Power Systems*, David Cebon (2020), The Centre for Sustainable Road Freight, "*Blog: Technologies for Large-Scale Electricity Storage*".

is expensive and not easily available in large quantities. What this graphic does not show, however, is the cost of such storage and, consequently, which type is most cost-effective for each type of use. Although some technologies are presently costly, in many cases the unit cost of storage will fall, perhaps dramatically, as technologies develop and as economies of scale and learning come into effect. It also does not show the improvement in flow battery technology, which now potentially enables over 100 hours of storage at low cost, nor some new 'heat battery' solutions.

At the bottom of Figure 11.1, very short-term reserves of energy are shown. They are useful for the provision of auxiliary services to the grid, such as voltage or frequency stabilisation. We will not cover short-term storage mechanisms but focus on those which potentially provide storage for hours rather than minutes or longer durations economically. Although we have categorised technologies this way, with technology moving so fast, a year from now storage durations, cost-effectiveness and technologies could all have changed. For this reason, we are not going to classify any particular technology as either daily balancing, longer duration balancing, or seasonal balancing. Some can already provide all types of services, while others may be able to do so within a few years.

11.1.1 *Revenue Stacking*

The key to the massive deployment of batteries is to make the investment profitable. To achieve that, so-called revenue stacking is necessary. Battery projects need diversified sources of income (different layers of revenue from providing diverse functions) and should be able to participate in any service that can generate income.

Price arbitrage in the day-ahead market and also in intraday markets is the basic function of storage, i.e. buy cheap and sell dear, as in financial markets, which is 'time arbitrage'. But in addition, there are so-called 'ancillary services' which need to be provided to the TSO, which makes payments for such services, as we covered above. But critically important in today's market is payment for providing 'capacity' to the grid. This so-called 'capacity market' ensures security of electricity supply by providing a payment for reliable sources of capacity to provide power when called upon by the TSO. This source of revenue is provided by the TSO simply for making power available. The actual provision of power, if required by the TSO, is paid for separately. The profitability of such a service may, however, be quite variable.

11.2 Electrochemical Energy and Grid-Scale

By far the best-known and most widely used power storage technology currently is the lithium-ion (Li-ion) battery, which can provide power that can be dispatched immediately. However, such storage is currently relatively expensive; therefore, the economic time for which it can provide power is generally limited to four hours. However, the development of lithium iron phosphate (LiFePO4) batteries, known as LFPs, may provide lower-cost and safer, large-scale, longer-duration batteries for grid storage.

The UK's largest battery storage project, at the time of writing, is located at DP World London Gateway on the Thames Estuary. This project, which will provide 320 MW/640 MWh of capacity (i.e. it can provide 320 MW of power for two hours or other combinations of power versus time).

11.2.1 *StorTera*

StorTera is a provider of energy storage systems. (We cover its AI system (Traicon), designed to take in data, analyse it, and make 'optimal' decisions, in Section 12.3.)

StorTera produces a non-flow battery using LFP technology. They also offer a flow battery which uses their lithium/sulphur technology. They predict capital costs of approximately £75/kWh, which would make it potentially an economically viable technology. The StorTera battery has received grants which have enabled it to develop the technology. They were also the prize winner in an international competition sponsored by the UK and the Canadian governments. The project took place in a town in Canada, and the statement below from a Canadian government minister was made at the end of August 2022 when the prize winner was announced:

> Our government challenged innovators to find breakthrough clean-tech solutions to help solve some of Canada's biggest problems — and they delivered. I'm proud to award the grand prize for the Power Forward Challenge to Equilibrium Engineering and their UK partner StorTera.
>
> The Honourable Jonathan Wilkinson
> Minister of Natural Resources Canada

11.2.2 *Fluence*

Fluence is a global provider of energy storage products, services, and digital applications for renewables and storage, originally launched and owned by AES and Siemens Energy and now publicly listed on the NASDAQ (ticker: FLNC). It is a large provider of these services, having deployed, contracted or managed 6.2 GW of storage and 225 energy storage projects across 44 markets and provided over 18 GW of AI-optimised renewables and storage. It provides an ecosystem including modular, scalable energy storage products such as Gridstack™. It does not manufacture battery cells but outsources this requirement to the most appropriate provider.

11.2.3 *Electrochemical Energy and Vehicle to Grid*

This potential source of storage is not yet fully developed but could prove important. The logic of this type of storage is that one day millions of vehicles around the world will be battery-powered and largely parked somewhere and therefore not using their battery power. That provides the potential to sell that power to the grid as 'peaking' power or as ancillary grid services (voltage and frequency regulation). At present, we already have a halfway solution to this in the concept of smart charging, which means the charging system will decide when there is the most surplus power available to the grid and will avoid charging vehicles during peak times. This would reduce the need for peaking generation. The next step is a two-way flow of power between a vehicle and the grid (V2G), whereby the charging unit would sell power to the grid at peak times (if the car was not being driven and had sufficient battery capacity for the next journey home). This would result in an even larger reduction in the need for peaking capacity, which remains the key problem for decarbonised grids.

11.2.4 *Electrochemical Energy Storage: Flow Batteries for Long-Duration Storage*

These use electrochemical storage as do Li-ion batteries but use different electrolytes. The interconversion of energy between electrical and stored chemical energy occurs in an electrochemical cell. A great advantage of flow batteries over more traditional batteries, such as Li-ion, is that while conventional electrode structures couple the energy storage capacity and the power rating of the battery, i.e. it is not possible to increase the duration of energy storage without also increasing its power rating, which requires more anodes and cathodes; this is not the case with flow batteries. By increasing the size of the separate electrolyte tanks in a flow battery, it is possible to achieve a long-duration battery without also having to increase its power output at the same time. Li-ion batteries currently have a limited ability to provide long-duration flexibility economically, as their storage capex increases in direct proportion to the required duration.

11.2.5 *ESS Inc.*

There are many flow battery technologies (including one of Stortera's technologies), but first we focus on ESS as one of the companies that has developed such systems over the last decade with financial backing from the Advanced Research Projects Agency – Energy (ARPA-E), a US government agency that advances high-potential, high-impact energy technologies that are too early for private-sector investment. ARPA-E provided grants to ESS Inc. in 2010 to enable it to develop its long-duration energy storage system. In May 2021, it announced that it had entered into a definitive agreement for a business combination (a SPAC) and thus became a publicly listed company on the New York Stock Exchange.

A flow battery is a system that stores the electrolyte as a liquid in external tanks. Currently, flow batteries account for less than 1% of the grid-scale energy storage market due to their high system costs to date. ESS uses iron in its innovative design, which increases power density and enables a smaller and, importantly, is less costly than other materials. ESS sells flow battery systems based on iron, salt, and water, without requiring the use of hazardous chemicals or rare-earth metals (it uses two chemical components dissolved in liquids that are pumped through the system on separate sides of a membrane).

Hugh McDermott, ESS senior vice president of business development and sales, has observed that ESS calculates that Patagonian utility Edelaysen, to which they have sold a system, will be able to save 12 years' worth of diesel emissions and $3 million in fuel and maintenance costs over 25 years by using a 300 kW, 2 MWhr *Energy Warehouse*™ flow battery. Flow batteries could last for 20 years with no major servicing, unlike lithium-ion units, which degrade over time. This could allow ESS to lease its batteries using *asset-backed financing* to cover the costs of battery manufacturing. ESS Inc. has also partnered with Munich RE to launch an industry-first insurance coverage for their flow batteries. The battery modules in their storage solutions come with up to a 10-year extended warranty backed by this global, investment-grade insurer.

11.2.6 *Form Energy*

Its first commercial product is an iron-air battery capable of storing electricity for 100 hours at system costs competitive with legacy power plants. It is made from iron, one of the most abundant minerals on Earth.

In its most recent equity fundraising in August 2021, Form raised $240 million. Form has now raised more than $360 million, including its earlier Series C round which raised $76 million. The investors are:

TPG Rise Climate and NGP Energy Technology Partners III;
Perry Creek Capital (a PE firm);
Coatue (an Investment manager);
Temasek (Singapore state investment group);
Energy Impact Partners, Breakthrough Energy Ventures, Prelude Ventures (impact VC funds);
MIT's The Engine;
Arcellor Mittal (iron and steel manufacturing group);
Capricorn Investment Group (mission-aligned investor);
Eni Next (the VC arm of Italian energy company Eni);
Macquarie Capital (part of Macquarie Group which has advisory, capital raising and financing functions).

11.3 Kinetic/Gravity, Compressed Air, Cryogenic, and Heat Energy Storage

11.3.1 *Pumped Hydro Storage*

This uses pumped hydro storage, also known as closed-loop pumped storage, in which water pumped into a high-level reservoir converts the potential energy (energy storage) in a water reservoir into mechanical/kinetic energy through the force of gravity, which then drives a turbine generator.

Thus, unlike 'flow-of-the-river' hydroelectric systems, which are used for the generation of firm, i.e. continuous, power, pumped storage can be used where a river has inadequate flow or where there is no river at all, but water can be made available. An upper and a lower reservoir can be created using traditional mining processes, enabling pumped hydro storage using low-cost off-peak nuclear or surplus solar or wind power to pump water from a lower reservoir to a higher reservoir. Such 'off-river storage' has a much lower environmental impact compared to a dam on a river, as it does not flood a valley and does not require a continuous supply of water; it uses the concept of water recycling, which has already run through a turbine many times before.

11.3.2 *Rye Development*

Such pumped hydro storage is best suited to operating as a 'peaker' but is less appropriate for intermittent, unscheduled, and unpredictable matching needs. The economics of this type of system depends on an arbitrage between the purchase price of low-cost power at night from nuclear or around midday from solar and the selling price of high-cost power to the Transition System Operator (TSO) at peak times. Note that pumped storage currently accounts for 95% of all utility-scale energy storage in the US. While the UK's pumped storage projects were originally for peaking and balancing to support fossil fuel power stations and nuclear, and with wind and solar replacing fossil fuels, these storage facilities can no longer work on the traditional daily cycle, i.e. pumping at night when power prices are low and generating when required for peaking and balancing. Instead, they now have to ramp and cycle frequently, sometimes as often as 50 times a day. This aims to 'fill in the gaps' caused by the loss of so much dispatchable power as a result of it being replaced by VRE, the supply of which will rise and fall many times during the day.

11.3.3 *Kinetic (Gravity) Energy Using Weights*

Pumped hydro storage uses low-cost power to raise water to a height when power is inexpensive and then releases it to fall by gravity and pass through turbines to generate power, which can be sold at a high price.

11.3.4 *Gravitricity*

GraviStore – which is Gravitricity's energy storage system – uses existing mineshafts to support thousands of tonnes of mass (e.g. concrete) to store energy. This can be converted from potential energy to electric power through a generator. Underground gravity energy storage technology uses the force of gravity to offer some of the best characteristics of both lithium batteries and pumped hydro storage. In practice, it would use abandoned deep mine shafts to allow the concrete block weights to fall a considerable distance, thus storing a substantial amount of energy. Gravitricity's energy storage system functions by suspending a large weight, up to 2,000 tonnes, in a vertical shaft and using low-cost, off-peak electricity to winch the weight to the top of the shaft. The kinetic energy can then be released when required.

11.3.5 *Advanced Compressed Air Energy Storage*

This uses some aspects of other compressed gas systems (which we cover in the following) and is also similar in some respects to pumped water storage, as it uses water pressure from gravity; however, they do not to directly provide power for turbines but provide 'back pressure' to the expanding air flowing through the turbine.

11.3.6 *Hydrostor*

This is a private company founded in 2010 and based in Toronto; it has developed a product which improves on existing mature compressed air energy storage (CAES) technology by eliminating the use of a gas turbine. It does this by storing the heat generated during the air compression process and then reusing it to heat the air on its way to the turbine.

Hydrostor is said to provide long-duration (up to 100 hours) energy storage, enabling direct replacement of fossil generation. Hydrostor is not a capital-intensive business but one which sells its technology skills to others. In January 2021, the Private Equity and Sustainable Investing business within Goldman Sachs Asset Management (GSAM) provided a preference share financing commitment of $250 million to Hydrostor.

11.3.7 *Highview Power Cryogenic Storage*

Air can be compressed as a storage medium, enabling it to store energy. When compressed and cooled to a low temperature, it turns to liquid, which can subsequently be heated (with waste heat if available) to vaporise and release the phase change energy (liquid to gas). This results in its volume increasing by 700 times. The gaseous air then powers a gas turbine connected to a generator. Highview provides an example of a company that has developed a liquid air system and has already sold its systems to a number of customers.

Air turns to liquid when cooled to −196°C (−320°F) and can then be stored very efficiently in insulated, low-pressure vessels. Exposure to ambient temperatures causes rapid re-gasification, which is then used to drive a turbine and generate electricity without combustion. This technology is said to allow storage from four hours to four weeks, unlike batteries which are intended for short-term storage only. It has also been described as having the same giga-scale, long-duration capabilities as pumped hydro but without geographical constraints, sometimes referred to as 'pumped hydro in a box'. In this respect, it is similar to Hydrostor.

CAES is also based on the gas turbine cycle. Power is used to compress air using a rotary compressor and then store it, often in an underground chamber. When power is required, air is released from the chamber and passed through a turbine that generates electricity from the flow of the high-pressure air. Output from the plant can be boosted by burning natural gas in high-pressure air before it enters the turbine, as would happen in a conventional gas turbine. However, this has the penalty of producing carbon dioxide emissions. More advanced plants can store heat during air compression and release it during the expansion phase (as in the Hydrostore system described above).

11.3.8 *Compressed CO_2 Energy Storage: Energy Dome*

One issue with cryogenic energy storage is the difficulty of liquefying air. An alternative medium for liquefaction and subsequent use, involving re-gasification and passage through a turbine, is the gas carbon dioxide. According to Energy Dome, which has developed such a storage system, CO_2 is the perfect fluid to store energy cost-effectively in a closed thermodynamic process, as it is one of the few gases that can be condensed and stored as a liquid under pressure at ambient temperature, allowing for high-density energy storage without the need to reach extreme cryogenic temperatures.

On 25 July 2025, Energy Dome announced a global commercial partnership with Google, utilising Energy Dome's CO_2 battery technology to enable carbon-free energy for the grids that power Google's operations. Alongside the commercial agreement, Google has made a strategic investment in Energy Dome. In contrast to Li-ion battery storage, which provides backup power for 4–6 or sometimes 8 hours, Energy Dome is capable of continuously dispatching energy for periods of 8–24 hours, providing enough firm electricity to meet both the baseload and flexibility requirements of large energy users. An additional advantage of this technology is that it uses a turbine to produce its electricity output. These turbines have 'inertia', just as those in power stations using coal or gas with a turbine, and as inertia has always been necessary to stabilise the grid (and some alternative methods are also being developed), it reduces one of the issues arising from large amounts of non-inertia-based power on a grid.

11.3.9 *Hydrogen as an Energy Storage Medium*

Hydrogen Storage – Electrical Energy Converted to Chemical Energy

Hydrogen can be stored in pressurised vessels, or solid metal hydrides or nanotubes can store hydrogen with a very high energy density but only up to a few MWh. Very large amounts of hydrogen can be stored in underground salt caverns at high pressures, which might be able to store 100 GWh of energy. Balancing seasonal variations might be possible with this system.

Power Generation: Chemical Energy Converted to Power

Hydrogen gas can be converted back into power using fuel cells, but there is an energy loss in the form of heat. This gives the possibility of creating a combined heat and power solution using fuel to utilise the waste heat from power generation for providing space heating in homes and offices. The efficiency of this conversion process from gas to heat and power is higher than that of a system which produces only power.

Hydrogen energy storage is shown in the top right-hand corner of the ETC graphic at the start of this chapter. This is the corner where a technology provides long discharge times (LDTs) and large storage size (a large amount of power). In terms of cost per kWh of storage, it is likely that the cost of hydrogen will likely need to continue falling to be economically viable, though in terms of the global climate crisis, cost is only one factor to be considered.

Hydrogen is likely to have the greatest potential among technologies for **seasonal** energy storage. In the 2050–2070 timeframe, hydrogen with as much as two weeks of stored energy is forecast to be a cost-effective storage method. Total global hydrogen use would grow substantially if it were to be used for energy storage in the way that natural gas has been used for this purpose.

11.3.10 *Rondo Heat Battery: Heat as an Energy Storage Medium*

The Rondo Heat Battery uses electric heating elements, similar to those in a toaster or oven, to convert available power into high-temperature heat at a low cost. Electrical heaters (so-called Joule heaters) convert electrical energy into heat at 100% efficiency and interact smoothly with grid and off-grid generation. Thousands of tonnes of brick are heated directly by this thermal radiation and store energy for hours or days with very low loss (less than 1% per day). When heat is wanted, air flows up through the brick stack and is heated to over 1000°C. The heat delivery rate is easily adjusted by changing airflow. Heat is delivered as either superheated steam or hot air.

We have already noted that one problem companies developing new technologies faces is persuading either venture capitalists or infrastructure investors to back their projects. In Rondo's case, they have persuaded the Breakthrough Fund (originated by Bill Gates) to provide the necessary funding to enable them to offer not just 'first of a kind' but second and third projects. If these are successful, Rondo will be well on its way to full commercialisation of its technology.

Heat energy can also be stored in molten salts. This is the mechanism by which the heat from TerraPower's nuclear reactor (see Sections 13.2 and 14.2.3.1) can be stored in its Natrium storage plant when the reactor output exceeds demand for electric power, and this heat can boost the power output of the generating system when the grid is most in need of it.

Part IV

Innovation in Systems/Networks: Digitalisation, Business Models, Market Design, and AI in Electricity and Heat Systems

Chapter 12

The Other Transition – Digitalisation: The Vital Requirement for the Green Transition

12.1 System Re-Design and Innovation Necessary for the Energy Transition

There are three key factors in the energy transition that are leading to the need for new business models among power producers, transmission and distribution network providers, and electricity retailers. These factors are the **decentralisation** of generation, the **decarbonisation** of generation, and the **digitalisation** of networks and retail distribution models, often known as the '3Ds'. In addition, there are constantly changing frameworks within which companies in these industries must operate as a result of changes in 'market design' regulatory change and innovation by 'disruptive' companies.

Much of the innovation in energy systems has been in energy storage, transmission, and distribution of electric power and 'demand-side management' systems, which enable demand to be reduced to better match supply. In addition, energy efficiency measures such as a reduction in the power required by individual appliances like vacuum cleaners, washing machines, and room lighting (LEDs) in the domestic context and decreasing power consumption in manufacturing processes can reduce the need for constructing new generation capacity. However, as power systems become more complex, they will increasingly be impossible for human beings to make all the necessary decisions to optimise 'the system', i.e. to achieve the best trade-off between energy security, affordability, and sustainability – or the 'Energy Trilemma', as it is known (which we covered in Section 10.1). Instead, it requires the abilities of an artificial intelligence (AI) system to take in thousands of pieces of data every second and process it to make instant decisions regarding system optimisation. In addition, digitalisation and AI systems provide the opportunity to create business models which generate more value both for the companies which have developed new models and for their customers.

The International Renewable Energy Agency (IRENA) sees the need for innovation not only in the 'hard asset' and 'soft asset' technologies but also in business models (private sector) and market design (state regulation), as shown in the upper half of Figure 12.1.[1] Business models and regulations that will enable these new technologies to be used in an economic, i.e. least-cost, manner are critical. There are numerous definitions of the term 'business model' but a useful one is *organizational structures designed to enact a commercial i.e. profitable, opportunity*. The business model can also be viewed as the way a company creates value, which is demonstrated through the rate of return on the investment that it is able to achieve.

A new and innovative market (re)design to reflect the different nature of the new power market and ensure that new technologies are attractive to investors is of key importance. The European Commission, for example, issued its proposals for the revision of the EU's 'Electricity Market Design (EMD)' rules in March 2023. These are intended to accelerate the energy transition, which is currently significantly off track. Reforming the current EMD, which is structured around pricing based on 'merit order', i.e. the sequence in which power plants are designated to deliver power to optimise the electricity supply in terms of the price of electricity, may not be achieving the lowest cost for consumers if they are unable to reap the benefits of low-cost renewables. The merit order is based on the lowest marginal cost producers being given priority in supplying power to the

---

[1] https://www.irena.org/Energy-Transition/Innovation/Systemic-Innovation and https://www.irena.org/-/media/Files/IRENA/Agency/Publication/2019/Feb/IRENA_Increasing_time_granularity_2019.pdf.

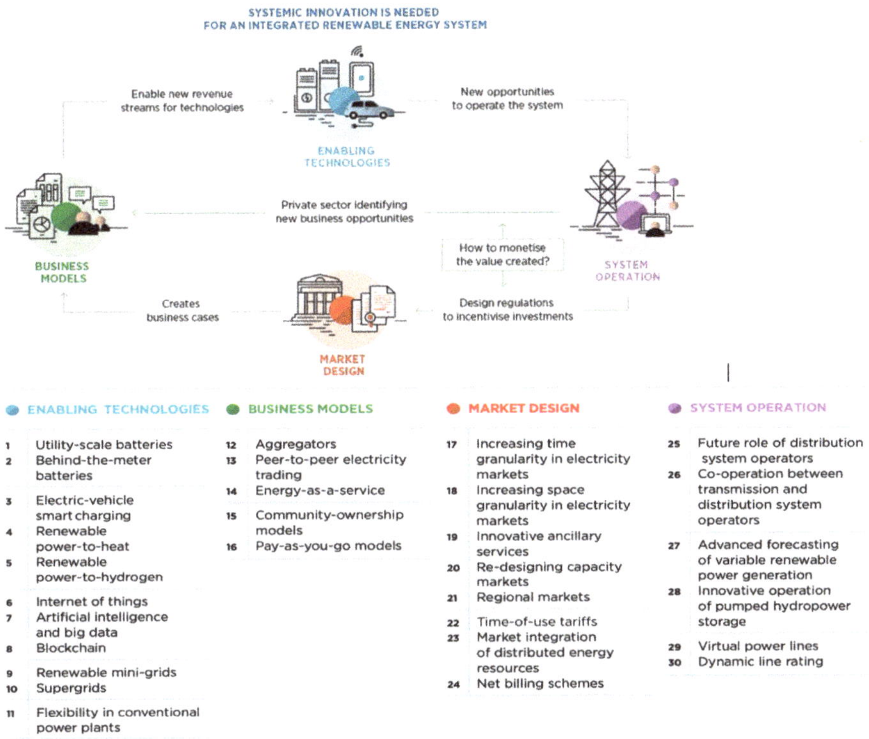

Figure 12.1. Systemic innovation in an energy system and what is necessary to achieve it.

grid. Another issue is whether a region (such as Scotland in the UK) which produces low-cost wind power should be allowed to have lower tariffs than a region that has to import power from such regions at a considerable cost in transmission fees.

In the bottom half of Figure 12.1, under the heading 'Market Design', is the expression 'net billing'. Net billing is a way to not only charge but also compensate retail customers who supply power to the grid based on the actual market value of electricity (rather than simply a notional figure), balancing what they consume against what they inject into the grid. It then raises a key issue for existing electricity utilities, what is known as 'Grid Defection'. This describes the phenomenon of retail customers leaving the grid (going off-grid) when the cost of photovoltaic (PV) panels and battery storage has decreased to a point where generating 'home-made' power is actually lower cost than purchasing electricity from a utility. In some countries, electricity prices vary substantially. For example, in the US, San Diego has a rate of $0.69/kWh, while in Nebraska it is only $0.06/kWh. Therefore, Nebraska utilities are much less likely to suffer defections from the grid than those in San Diego. For many countries, this is likely to be a key issue for regulators and the design of their electricity markets.

In the graphic, the technologies and new regulatory frameworks are predicated on an increasing proportion of renewable energy in the electricity system and a much more 'distributed' system. At the top of the graphic are the new technologies and expanded and improved enabling technologies we cover in subsequent chapters. To the right of the graphic is 'System Operation', which needs to be updated to reflect the new technologies. In turn, this requires innovation from transmission system operators (TSOs) and innovation in business models from power companies if investors who finance innovations and expansion in existing solar and wind technologies are to be satisfied with the expected return from such new assets at the level of risk associated with those expected returns – the traditional, risk/required return investment framework. At this point, we are focusing on 'Enabling Technologies' 6, 7, and 8, which are principally digitalisation and IT. But as can be seen in the graphic, every aspect of the energy system requires innovation. Note that under Market Design, 21 and 22 refer to a consumer pricing system where location and time of day (or grid imbalance) can determine the price paid for electricity or paid by the distributor.

Electricity market redesign in any country will need to be able to best integrate variable renewable energy (VRE) projects into the national power system of the country without risking brownouts or blackouts. It also needs to be compatible with, for example, an ability for companies to develop a business model for pricing longer-duration storage at a rate which provides an acceptable return to investors and can be used within the revised market structure. AI will be key to optimising systems. To supply the AI system with the necessary data, all components of the physical system must transmit information to the 'central brain', and this requires the Internet of Things (IoT), i.e. sensors which can pick up the required information from physical assets, including solar and wind generators and energy storage assets, and transmit it to the centre. In addition, there needs to be a system to request, receive, and transmit information from satellite observations, prices from electricity markets, data from the TSOs, and weather data from providers such as the **Met Office** in the UK and the **European Centre for Medium-Range Weather Forecasts (ECMWF)**, which is based in Reading, UK, and will shortly move to the University of Reading campus near the university's meteorology department. These data must then be transmitted to the AI computer system.

12.2 EU Green and Digital Transitions: Green Power and Digitalisation for 24/7 CFE

Digitalisation is considered so important by the European Commission that it devotes a key section of its 2022 *Strategic Foresight Report* to energy markets and digitalisation.

Going back a little in history, at one time, power generation was centred in cities with one large power station to meet the needs of that city (often owned by a local authority rather than a private company). Gradually, as power stations grew in size and number, they were increasingly located outside cities and were later connected to a national grid. Today, **the major shift is to decentralised power** with numerous small generating facilities (solar, wind, and storage). By 2035, there are also likely to be many small modular reactors (SMRs) and advanced nuclear technology (ANT) providing power directly to users (rather than to the grid) or on the grid.

The problem this raises is the need to integrate renewables, particularly from small-scale generation such as from rooftops and also SMRs/AMRs, into the power system. The issue becomes more challenging to resolve as the number of assets on the grid (solar farms, rooftop systems, wind farms, SMRs, AMRs, and energy storage) grows. As a result, the need for systems comprising power generators, storage systems, transmission lines (the physical assets), and cyber components (including sensors, IoT, and connections to external data sources) that control the operation of the grid has grown dramatically. This, of course, also provides investment opportunities. Controlling the interplay between so many supply sources (generation), demand management systems (DMSs), and storage would be virtually impossible for humans to undertake without these intelligent control systems.

What is happening in the energy field is one example (and a particularly important one) of what has become known as the Fourth Industrial Revolution. In Figure 12.2,[2] we are focused on the new technologies of

[2] World Economic Forum (WEF) '3 ways to be a good leader in the Fourth Industrial Revolution' https://www.weforum.org/agenda/2019/01/the-fourth-industrial-revolution-needs-new-forms-of-leadership/. Adapted from: Shockoe.

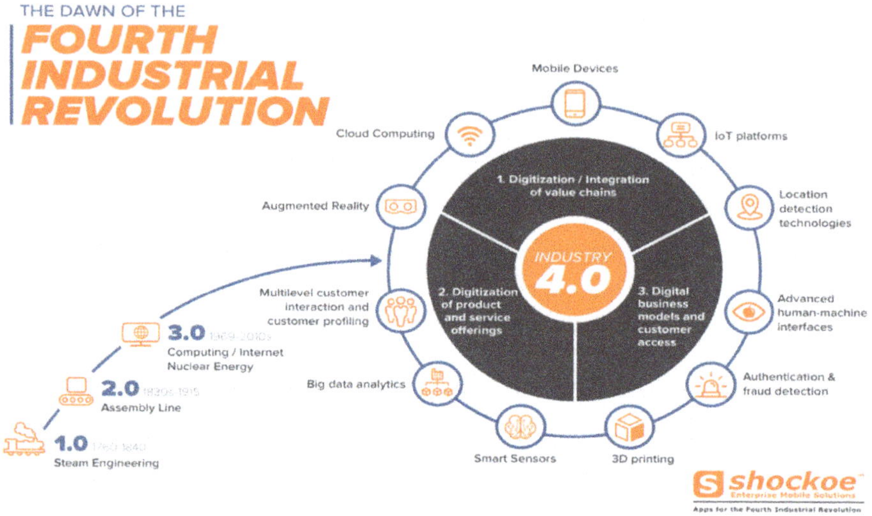

Figure 12.2. Where we are today – the Fourth Industrial Revolution.

today. The best way of looking at what is happening in the power industry (but also in many other industries) today is the implementation of what has become known as **cyber-physical systems (CPSs)**.

These are systems that integrate sensing (for example, temperature, wind strength, and power flows) and transmission of that sensor data through the internet (IoT). This then enables the computation, control, and networking of physical objects and infrastructure via their internet connections to each other. Such systems can also take in other information through the internet, i.e. from sources other than sensors, and can thus also integrate weather data (including forecasts), power market price data, data from TSOs, cloud cover images from satellites, etc. to enable decision-making. Humans could, in theory, make these decisions, but in practice, they have to be made by AI. The new technologies involved are listed under Industry 4.0 later.

Machine learning (ML) is often used to identify patterns in data and provide specific results. ML has been improving by leaps and bounds, not only because of the skills of those designing ML systems but also because of the falling cost and increasing power of computer systems. This has led to the cost of 'compute' (to use the industry jargon for computing power) falling substantially and thus enabling much greater compute to be applied to decision-making in, for example, the power industry.

Energy management systems have traditionally been rule- or logic-based and specialised for the function of managing a collection of assets and their links to other systems. Recently, some energy companies and project managers have been experimenting with the use of more advanced ML systems – Generative Pre-trained Transformers (GPT) such as Chat GPT, which we covered above. However, it is important to understand that the use of GPT in the energy field (apart from its use in call centres) is at the research end of the spectrum and is not at this time being let loose, except experimentally, for decision-making in operating energy networks.

12.3 Case Studies of Software Companies Contributing to the Energy Transition

This section presents a selection of relevant companies. Many others are also making a substantial contribution.

12.3.1 *StorTera*

StorTera is an energy storage innovator in Edinburgh, which was awarded the $1 million prize along with its partner, Equilibrium Engineering, in the highly competitive Canada-UK Power Forward Challenge. One of the key issues in the competition was how to bring together multiple stakeholders, such as grid network operators and consumers, to share both the costs and benefits of upgrading an energy system to handle the integration of renewable energy, electrification of heating, and charging of electric vehicles.

StorTera's innovative solution utilised an AI platform to optimise the performance of the smart grid in real time by having access to **accurate forecasting of weather and energy demand for the day ahead**. The installation included intelligent energy storage, energy efficiency upgrades to buildings, and the ability to control electrical loads such as hot water heating (demand-side response).

Figure 12.3 shows a graphic of StorTera's Traicon[3] (tri-layer artificial intelligence controller) platform, which enables AI control and optimisation. The Traicon system utilises neural network-based machine learning to analyse large datasets, monitor, optimise, and proactively control all

[3] www.stortera.com.

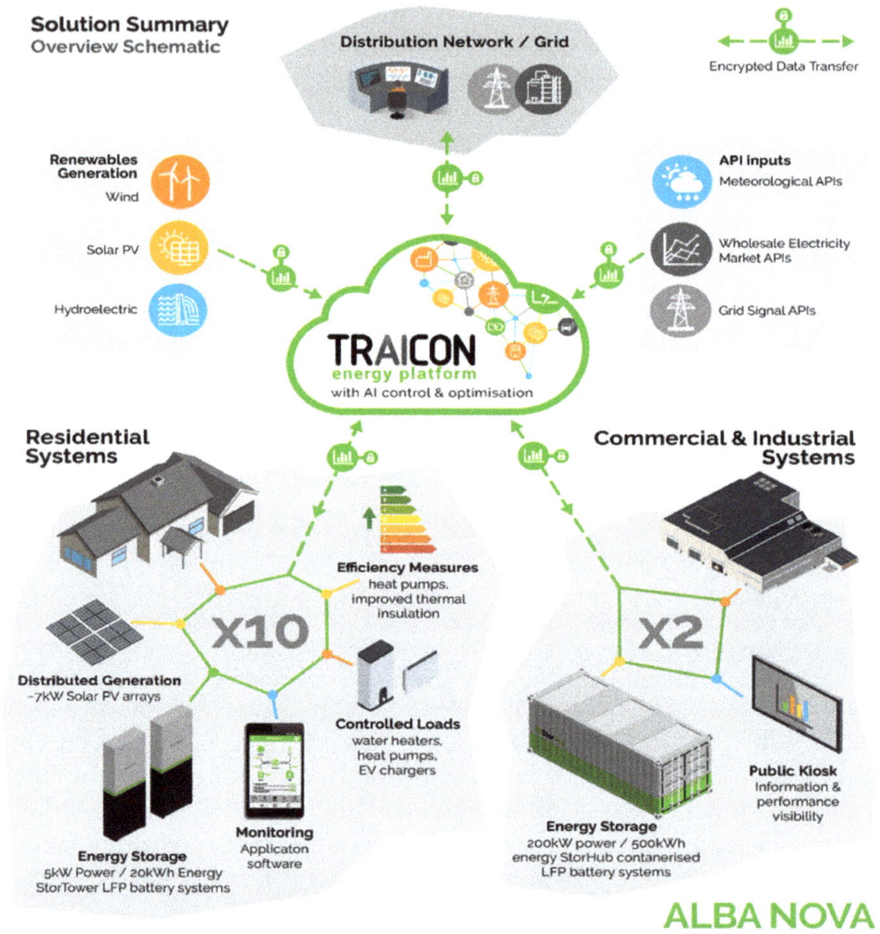

Figure 12.3. AI control and optimisation of an energy system.

connected assets. The controller learns local energy usage and storage patterns and uses cloud-based ML to integrate weather forecasting and other available APIs, allowing networked synchronisation of multiple systems and continually optimised energy performance. The key innovation is the use of the AI platform that manages data and control, ranging from battery cell information to sophisticated algorithms that predict solar energy and customer energy demand for the day ahead. The figure also shows the data inputs and associated APIs required for optimal AI decision-making and the links to power users.

12.3.2 *Fluence IQ and Nispera*

Fluence IQ delivers AI-enabled digital applications for managing and optimising renewables and storage *from any provider*, i.e. it is technology and storage cell provider agnostic. In 2022, it acquired Nispera AG, an advanced technology which enables customers to monitor, analyse, forecast, and optimise the performance and value of renewable energy assets. Its flagship offering is an AI-driven utility-scale asset performance management platform.

In combination with Fluence IQ's digital application offerings, Nispera's suite of five applications enables the combined firm to offer its customers a user-centric 'manage app'. Nispera's technology portfolio includes a predictive maintenance app, an operations and maintenance app, and a portfolio management app. The company also offers a fifth application that provides enhanced power generation forecasting services for energy assets, which is particularly attractive to energy traders, large asset managers, grid operators, and financial institutions.

12.3.3 *Virtual Power Plants*

12.3.3.1 *ABB*

AI systems can control a wide variety of decentralised assets (demand side and supply side) of different types over large areas to aggregate them into a 'virtual power plant' (VPP). The collection of assets owned by different people and companies can then function as if it is a 'power station' providing continuous power.

In more detail, VPPs are cloud-based data control centres that aggregate data from distributed energy resources (DERs). Examples include solar PV plants, battery storage facilities, residential storage systems, electric battery-powered road vehicles, and other sources of power demand and supply connected to the grid. The data centre uses the internet and 5G mobile services, along with information supplied by sensors connected to IoT sensors, to gather data. An operator can then monitor and control the production of each plant to optimise security of supply and costs. Alternatively, this optimising will be undertaken by AI algorithms. A number of different power sources, such as solar, wind, and small hydro, supported by battery storage and also by demand-side management systems, can create a VPP. ABB is one of the companies which provides this type of software. A VPP can overcome pressure on the grid at times when demand would otherwise exceed supply by aggregating various energy sources (and enabling a reduction in demand through demand management) into a VPP coming on to the grid almost instantly when required.

12.3.3.2 *NextKraftwerke*

Another provider of the services needed to provide VPP services is Next Kraftwerke. The strapline under its name on the logo succinctly explains what it is and hence the service provided by such a company: 'THE POWER OF MANY', as it enables many types of assets to be connected to the network. That network can respond either to the grid operator when it demands flexibility by supplying it or (and) it can supply that flexibility to the power system by trading its ability to provide flexibility on an energy exchange.

12.3.3.3 *Piclo*

This company is a 'flexibility marketplace provider'. It offers two platforms. The first is Piclo Max, which is for flexibility service providers to enable them to maximise revenue stacking opportunities for flexibility. It offers buyers and sellers a 'one-stop shop' for all electricity markets globally. For transmission system operators, it offers Piclo Flex, which is a marketplace for local flexibility currently in six countries. Piclo Flex provides an end-to-end solution for flex buyers (system operators) worldwide. Piclo is headquartered in London.

12.3.3.4 *GridBeyond*

GridBeyond defines its role as intelligently dispatching flexibility into the right market at the right time. By achieving this, it supports asset owners and energy consumers to unlock new revenues and savings, achieve resilience, and manage price volatility, helping support the transition to a net-zero future. GridBeyond connects distributed energy resources and electricity markets. It bridges the gap between distributed energy resources around the grid and the requirements of the grid at any point in time.

The need for such optimisation solutions arises from the ever-increasing proportion of renewables on the grid, whose power outputs vary continuously, creating an ever-increasing need for grid balancing services. This, in turn, requires real-time solutions to manage this new power market structure. A technology platform is needed to connect distributed energy resources (DERs) and grid-connected technologies that consume, generate, or store electricity. When optimised across traded power markets and flexibility programmes, DERs can lead to significant cost savings for grid operators and utilities.

GridBeyond provides a link between renewable energy generators and energy buyers, enabling them to hedge against risk and gain benefits through corporate PPAs. By accurately forecasting forward power prices and a company's business demand, its AI-powered optimisation solution runs Monte Carlo simulations to match site demand with managed renewable energy generation and battery storage assets, helping manage price risks and generate savings per MW.

The company works with existing and soon-to-be-operating renewable energy projects to structure long-term offtake agreements that match a customer's demand and renewable energy goals. While there is no direct physical connection between the power generator and the customer, the PPA guarantees that the energy the customer offtakes from the grid can be traced back to a specific wind or solar farm that feeds equivalent power into the grid. This is evidenced by certificates that GridBeyond issues for every MWh generated. The certificates prove that the electricity purchased is renewable and originates from a specific renewable energy asset, such as a wind farm or solar farm, to support a company's Scope 2 GHG reporting.

12.3.3.5 *GraphCast: A New Way of Forecasting Solar Power Generation*

Traditional forecasts, such as those from the Met Office in the UK, have relied on physics-based models. They solve equations based on airflow, moisture, and atmospheric thermodynamics. This new approach to short-term (one or two hours ahead) is quite different. It makes predictions based on gigabytes of previous data on similar weather conditions. It is said to take only seconds or minutes to come up with a forecast, rather than the hours taken by the traditional computing approach. The new approach uses 'prediction engines', which feed historical data into Google's AI DeepMind weather simulator known as GraphCast. In weather forecasting, ML systems such as this are beginning to improve upon traditional numerical weather prediction (NWP)-based forecasting models while also being lower cost, faster, and therefore more useful.

A 'hoped-for' use of this new technology is to provide more accurate and much faster, up-to-date forecasts of wind strength and cloud cover, which reduce solar output. To cope with the frequently significant fluctuations in the output of wind and solar power entering the grid, grid operators (TSOs) schedule and request power-up from natural gas turbines well in advance if solar power production falls short of power demand. Thus, if the approach proves effective, it will make a significant impact on GHG emissions from power generation by reducing the number of 'false' gas switch-ons. The alternative to natural gas for load-balancing flexibility is, of course, on-grid battery storage, which is immediately dispatchable and would also benefit from more accurate weather forecasts.

12.3.3.6 *Iberdrola as an Exemplar of Web 3.0 and Blockchain in Genuine 100% Clean Energy*

Web 3.0 is a term used to describe the next iteration of the internet built on blockchain technology. A blockchain is a digitally distributed, decentralised ledger across a computer network to record transactions. As new data are added to a network, a new block is created and permanently appended to the chain. All nodes on the blockchain are then updated to reflect the change. Consequently, the system is not subject to a single point of control or failure. Web 3.0 refers to decentralised databases and systems architecture.

Iberdrola Group has a pilot project based on using blockchain to guarantee, in real time, that the energy supplied and consumed is 100% renewable, i.e. 24/7 carbon-free energy (CFE). Using this technology, they have managed to link plants where electricity is produced to specific points of consumption, allowing the source of the energy to be traced. This increases transparency and ultimately encourages the use of renewable energy. These systems are used by energy companies to enable them to achieve the best rate of return on capital from their portfolio of energy generation assets (wind, solar, and storage), given time-of-day power pricing and customer demand (as well as demand management) patterns.

12.3.3.7 *Flexi DAO*

This is a European climate tech startup which provides end-to-end CFE solutions for energy buyers and sellers. Its software as a service (SaaS) platform enables organisations to monitor, quantify, and demonstrate to society their contributions to decarbonisation. Their vision is that we should be living in a 24/7 CFE world. They envisage this happening by using their Energy and Emissions Management Platform, which enables corporates to accurately measure, report, and reduce their energy-related emissions. Iberdrola and Red Eléctrica de España are two of the energy companies that use FlexiDAOs SAAS.

12.3.3.8 *ACCIONA Energía*

In Continental Europe, ACCIONA Energía, said to be the world's largest fossil-free utility operating exclusively in renewable technologies, has become the first utility to use blockchain technology to certify the origin of its energy to consumers. It has created a commercial demonstration to trace renewable generation from five wind and hydro facilities in Spain through to four corporate clients in Portugal. The company used a platform and software developed by the Energy Web Foundation (EWF). EWF believes that blockchain technology has the potential to reduce transaction costs, enhance data security and transparency, open markets to a larger number of active participants (consumers and devices), and accelerate the transition towards a digitalised, decarbonised, decentralised, democratised, and resilient energy system.

12.3.3.9 *Energy Trading: The Shift from Manual Trading to Algorithmic Trading*

We noted that controlling energy systems is only practically possible using AI assistance. Figure 12.3 provided an example in a graphic of the Traicon system, which uses AI as a control and optimisation mechanism. While such systems are used for controlling **physical** assets in an energy (electricity) network, there are also energy traders who may or may not work for an electricity utility and simply want to trade electricity in both **spot and forward markets**.

Many of the large investment banks have energy trading desks, as they do with interest rate trading desks. Such traders will arbitrage markets to take advantage of large price swings over short and longer periods of time. As with arbitrage more generally, this should reduce price volatility and generate an acceptable rate of return on their investment in trading.

As an example of available opportunities, electricity spot prices in Germany can switch from negative on days when there is too much power to more than €600 per megawatt hour on days of *Dunkelflaute* (windless and sunless periods). As renewables provide an ever-higher proportion of electricity in a system, the need for real-time balancing of energy markets is increasing. In addition, energy markets and thus the price of electricity respond quickly and sharply to geopolitical tensions, macroeconomic news, regulatory frameworks, new technologies and their expected impact, the extent to which demand flexibility is available in a system, and, of course, the weather conditions.

12.4 Developing Data Standards to Enable Inter-Operability

12.4.1 *BRIDGE*

This is an EU initiative to gather all the varied Horizon 2020 smart grid innovation projects. It is designed to foster collaboration and consolidate results among research and innovation projects in smart grids, energy storage, islands, and the digitalisation of the energy system funded under the Horizon 2020 programme. Since its creation, it has gathered more than 80 projects involving over 700 organisations from 38 countries for a total EU funding of €750 million. BRIDGE is structured around four working groups: Business Models, Regulation, Customer & Citizen Engagement, and **Data Management**. The Working Group on Data Management is focused on the following:

- **Communication Infrastructure**, the technical and non-technical aspects of the infrastructure needed to exchange data and related requirements, such as issues faced by TSOs and DSOs.

- **Cybersecurity and Data Privacy**, entailing data integrity, customer privacy, and protection.
- **Data Handling**, including the framework for data exchange and related roles and responsibilities, together with the technical issues supporting the exchange of data in a secure and interoperable manner, and data analytics techniques for data processing.

12.4.2 *Trusted Energy Interoperability Alliance (TEIA)*

In the US, several organisations have come together to try to tackle the issue of standards and interoperability. The organisations involved include E.ON SE, Intertrust, JERA, and Origin Energy, which in 2023 were the founders of the Trusted Energy Interoperability Alliance (TEIA), set up to develop and promote open technology standards for energy devices, data interoperability, and security for the global energy industry ecosystem.

12.4.3 *UK National Grid ESO (Now Called NESO) and the Energy Data Taskforce*

The UK national energy regulator (Ofgem), along with the government, established the Energy Data Taskforce to provide a set of recommendations on how data can assist with unlocking the opportunities provided by a modern, decarbonised, and decentralised energy system at the best value for consumers. This work is part of the smart system strategy being developed by government.

The Energy Data Task Force's aim is to provide a set of recommendations that will support the delivery of a modern, digitalised energy system, unlocking the benefits of decarbonisation and decentralisation through better use of data. The taskforce believes that digitalisation releases value, opportunity, and resilience, enabling decarbonisation and decentralisation to be delivered at optimal cost for the benefit of consumers. As the system becomes more disparate, diverse, and decentralised, data sharing will be crucial to coordinate the wide range of actors undertaking new roles across the sector and ensure system stability.

Part V

NUCLEAR – A New Approach to Decarbonisation: Innovation in Technology, Co-Location of Supply & Demand, and Nuclear Financing

Chapter 13

Nuclear and Renewables Have Quite Different Characteristics

It is likely that wind and solar power, along with energy storage and demand management, will eventually provide more than 50% of the generation capacity in most countries. What percentage of renewables will in practice be possible or optimal, however, in the sense of not only ***minimising the cost of electricity and heat*** *and ensuring* ***supply security to customers*** *but also enabling* ***supply security from 'bad actor' actions from Russia, China, or Iran*** *(the current actual and potential bad actors) is not yet clear. A clear risk in the case of offshore wind is 'cable cutting' using unmanned small submarines. Not only could such attacks give rise to power brownouts and blackouts but rejoining cables in deep water offshore is not a simple task.*

Some proportion of the balance of generation capacity will almost certainly be made up by using 'dispatchable' 24/7 power from nuclear, hydro, and geothermal, with a proportion of natural gas generation, which may be required to cover long periods of 'dunkelflaute'. This is a German word meaning 'dark doldrums', referring to a period of low or no wind and minimal sunlight, which impacts renewable energy generation. It is also known as 'anticyclonic gloom' and can continue for many weeks over large geographical areas, such as most of the European Union.

Another phenomenon – this time one which affects hydro – is a lack of rainfall, which lowers electricity output as the water levels in reservoirs fall. This may also require the installation of genuinely 'dispatchable' generation capacity, of which nuclear is likely to be the key technology if a country is taking decarbonisation seriously and minimising the need for backup natural gas generation.

A good example of the type of problem we cover above is demonstrated by developments in Italy. In January 2025, renewable energy production fell compared with the previous year, even though installed renewable capacity grew. This decline was mainly driven by a drop in wind (−8.7%), hydroelectric (−5.1%), and photovoltaic (−0.5%) production. In contrast, generation from coal and gas increased by 11.5%.[1] *The problem of a massive blackout in Iberia, which we cover later, also suggests that electricity systems in Europe more generally need modernising, as ever more power from renewables needs 'integrating' into the grid system while avoiding the risk of electricity shortages and blackouts.*

We have not given extensive coverage in this book to solar PV generation technology or to wind turbine technology, as these technologies are well known and well developed, while 'new nuclear' employs new technologies which are not yet widely understood. For this reason, Part V on nuclear is the most extensive part of this book.

In this chapter, we focus on Europe as a region that has been trying to decarbonise the economy rapidly but is falling behind targets despite the high degree of effort that has been expended. However, the same logic applies worldwide. What we expect to happen, which will make a contribution to meeting decarbonisation targets in many countries, is the use of the new nuclear technology that is about to come on-stream to overcome some of the problems that are becoming apparent in trying to decarbonise an economy **at the pace required to keep the global temperature rise below 2°C while focusing only on renewables. It will also be necessary to** meet the electricity 'demands' of the hyperscalers' datacentres, which become more vocal every month.

[1] Statistics from Terna, the Italian grid operator.

13.1 Embedded Nuclear, Hyperscalers, and Five Key Differences Between Renewables and Small and Micro Advanced Nuclear Technology Reactors

Nuclear **can facilitate the integration of renewable electricity into a grid system** (filling in the gaps from dark nights and light winds), can provide baseload power (24/7/365 dispatchable electricity), can provide industrial heat and space heating (which critically and urgently require decarbonisation), and can be used as a heat generator for energy storage in a heat battery for use as industrial heat or for subsequent conversion to power. It can also provide portable power from micro-reactors, which can be easily transported to where heat and/or power is needed, embedded within an industrial complex either as a single unit or in multiple units, or used to generate heat or heat and power. The micro-reactors we cover can also be very attractive for military use to generate electricity and heat as an alternative to diesel-generated power, which is noisy and emits fumes and is therefore easily detected by an enemy. Micro and small reactors can also have a role to play in maritime transport, particularly for container ships. There are already many submarines using nuclear propulsion and also icebreakers which have not caused any dangerous incidents over many decades. In the future, nuclear power is likely to be used much more widely at sea as its safety comes to be understood better.

As we noted above, the European Commission is one of the most active organisations striving to ensure the maximum contribution to global decarbonisation from European countries. On 26 February 2025, the European Commission issued a document: the **Clean Industrial Deal** (CID).

> The EU aims to become climate-neutral by 2050, with the Clean Industrial Deal and the Green Deal Industrial Plan as key strategies to support this, focusing on **decarbonizing industries** and boosting clean-tech sectors through increased investment and simplified regulations.

While the objective of the CID is to decarbonise industry, a well-considered plan to decarbonise space heating in domestic residences and commercial premises and in industry is also required.

Much of this decarbonisation will involve **energy in the form of heat rather than electricity.** It will also require the production of much more green hydrogen (some produced using low-cost nuclear **heat** and not

100% electricity) rather than hydrogen made from natural gas. In fact, electricity accounts for only 20% of the total energy market, while heat and transport each account for 40%. Depending on the temperature of carbon-free heat required, it can be provided by electricity (power to heat) using resistance heating, induction heating, electric furnaces, electric boilers, industrial heat pump, or from nuclear heat.

The key problem in Europe's conversion from fossil to carbon-free energy (CFE) is the difficulty of integrating CFE **in ever higher proportions into an existing transmission and distribution system designed for very different energy sources**. The traditional energy system built many decades ago was very different from the one it is hoped to have later this century – and certainly by 2050 – but might benefit from a reversion to some aspects of that system (community energy). As the International Energy Agency notes in a recent report[2]:

> This report emphasises that while systems at early phases of VRE integration can accelerate deployment with relatively low system impact, those in high phases face more complex challenges related to **stability and flexibility, which call for a transformation of how power systems are operated, planned and financed**.

Traditionally, electricity was generated in power stations relatively close to demand, which is very different from today when an ever-increasing amount of electricity for European customers is generated in the North Sea or from distant solar farms. On the other hand, in Europe, users of electricity are mostly hundreds of kilometres south of the North Sea and distant from good solar or wind locations. When existing legacy transmission systems in many countries were constructed many decades ago using pylons with overhead cables, it was relatively straightforward to connect electricity producers with electricity users, and these were generally not very far apart. In those days, it was also possible to overcome resistance from people living near intended transmission routes. Today the distances required for transmission are so much greater and the opposition by local communities is so much stronger, which will likely take many years to connect all generating facilities to users, if it is possible at all.

---

[2] IEA *Integrating Solar and Wind September 2024* https://www.iea.org/reports/integrating-solar-and-wind.

The cost may increase sharply if opposition to pylons and overhead cables is such that energy networks must be installed underground.

Although expanding and strengthening the network is necessary and subject to difficulties and delays, ENTSO-E's cost–benefit analysis suggests that implementing the whole portfolio of planned transmission projects would create up to €33 billion increase in socio-economic welfare for EU citizens in 2040 (i.e. socio-economic welfare generated exclusively in that single year). However, what is planned across the EU is, according to ENTSO-E, **unlikely to be sufficient to achieve the hoped-for outcomes**.

One solution to the transmission problem, though a somewhat limited one, is greater use of rooftop solar, though this would need to be connected to adequate energy storage; this could be in a community energy storage facility rather than individual homes, which is a less efficient solution. Such a system can provide grid benefits beyond simply the power that it produces. It generates electricity exactly where it will be used and therefore does not require additional transmission capacity (unless also connected to the grid). As well as reducing transmission costs, it can also help reduce line losses and congestion without requiring new land for solar farms.

What we are suggesting for nuclear would be on a much larger scale and would have a much larger impact in overcoming the problem of lack of transmission capacity. It would also bring benefits in terms of decarbonisation particularly through its ability to decarbonise heat. To achieve these benefits, there is a need to revert to some aspects of the **model of yesteryear**, when electricity generating capacity was sited near domestic and industrial users in cities or regions. Today, it can also be **embedded within an industrial facility** or, as we cover elsewhere, next to a **hyperscale datacentre**. We propose that the best solution would be to further facilitate the use of small modular reactors (SMRs) but, more particularly, **advanced nuclear technology (ANT)** high-temperature **small and micro** reactors **embedded in communities** (somewhat parallel to the EU's 'Community Energy' concept) and industrial zones. A key advantage of nuclear (and also geothermal) is that, unlike renewables, it provides continuous power 24/7/365 (**dispatchable power/baseload**) and other necessary grid services, which it is difficult, and can be expensive, for renewables alone to provide as the proportion of renewables on the grid increases.

In Chapter 10, we noted the issues involved in trying to 'convert' renewable intermittent power into dispatchable 24/7/365 electricity. This is a 'system' issue. When designing an electricity system with a high proportion of renewables, a key consideration is ensuring that a sufficient proportion of the assets on the system are 'system-beneficial' assets. Battery storage clearly is, but such systems do not generate power though they are of critical importance in providing instant availability of electricity during short-term system imbalances but are less useful for longer periods when actual **generation capacity** falls short of demand.

Interconnectors to other countries are also invaluable day-to-day to match supply and demand over a much wider geographical area, as renewables generation varies by geography, and the time of peak demand may also vary. There thus is a 'portfolio' diversification benefit of reducing risk (in this case, the risk of short-term fluctuations in electricity supply, brownouts, or blackouts). However, if there is a long *dunkelflaute* (using the German term), this means it is both dark (which increases electricity demand and reduces solar supply) and a fall in wind force (also resulting in less power being generated). In this situation, if wind strength is down across Europe for days at a time, interconnectors are less useful.

S&P Global Commodity Insights held a webinar in late March 2025, which, among other things, provided a graphic (Table 13.1) illustrating the misalignment of data centre future requirements in many countries. What this figure notes is that by bringing demand and supply for power closer together geographically, current and future grid constraints will be eased. It then goes on to note that offshore wind and existing nuclear power stations are often located far from demand. On the other hand, the new designs of reactor, which we cover in the following and which are currently coming to fruition, need not suffer from this problem.

The first point in the 'wish list' below is that data centres do not want to locate near renewable energy facilities if they are far distant from their customers. They also need to be in a location where there is **adequate data infrastructure, i.e. capacity for their huge bandwidth demands**. Hyperscalers cannot move to a distant location from users unless they also have the necessary data capacity to meet their customer needs. The second point is that they need to have low-carbon electricity for regulatory reasons and power which is additional (**'additionality'** in EU parlance) to existing supplies as many countries do not want data centres monopolising existing and planned future low-carbon power.

The third point is that their 'operating profile' requires 24/7/365 clean power **on an hourly (or 15-minute) basis**. The final point – 'Delivery

Table 13.1. Data centre needs and how the power sector might adapt to meet them.

	Data Centres	Power Sector
Location	• Want to be close to their customer base and to data infrastructure – typically large cities with high concentration of power demand	• Locating demand close to supply limits impact of grid constraints and lowers overall system costs
Low carbon supply	• Want certified (ideally 24/7) low carbon power to fulfil their sustainability targets	• Low carbon sources (e.g. offshore wind and nuclear) are often located some distance away from demand centres. Renewable generation is highly variable
Operating profile	• Want reliable 24/7 baseload power for their operations	• Increasingly values (and is willing to pay for) flexibility of grid offtakes, either through actual demand turndown or through on-site storage/generation
Delivery timelines	• Relatively short time to market, 2–3 years • Long-term demand trends are uncertain, hinged on the growth in AI usage vs efficiency gains	• New power projects, esp. low carbon baseload ones like nuclear or BECCS, take at least 5 years; upgrades to grid infrastructure can take 10+ years • Values long-term certainty on demand to support investment decisions

Timelines' – notes that new generation assets are not yet ready for deployment and are expected to be at least five years away. To meet all the requirements of these growing customers, the only two sources of clean **baseload** (24/7) electricity are nuclear and, in some countries, geothermal generation. Bioenergy with carbon capture and storage (BCCS) is a potential source of such electricity, but in practice it may not be low carbon and is unlikely to be available at the scale required.

The graphic also notes that updates to grid infrastructure can take 10+ years (and this is said to be the case in many countries), so the fact that advanced nuclear is probably five years away is much less of an issue than might be thought, as it may not require the same use of the transmission network as renewables to enable it to supply power in the required locations. It can be located close to demand in an SMR campus or an industrial zone, or even be embedded within a datacentre itself, though it might also require a connection to the grid.

We summarise in the following why today's and tomorrow's nuclear 'fits the bill' for data centres, while it can also enable the necessary integration of higher proportions of renewable power into the grid system. Nuclear, as well as supporting renewables grid integration, is also ideal for providing baseload power to the system. Finally, we must also mention the need for heat, both in industry and in space heating, where nuclear can supply relatively low-cost heat for district heating systems from its 'waste' heat.

There are five key differences between renewables and small and micro ANT reactors:

1. Nuclear power is **dispatchable** 24/7/365, unlike variable output renewables which require ever-increasing storage, demand management, and backup gas generation.
2. ANT nuclear, in addition to providing electricity, can provide **high-temperature heat for industrial processes** (as well as low-temperature heat for district heating networks and thus can replace coal and gas which renewables cannot achieve so easily.
3. It can help **reduce the problem of how to transmit ever-larger amounts of power**, requiring ever-more transmission capacity, for example between the North Sea wind farms and industrial centres in Europe, as it enables the source of electricity to be located close to demand in built-up areas, as it was in the early days of coal power stations and also town gas manufacture from coal.
4. Micro (high-temperature or low-temperature) reactors, much smaller than SMRs, can be embedded within an industrial complex, which reduces some of the transmission and distribution problems that countries face today and thus should also enable the power to be provided to industry below the 'zonal price' laid down by the energy regulator. This could substantially reduce the price of power for industrial use.

5. Nuclear can provide 'inertia' services and voltage stabilisation. Generators using steam turbines have spinning rotors rotate at the required frequency to help balance supply and demand, and can spin faster or slower as needed. Also, the kinetic energy 'stored' in these spinning turbines provides system inertia. Turbines continue spinning even if the generator itself has lost power, thus slowing the rate of change of frequency (normally 50 cycles per second) and allowing the central control room more time to restore balance. This provides a dampening effect on the system. While new ways of trying to achieve 'synthetic' inertia are being trialled, having baseload nuclear mostly overcomes such problems.

13.2 The Convertibility of Heat to Power and Power to Heat

Much of the urgently required decarbonisation across the world will involve **energy in the form of heat as well as the use of electricity**. The use of heat can also enable the production of lower-cost green hydrogen using some proportion of low-cost nuclear heat in electrolysers rather than 100% electricity. In fact, as we have already noted, electricity accounts for approximately 20% of the total energy market, while heat accounts for 40% and transport also accounts for 40%. Renewables can be a source of heat, but they can only produce heat by passing an electric current through a resistance. Nuclear is likely to be ever more necessary in its role in providing industry with high- and low-temperature heat, electricity, and hydrogen.

Heat has three principal functions in the energy system:

1. Low-temperature heat is used for space heating in domestic and commercial premises using either heat pumps or district heating systems.
2. High-temperature heat is required for industrial processes, mainly directly as 'dry heat' or as steam.
3. Heat is also used to produce steam for **dispatchable** generation and this heat can be from nuclear, geothermal or concentrating solar, all of which can provide CFE using a turbogenerator set. Heat also enables relatively low-cost energy storage, with that stored heat later being converted to electricity.

Almost half of the final energy consumed globally is for the provision of heat itself (1 and 2 above). More than half of this heat goes towards meeting the space heating and water heating requirements of our homes, offices, and commercial activities. The balance primarily covers food processing, general manufacturing, and the 'hard-to-decarbonise' industries – iron and steel, cement, glass, aluminium, chemicals, and paper. In the future, new processes may be created for these using CFE or hydrogen as a fuel and reagent and from dry nuclear or nuclear steam or concentrated solar heat using mirrors in desert sites (such solar is discussed in Chapter 14). In addition, heat can be used for the thermochemical or electrolytic splitting of water into hydrogen and oxygen, with the hydrogen being stored for subsequent use. It can then, if required, be used as an energy storage medium (to be converted back to electricity through a fuel cell on demand) though the efficiency of this process is low.

Globally, most heat is produced by coal and natural gas (and wood). But while natural gas may emit only half as much CO_2 as coal, there are also frequently high levels of 'fugitive' emissions of methane (itself a potent greenhouse gas) from the extraction and transport of natural gas. It can thus only be a very short-term expedient to help decarbonisation. Today, the role of natural gas in Europe has taken centre stage after the Russian invasion of Ukraine, as Russia was providing up to 40% of Europe's natural gas and a higher proportion in the case of Germany. Today, much of Europe's natural gas is imported from the United States as Liquefied Natural Gas (LNG), though unfortunately its liquefaction, transport, and regasification are energy intensive. Three products – **carbon-free heat, carbon-free green electric power, and carbon-free green hydrogen for industrial heat and chemicals** – can all be produced using nuclear, as can desalinated water. This latter point may become more important in Europe as sources of clean water become scarce due to droughts.

Heat itself can be stored, unlike electricity which can only be stored as other forms of energy, such as electrochemical (battery) storage. Heat stores in district heating systems typically have insulated hot water tanks holding many thousands of litres as a simple and cheap way to enable low-temperature heat supply to be matched to demand. High-temperature heat storage, such as that used in the TerraPower Natrium reactor, usually takes the form of molten salts, where the latent heat of phase change (from solid to liquid) is used to keep down the size of expensive thermally insulated storage vessels. Some Generation IV high-temperature nuclear reactors use molten salt both as a heat transfer medium and as the carrier of fissile material (nuclear fuel).

Heat can be classified as high temperature (>600°C) and lower temperature (<250°C). In the absence of fossil fuels, only nuclear and electricity can produce high-grade heat from which industrial value can be gained. Low-temperature heat is useful for space heating and horticulture. Very-high-temperature heat (above 750°C) can also energise the direct thermochemical decomposition of water (H_2O) into hydrogen and oxygen more efficiently, i.e. at a lower cost, than by electrolysis. This thermochemical process has driven the ongoing development work on advanced Generation IV nuclear reactors, which may be able to operate at 800°C and above and also opens other essential clean energy needs such as green steel.

Heat, unlike electricity, is expensive to transport, generally needing large-diameter pipework for outbound and returning fluid, usually superheated steam or water. This limits heat distribution to a few kilometres from the source, which **favours small, local energy plants typically owned by the community, municipality, or city**. Hydrogen as a gaseous fuel is easier to transport than high-temperature heat, although some parts of the existing gas pipe network in a post-fossil world could become so-called 'stranded assets', i.e. of no economic value. Some of the gas network may be suitable for hydrogen transmission, but other parts may not be suitable for metallurgical and safety reasons.

13.3 Nuclear Desalination Using Heat and Power, Nuclear Hydrogen, and Regional Economic Growth

In Europe, Spain is by far the country with the largest desalination capacity, totalling nearly three-quarters of total capacity. It is followed by Italy, Cyprus, Malta, and Greece. But the need for fresh water supplies is rapidly increasing. Cyprus would seem to be particularly exposed to water shortages, not only due to drought but also to underinvestment in water systems. It has been reported that in June 2025, the country's reservoirs were only around 22% full compared to the situation in 2020 when they were at 97% capacity in the same month of June.

Another region which may require desalination of seawater is the Oxford/Milton Keynes/Cambridge Arc, where the government aims to massively increase housebuilding and science research facilities and other infrastructure, such as electric rail lines. Oxford and Cambridge have both suffered from shortages of water and electricity, which have led to environmental regulators temporarily blocking housing and science developments. Generation IV nuclear might be the optimal solution to enabling further economic growth in this region by increasing the supply of electricity, heat, freshwater, and, if required, hydrogen (as covered in the following paragraph).

According to the International Atomic Energy Agency (IAEA),[3] nuclear hydrogen production technologies have great potential and advantages over other sources that might be considered for growing the hydrogen share of energy in a future world energy economy. The selection of hydrogen technologies (to be coupled with nuclear power reactors) greatly depends on the type of the nuclear power plant itself. Some hydrogen production technologies, such as conventional electrolysis, require only electric power. Others, such as thermochemical production, may require only process heat. High-temperature steam electrolysis (HTSE) and hybrid thermochemical cycles can also be used, and these require both heat and electricity. While at present the only nuclear power available is generally low temperature (around 300°C), once Generation IV reactors are available at scale and commissioned, they will be able to provide heat at 600°C and above, which may eliminate the need to use electricity as well as heat in the production of hydrogen.

13.4 But Is Nuclear 'Safe'? Lifetime and End of Life Risk Management of Nuclear, Renewables, and Battery Storage

Nuclear is one of the few carbon-free sources of heat. It can also, of course, produce electricity. *Decarbonising both heat and electricity (and hence domestic and commercial heating, transport and industry) is the basic requirement for mitigating climate change.*

> Do you believe humanity can invent something to stabilise the climate?
> *I always advocated nuclear as a good and cheap and sensible way of getting energy.... But too many people hate it.*

The above quotation, published by *The Observer* newspaper in July 2020, came in an interview with the creator of the Gaia concept, James

[3] https://www.iaea.org/topics/non-electric-applications/nuclear-hydrogen-production.

Lovelock, when he was 101 (someone who was certainly 'green' and very aware of the global risk from climate change). Like him, we believe that nuclear is likely to be essential to stabilising Earth's temperature – a task which requires a rapid increase in all forms of clean energy production. New reactor technologies, providing both direct heat and electricity, include Gen III+ (both gigawatt-scale and SMRs) and small and micro-sized Gen IV Advanced Modular Reactors (AMRs **Exeltium**). AMRs are very different from previous generations of nuclear and utilise completely new, inherently safe designs based on physical principles and not engineering add-ons, as was the case with legacy reactors.

The principal author of this book has a similar view to James Lovelock, namely that nuclear is a relatively low-cost and sensible way of providing carbon-free heat and electricity. The principal author was first introduced to nuclear as a schoolboy when the chief engineer of the first Hunterston nuclear power project took him on a guided tour of this bottom-loading reactor before its fuel had been inserted, enabling him to climb a ladder and walk around on the floor of the pressure vessel. This tour piqued his interest in nuclear and also his decision to start a degree in mechanical engineering. He is also a former member of the Advisory Board of Penultimate Power UK (at one time a potential supplier of UK-manufactured Chinese nuclear ANT), and thus the technology and financing issues related to advanced nuclear are both a special interest.

The UKs National Energy System Operator in its December 2024 report entitled 'Clean Power 2030', states (in relation to electricity alone) that:

> "Nuclear power will play an important role in achieving a clean power system by 2030 and beyond into the 2030s". On SMRs and AMRs specifically, the Report states "Government could consider the value of bringing this rollout forward".

Nuclear fission as a source of heat and power remains controversial but decreasingly so in Europe. Factually, given the number of reactors in the world (over 400 in 32 countries), there have been few accidents and very few deaths compared with most human activities of comparable scale and purpose. Three – Fukushima (2011), Chernobyl (1986), and Three Mile Island (1979) have had 'incidents'. Those reactors which had incidents used what are now old and long since abandoned technologies developed at a time when knowledge levels were very much less then today, and all reactors designed since have taken note of the root causes of these accidents and have achieved design and operational standards to control risks to much higher standards.

Charles Perrow's *Normal Accident Theory* describes 'unanticipated interactions of multiple failures in a complex system'. Perrow believes that the failure at Three Mile Island was a consequence of the system's immense complexity. Such systems are at high risk, particularly if not well managed. He then suggested the need for a radical redesign of reactors. In fact, designers of Generation III+ and *a fortiori* Generation IV have taken note of his observations. Generation IV reactors are generally designed **without using water for cooling and heat transfer**. Generation IV reactors are specifically smaller in size and their resultant inherent physics enables them to function safely through their inherent design without human intervention or external power in all physically possible internal and external conditions. *This was not possible for the larger reactors of the past due to their being sized far above a critical geometrical limit for the self-dissipation of their own decay heat.*

Italy had Gen II and III nuclear power plants in the 1960s and 1970s, but after the Fukushima tsunami, a referendum resulted in citizens voting against further expansion of the nuclear fleet. Following subsequent pressure from activists, the government decided to shut down its existing plants. In practice, however, the exposure doses from radioactive iodine-131 released from the Fukushima nuclear accident were less than one thousandth of those from natural radionuclides. The incident did, however, create a high level of fear among the public across Europe.

More recently, the energy security minister in the Italian government disclosed that Italy is now planning to have Generation IV small modular reactors operating within 10 years. He believes it is important to reduce dependence on fossil fuels (which includes Russian gas and oil). Italy has now set a target of at least 11% of electricity generation being from nuclear by 2050. The Italian government has also recently imposed restrictions on new solar power projects in the belief that they threaten food security by using cultivatable land. **So, for the Italian government, there is a trade-off between energy security and food security.** Also in Italy, as we note elsewhere, the Italian government has been concerned about the growth of solar farms reducing agricultural output.

Some people, unsurprisingly, are concerned about the risk of having a nuclear reactor in their near vicinity. Indeed, many people see Chernobyl as the worst energy disaster in history. But deaths as a result of nuclear are lower than for other types of energy generation. For example, while the UN has registered only 43 deaths as a result of Chernobyl, in comparison 240,000 people lost their lives when the Banqiao Dam in China (which provided hydropower) collapsed. Another statistic would compare the 43 deaths associated with Chernobyl with 42,514 deaths from vehicle crashes in 2022 in the United States alone. Erlend Hagen[4] suggests that we are moving quite rapidly towards technologies that will further reduce any radiation or environmental impacts of nuclear power generation that may or may not exist at present.

Understanding the full scale of potential impacts from all current and future electricity generation systems involves life cycle assessments across a wide range of environmental indicators, including risks from radiation, CO_2 emissions, the recycling costs of the technology, toxic materials in landfill leading to water table pollution issues, and many other considerations. Nuclear has always been thought by some to be the 'most risky' technology. But given the current very high recycling costs of renewables waste, of which there will be many thousands of tons, it seems likely that most used panels and wind turbine blades will still go straight to landfill. This may then cause leaching of dangerous chemicals into the water table. The International Energy Agency (IEA) has noted that only 3% of solar panels manufactured prior to 2009 have been recycled. Battery manufacture is highly CO_2 emissive due to the electricity used in manufacture frequently produced from coal. In addition, every tonne of lithium extracted through mining is said to lead to the release of approximately 15 tonnes of carbon dioxide.

In the case of solar panels, the Harvard Business Review estimates[5] that by 2035, discarded panels would outweigh new units sold by 2.56 times. In turn, **this would catapult the LCOE (levelised cost of energy, a measure of the overall cost of an energy-producing asset over its lifetime) to four times** the current projection due to the high recycling costs that one-day someone must bear. The economics of solar would darken quickly as the industry sinks under the weight of its own trash.

---

[4] https://www.linkedin.com/pulse/from-waste-watts-europes-fast-reactor-roadmap-net-zero-erlend-hagen-v3g9f/?trackingId=BUdtf1wnSh%2B%2FKzqkCiBGGA%3D%3D.

[5] *Harvard Business Review* The Dark Side of Solar Power https://hbr.org/2021/06/the-dark-side-of-solar-power June 18, 2021.

The view that the European Commission has taken, on the basis of its Joint Research Centre (JRC), is that nuclear is no more dangerous than other forms of energy generation and may indeed be safer, as explained in the following.

Technical assessment of nuclear energy with respect to the **'do no significant harm' criteria** of Regulation (EU) 2020/852 ('Taxonomy Regulation') by the EU JRC.[6]

The inclusion or exclusion of nuclear energy in the EU taxonomy was a subject of debate throughout the negotiations on the EU Taxonomy Regulation. While there are indirect references in regulations to the issue of nuclear energy (including on radioactive waste), co-legislators ultimately left the assessment of nuclear energy to the Commission as part of its work on the delegated acts establishing the technical screening criteria.

As the in-house science and knowledge service of the Commission, the JRC – with extensive technical expertise on nuclear energy and technology – was invited to carry out such analysis and to draft a technical assessment report on the 'do no significant harm' (DNSH) aspects of nuclear energy, including aspects related to the long-term management of high-level radioactive waste and spent nuclear fuel, consistent with the specifications of Articles 17 and 19 of the Taxonomy Regulation. This report is the result of that JRC analysis.

- The analyses did not reveal any science-based evidence that nuclear energy does more harm to human health or to the environment than other electricity production technologies already included in the Taxonomy as activities supporting climate change mitigation.
- The comparison of the impacts of various electricity generation technologies (e.g. oil, gas, renewables, and nuclear energy) on human health and the environment, based on recent lifecycle analyses (LCAs) shows that the impacts of nuclear energy are largely comparable with hydropower and renewables, with regard to non-radiological effects.
- The management of radioactive waste and its safe and secure disposal is a necessary step in the lifecycle of all applications of nuclear science

[6] The Joint Research Centre (JRC) is the European Commission's science and knowledge service. It aims to provide evidence-based scientific support to the European policymaking process. The text in this section is from a 387 page JRC Science for Policy Report published by the European Commission in 2021 under the title of the Chapter heading above. Available at https://publications.jrc.ec.europa.eu/repository/handle/JRC125953.

and technology (nuclear energy, research, industry, education, medical, and other uses). Radioactive waste is therefore generated in practically every country, with the largest contribution coming from the nuclear energy lifecycle in countries operating nuclear power plants. Currently, there is broad scientific and technical consensus that the disposal of high-level, long-lived radioactive waste in deep geological formations is, to the best of current knowledge, considered an appropriate and safe means of isolating it from the biosphere for very long time scales.

- Measures to ensure that radioactive waste does not harm the public and the environment include a combination of technical solutions and an appropriate administrative, legal and regulatory framework. Although there remain contrasting views, it is generally acknowledged that the necessary technologies for geological disposal are now available and can be deployed when public and political conditions are favourable. No long-term operational experience is presently available as technologies and solutions are still in demonstration and testing phase moving towards the first stage of operational implementation. Finland, Sweden, and France are in an advanced stage of implementing their national deep geological disposal facilities, which are expected to start operation within this decade.

The comparison of environmental impacts of various electricity generation technologies on human health and the environment leads to the following main findings:

> Average lifecycle GHG emissions determined for electricity production from nuclear energy are comparable to the values characteristic of hydropower and wind.
>
> Nuclear energy produces very low NOx (nitrogen oxides), SO_2 (sulphur dioxide), PM (particulate matter), and NMVOC (non-methane volatile organic compounds) emissions. The values are comparable to or better than the corresponding emissions from the solar PV and wind energy chains.
>
> Land occupation of nuclear energy generation is about the same as for an equivalent-capacity gas-fired power plant but significantly lower than for wind or solar PV.

In terms of cost, some forecasters report that once the new generation of ANT reactors has production lines operating in a mass production system similar to that for car manufacture, they should be able to generate power 'at the farm gate' at a lower cost per unit of output than gigawatt power stations such as Hinkley Point C in Somerset, England. Furthermore, small nuclear once standardised and manufactured at a mass-production scale is likely to be able to supply electricity and heat at a lower cost to the end customer without the high system costs of transmission involved when power is generated by distant wind and solar farms or legacy nuclear. It should be possible, once a mass production scale of manufacture is achieved, to provide electricity or heat without subsidy **while commanding a commercial rate of return for the investor**.

Two new factors now favour nuclear instead of natural gas. First, the passive safety systems of new nuclear technologies make them more widely deployable, responsibly, across the developed and developing world. **Second, the realisation that the world is not in practice decarbonising at the required rate of progress because heat and transport, which are the dominant carbon emitters will require a massive increase in carbon-free heat and electricity production, transmission, and distribution to become carbon-free. Unfortunately, these needs are unlikely to be addressable by the methods adopted based on intermittent renewable power alone.**

Solar and wind are not producing more than a small fraction (just under 14% in 2023) of electricity worldwide or even in large economies, despite sizable capital flows into these technologies. It is now appreciated that **a massive expansion of solar and wind would require a huge investment in energy storage to overcome intermittency and a huge investment in transmission capacity to transport power from where it is produced to where it is required. It also requires the use of large land areas, which might alternatively have been used for food production, as the Italian government has already realised.** Additionally, the quantities of material needing to be mined to build a global low-carbon electricity network using only wind, solar, and energy storage required for electricity alone (without considering the absolute need for carbon-free heat) would likely wreak intolerable environmental damage. According to EU studies, such material resources may not even be geologically available. **Accordingly, the EU has now positioned nuclear as a 'transition fuel', making it effectively a 'green' fuel in the EU Taxonomy.**

Chapter 14

NEW Nuclear and 'LEGACY' Nuclear Have Quite Different Characteristics

There have been several trends in nuclear design in recent years. The main ones have been the following:

- a great simplifying of designs to reduce the complexity of all types of reactor;
- a gradual shift currently taking place to smaller designs, which should overcome the complexity of attempting to construct 'giga'-scale reactors and the time taken for their construction;
- the use of passive safety systems, which use the principles of physics to minimise risk;
- the ongoing design work not only on 'small' reactors (usually 300 MWe or less) but also on much smaller (micro) designs and on low-temperature reactors to produce heat alone at 100–150°C;
- the expected eventual shift from PWRs operating at around 300°C to AMRs operating at 600°C or above, which results in a higher conversion rate of heat to power and also provides the high temperatures required for some industrial processes (however, for immediate availability for construction, currently mostly developers must use Generation III+ pressurised water reactors (PWRs), giga reactors, and small modular reactors (SMRs));
- the development of 'heat-only' reactors for district heating, industry, and desalination.

Generation I
Now obsolete and mostly decommissioned: legacy technology.

Generation II
Now obsolete and being decommissioned or currently having life extension: legacy technology.

Generation III
The mainstay of the current generation of nuclear operating worldwide.

Generation III+ (Gen III+) using water for cooling and heat transfer:
1. Gigawatt (GW) Generation III+,
2. SMRs using Gen III+,
3. Micro Modular Reactors (MMRs).

Generation IV (Gen IV) using liquid metals, liquid salts, or gases such as helium for cooling and heat transfer:
1. SMRs using Advanced Nuclear Technology (ANT),
2. MMRs using ANT.

We do not cover Generations I, II, or III, as these are no longer being manufactured. We start with Generation III+.

14.1 Generation III+ Pressurised Light Water Reactors: Some Examples

1. **Gigascale:** Westinghouse AP1000, EPR (PWR).
2. **Small Modular Reactors:** Westinghouse AP300 (PWR), General Electric/Hitachi BWRX300 (PWR), Holtec SMR300 (PWR), NextEnergy (PWR), Rolls-Royce SMR (PWR), NewScale (PWR), and Neward (PWR) Rosatom (PWR).
3. **Micro Modular:** Deep Atomic (PWR), Steady Energy (PWR), Last Energy (PWR), and Calogena (very low pressure LWR).

There are currently many nuclear technologies in use or in development. To date, 75% of all operating reactors have used a technology known as 'light water'. This is essentially treated tap water, in which the water not only serves as the coolant and heat transfer fluid but also functions as the 'moderator' to slow down neutrons and sustain the fission process which produces the required heat. The other 25% includes, for example, gas-cooled (CO_2 or helium) reactors, which can operate at a higher temperature and are, therefore, more thermodynamically efficient; this type of reactors predominate in the UK's legacy fleet. Also, heavy water technologies (where heavy water contains a proportion of deuterium atoms, which are an isotope of H_2), such as CANDU and in most of India's 22 current production plants. More recent water-cooled reactors are the Westinghouse AP1000, Korean APR1400, Russian VVER, Chinese Jualong 1000, and the French EPR, originally standing for European Pressurised Reactor or Evolutionary Power Reactor but now simply EPR. Most of these are known as Generation III+ technology reactors. They mostly have an electrical power output of 1 GW or more and can be built as two or more units combined to generate a multiple of the unit size.

Some GW LWR reactors have been scaled down to become SMRs, such as the Westinghouse AP300 scaled down from the AP1000. NuScale manufactures small LWRs, with the individual reactors generating only 77 MW of power but which can be assembled in groups of 12 to provide almost 1 GW. NuScale also uses passive safety systems, which can function without power to enable a reactor to cool down after an incident. When we refer in the text to SMRs, by this abbreviation we mean small LWR/PWRs.

14.1.1 *Gigawatt Reactors: Some Examples of Gen III+ LWR/PWR*

14.1.1.1 *Westinghouse AP1000™*

This design is an 'advanced' (Gen III+) PWR with improved use of advanced passive (AP) nuclear safety and many design features intended to lower its capital cost and improve its economics. It can provide 1.1 GWe. Twelve AP1000s are currently in operation or under construction – four at two sites in China. Six more are under construction. Nineteen more AP1000s are planned – six in India, nine in Ukraine, three in Poland, and one in Bulgaria.

Simplification was a major design objective of the **AP1000™**. Simplifications in overall safety systems, normal operating systems, the control room, construction techniques, and instrumentation and control systems provide a technology that is easier and less expensive to build, operate, and maintain. These simplifications yield fewer components, cables, and building volume, all of which contribute to considerable savings in capital investment, as well as lower operation and maintenance costs. At the same time, the safety margins for the reactor have been increased dramatically over currently operating reactors. In December 2005, the US Nuclear Regulatory Commission (NRC) approved the final design certification for the AP1000. Since then, many have been constructed, are under construction, or are planned. This information is based on the Westinghouse website: https://westinghousenuclear.com/energy-systems/ap1000-pwr/.

14.1.1.2 *EPR (Also Known as European Pressurised Reactor or Evolutionary Power Reactor)*

This reactor was designed and developed jointly by Framatome and EDF (the French nationalised power generation company). Two such reactors are operating in China, and both had commenced operation by 2019. EDF has admitted that there have been considerable difficulties outside of China in constructing these reactors. One has been under construction in France (Flamanville 3) since 2007. In 2024, fuel loading commenced, and in September that year, the first very low level of power was generated in test circumstances. It is hoped that the reactor can be ramped up to full power in 2025.

In the UK, the Hinkley Point C power station in Somerset currently being constructed is an EPR with a rated output of 3.2 GW generated from two reactors. Although planning for this project commenced in 2013, it is not expected to be generating its full power output until 2030.

14.1.1.3 *South Korean APR 1400*

Korea is an important country for the research, design, and manufacture of nuclear reactors and components. The Korea Atomic Energy Research Institute (KAERI) has been at the forefront of innovative reactor design. The APR 1400 is a light water reactor (LWR) based on earlier successful designs but incorporating several innovative safety and control features and digital instrumentation and control. The design was licensed in Korea in 2002 and in the US in 2019. To achieve its goal of expanding clean power generation in the country, South Korea will build two new APR

1400 nuclear power plants with a combined generation capacity of 2.8 GW and start operations before the end of 2038. The new plants will be constructed by the state-run Korea Hydro & Nuclear Power Co. Korea is also developing an SMR and an AMR, which we cover in subsequent sections.

14.1.1.4 *CANDU*

Canadian nuclear power reactors, CANDU, are heavy-water reactors developed by Canadian scientists and engineers. CANDU stands for **Canada Deuterium Uranium** because it uses deuterium oxide (heavy water) as a moderator and coolant and utilises natural (not enriched) uranium as a fuel. There are currently 30 CANDU reactors operating worldwide. Following publication of four executive orders by the Trump administration in May 2025, which set targets to quadruple nuclear energy capacity in the country by 2050, Atkins-Realis plans to start work on 10 large reactors by 2030 and accelerate regulatory approvals. Atkins Réalis Engineering is a services and nuclear organisation which operates existing CANDU reactors in Canada. These new reactors are a modified version of existing reactors, using a lesser amount of heavy water (thus saving cost) and are larger than the earlier generation. The new reactor is a Gen III+ with an output of 1 GWe and is known as the CANDU MONARCH.

14.1.2 *Small Modular Reactors (SMRs): Examples of Gen III+ LWR/PWR*

14.1.2.1 *Westinghouse Nuclear AP300™*

In contrast to Generation IV reactors which are mostly still in development, this Gen III+ design is based on the existing, operational reactor the AP1000™ (approximately 1 GW). The same technology that is in use for the AP1000 will be used in the AP300.

The AP300 SMR is the only SMR based on an advanced, large Generation III+ reactor already in operation globally – Westinghouse's proven in-use AP1000 technology. Unlike AMRs under development with first-of-a-kind technologies and risks, Westinghouse's AP300 SMR utilises the AP1000 engineering, components, and supply chain, enabling quicker licencing and the ability to leverage already available technical skills. Together, these suggest that the first operating unit may be available in the early 2030s.

In August 2024, the company received formal approval from the UK Department of Energy Security and Net Zero (DESNZ) to enter the Generic Design Assessment (GDA) for the AP300™ SMR. This is the first step in the two-stage GDA. In September 2024, the AP300 SMR was one of four reactor designs selected for the current phase of Great British Nuclear competition in October 2024. Earlier that year, in February 2024, Westinghouse signed an agreement with Community Nuclear Power Ltd (CNP) to build four AP300 SMRs in Northeast England (see Section 15.4.3).

14.1.2.2 *General Electric Hitachi BWRX-300*

This reactor is of a type known as 'boiling water reactor' (BWR). It uses light water (ordinary water) as the coolant (heat transfer medium and moderator) but rather than employing a separate steam generator, it produces steam within the reactor. The technology is simpler than that of traditional light water reactors (LWRs), which reduces its capital cost. It also has a smaller footprint. The estimated construction time is 24–36 months while achieving approximately a 90% volume reduction in plant layout. Reducing the building volume by about 50% per MW should also enable them to become operational with the use of only 50% of the concrete per MW compared to earlier reactor designs—a significant improvement in both affordability and size (and carbon emissions from the production of cement). It has a design life of 60 years. GE Hitachi hope to have the first commercial installation operating by 2029.

14.1.2.3 *Holtec SMR-300*

Like all small water-cooled reactors, Holtec's SMR-300 is designed to deliver high performance but with a smaller footprint. Its design requires fewer components, thus reducing the need for substantial on-site construction and lowering operational and maintenance costs. It is expected to have an 80-year service life, which is considerably longer than traditional nuclear power plants. Maintenance costs are also expected to be lower. The heat from fission is used to boil water and uses the resulting steam to make power in a conventional Rankine power cycle via a turbine. This next-generation reactor design incorporates passive features to ensure safety of the reactor without any human intervention and to simplify its operation. It can use air cooling (with cooling fans driven by some of the power being generated), thus avoiding the problem of obtaining water use permits where that is an issue. The plant is quite compact requiring ~30 acres for two co-located units, though it uses more land than Generation IV reactors are expected to require. In the UK, it has been supported by £60 million of investment, of which half came from the UK government's Future Nuclear Enabling Fund award last year.

It works with Hyundai Engineering and Construction Co. of Korea and Mitsubishi Electric Co. of Japan. Holtec's plans in Europe include the establishment of a manufacturing plant in the UK to build SMR-300 components. It has already secured regulatory Step 1 endorsement in record time.

14.1.2.4 *Rolls-Royce (RR) 470 MWe*

This reactor does not easily fit into the usual categorisation of an SMR, being 300 MWe we or less. The reactors considered above are all 300 Mwe, while this one has an output more than 50% larger. Thus, two such reactors in tandem generate almost a gigawatt of power.

This company has manufactured LWRs for UK atomic powered submarines for some decades, though they differed in many ways from those being designed today. For example, they had a higher level of uranium enrichment. They have used their experience of developing and maintaining such reactors to design a 'small' LWR for civilian use. Because of its size and high output, this reactor addresses a different market to the embedded SMRs or AMRs which other companies are developing. Furthermore, the thermal output of the reactor exceeds that of most industrial heat loads and is at a temperature below that classified as high-temperature heat. Like all LWRs, it is less suitable for industrial heat provision. Of course, an LWR brings some advantages such as regulatory familiarity from legacy technologies and fuel types, but it is very different in design, manufacture, and construction from existing LWRs. RR themselves claim that:

> The Rolls-Royce SMR offers a radically different approach to delivering nuclear power, we have drastically reduced the amount of construction

activities and transformed the delivery environment from a large complex infrastructure programme into a factory-built commoditised product.

We have already noted the need for mass production of SMRs/AMRs in order to make them competitive in the electricity and heat generation market. In 2022, Rolls-Royce CEO SMR is reported as saying that the investment case for setting up an SMR business was based on selling many hundreds of SMRs by 2050. In terms of when it might be expected that reactors would be 'rolling off the production line', one constraint is the time taken for the regulatory GDA process. The process for the RR SMR commenced in 2022 but is expected to take four years.

The development cost of this reactor has been considerable and to help finance it, RR has brought in minority investors, which have reduced its ownership of the company – Rolls-Royce SMR Limited – to 76%. Other investors now include Qatar and Constellation Energy, a US energy utility.

14.1.2.5 *NuScale 77 MWe SMR VOYGR*

This company was the first to obtain US certification for its small LWR module. Rather than providing clients with a standard 300 MWe reactor, as many Gen III+ manufacturers do, this company has designed smaller modules intended to be grouped to provide the required power output. An expected configuration for initial plants is 12 modules providing almost 1 GW of power. The first overseas installation of this configuration is expected to be in Romania. While the company has had financial difficulties at various stages in its development, it now seems well supported by Fluor Corporation and other investors. It is now a public company whose shares performed extremely well in 2024.

14.1.2.6 *Nuward*

Nuward is a project commenced 2019 by the French Alternative Energies and Atomic Energy Commission, EDF, Naval Group, and TechnicAtome. The small modular SMR will generate 340 MWe using two PWRs of 170 MWe each. Its design is based on EDFs experience in PWRs. In mid-July 2024, the company changed its intended design and abandoned some of the technologies it was developing. After considering customer needs, including the earliest possible availability, EDF decided to change the approach to development by shifting to proven technology only. They believe that this will provide a higher level of success by facilitating technical feasibility.

14.1.2.7 *Korean SMART 100 SMR*

This is a water-cooled reactor with many new safety features which eliminate, for example, the possibility of a coolant pipe fracture.

14.1.2.8 *Rosatom SMR RITM-200N*

Russia's first small reactor is based on one which is employed in icebreakers. It is a light water-cooled reactor. Its output is 55 MWe. It has both active and passive safety systems. Work has started on a power plant in Uzbekistan, which will employ six units, giving a total power output of 330 MW. Rosatom is also reportedly aiming to build much larger PWRs. It is also reported that Rosatom is in discussion with Myanmar for a PWR and with India for a lead-cooled fast neutron BREST-300.OD. The role of the proposed **International Bank for Nuclear Infrastructure** in relation to creating strong competition to Rosatom in international markets is covered in Section 15.6.3.

14.1.3 *Micro Modular Reactors*

14.1.3.1 *Deep Atomic MK60 LWR: Purpose-Built for Datacentres*

Deep Atomic is a relatively new company based in Switzerland, which has as its tagline **'Compact Nuclear Reactors, Purpose-Built for Compute'.** It is being designed with data centre requirements at the forefront – both for power and cooling. The design is for a LWR with an output of 60 MWe + 60 MW cooling. It is described as 'plug and play' in the sense that generation can start small and be increased as necessary by adding more units and thus able to provide the generating capacity for a megawatt power station, as is the case with many designs of small reactors. The three services that this design of small reactor can provide – cooling, electricity, and heat – make it stand out from other designs with cooling being the additional service, which others have not yet focussed on but which is becoming increasingly necessary.

Deep Atomic says it is 'compact, scalable, and built on a foundation of proven technology'. This last point is also a critical one at this point. While LWRs are not 'the latest thing', their basic underlying technology has been in use for decades. Given the small size of individual units (giving a high surface area to volume ratio), the ability to cool down naturally in the event of an incident should be assured which is critical for a project intended to embed reactors within industrial facilities.

14.1.3.2 *Steady Energy LDR 50*

This is a Finnish design which will generate 50 MWth. It is not designed to generate electricity but if it were it would be around 20 MWe. It will provide energy at a temperature of up to 150°C. Its main use initially is likely to be for district heating, as this is the main form of heating in Finnish cities. The first operational plant is expected to start construction by 2028 and to be operational by 2030. In June 2024, Steady Energy and Swedish project developer Kärnfull Next (see later) announced a strategic partnership to use the LDR 50 SMR in projects. According to Steady Energy:

> With heating using 40% of all energy and the single biggest source of CO2 and in Europe, fossil fuels being the main source of district heat, the LDR-50 offers the possibility of replacement with heat to be delivered at an estimated cost below €40/MWh.

14.1.3.3 *Last Energy PWR 20*

This company is set to produce an air-cooled light water 20 MWe reactor, with the 'nuclear island' being installed underground. The reactor operates at a pressure comparable to that of a household espresso machine. As expresso machines are safe for use in the home, this reactor should be much safer even than the Gen III+ reactors, which operate at very much higher pressures. As the technology uses a standardised balance of plant and modular construction, the company claims its reactors could be fully factory fabricated and installed on-site in as little as three months. Factory fabrication simplifies the regulatory process and the company believes it can keep project costs below $100 million. This should also make it privately financeable. Already, the company has secured power purchase agreements (PPAs) for 34 PWR-20 SMR units with four industrial partners in the UK and Poland. Securing PPAs with customers is a key factor in enabling nuclear to be privately financed.

One of its first projects will be in the UK, where it is planning four 20 MW micro reactors at a site in Bridgend in Wales, which they say could be up and running by 2027. It is said to have a provisional offer of over $100 million from the US Export-Import bank to enable this project. The location in Wales was previously occupied by a coal-fired power station. But it sees its main market as owners of server farms and hopes to build around 10,000 units of its reactor.

14.1.3.4 *Calogena Low Temperature Heat Project for District Heating*

According to Calogena, heat accounts for 40% of energy consumption in Europe. Nearly 70% of this is produced by fossil fuels. And 80% of this heat is for hot water and heating at less than 100°C. The two major low-carbon solutions for the transition are heat pumps and heat networks. The Nuward project is aiming at enabling existing and new heat networks to be powered by low-temperature nuclear instead of natural gas or coal. Given the extent of domestic and commercial use of natural gas boilers for space heating, there is considerable scope for decarbonising this part of the economy – difficult though this is.

Gas-fired boilers emit just over 220 g of CO_2 per kWh, while coal-fired power stations emit around twice as much. Each Calogena module could decarbonise district heating by 30,000 tonnes of CO_2 a year if it replaces a gas-fired installation and twice as much if it uses coal. Gas boilers also emit other pollutants in addition to CO_2 and CH_4 (methane) from natural gas production. A particular source of pollutants are nitrogen oxides (NOx), which have harmful effects on human health and the environment. By comparison, lifecycle studies carried out by Ademe for large reactors (SMRs and AMRs would be similar) show that these have a carbon footprint of around 6 g of CO_2 per kWh of electricity, and 2 g per kWh of heat.

The Calogena reactor operates at a pressure of only 5 bar, which is less than the pressure of a domestic water heater and is therefore effectively explosion proof. It also runs at a low temperature of just under 100°C. It will use 3.4% low-enriched uranium pellets. Because of its simplicity of its design, it requires far fewer auxiliary systems and is even safer than any reactor currently in operation or planned. Its heat output is only 30 MWth (it does not produce any electricity). It is thus a design that can safely be installed close to residential or commercial premises.

In Section 15.6.1, we cover a potentially key initiative by the European Commission to encourage the creation of a European Industrial Alliance on SMRs and AMRs. In June 2024, the Alliance launched a call for SMR projects, wishing to be considered for the Alliance's Project Working Groups (PWGs). Subsequently, the governing board, with the assistance of the Alliance secretariat, conducted a review and assessment of the 22 applications received. The CityHeat project (Calogena, Neward/Steady Energy) was one of those accepted in the initial batch, which should be helpful in ensuring that Calogena can obtain further funding for the project.

14.2 Generation IV (Gen IV) Reactors Have Quite Different Characteristics to Generation III+ (Gen III+)

14.2.1 *Generation IV (Gen IV) AMRs: Two Sizes*

1. SMRs using ANT known as advanced modular reactors (AMRs), usually around 300 MWe;
2. Micro modular reactors using Advanced Nuclear Technology (ANT MMRs), usually 1–45 MWt.

The new ANT reactors generally described as Generation IV AMRs are different from water-cooled reactors in many ways. None has yet become commercially operational, except in China and Russia, but test AMRs have been operated for many decades in many countries. Some are now close to being licensed and expected to be operational by the early 2030s. However, many will never achieve the large-scale production which is necessary to bring their costs down to enable unsubsidised operation under truly commercial financing, but in our view some will, and for shareholders in such companies, we expect the rewards to be substantial.

Generation IV, ANT high-temperature modular reactors (HTMRs) are of quite a different design to Gen III+ and use quite different physical features and technologies. Almost the only thing they have in common is that they both use nuclear fission to create heat. They are also physically much smaller, do not require a large high-pressure vessel, which is technically difficult and also very costly, have fail safe characteristics, and can be located close to a customer. The most important distinction between Gen III+ (SMRs) and Gen IV (AMRs) is that the former generally use 'light water', whereas the latter use other coolant/heat transfer agents, such as **helium gas, molten sodium, molten lead, or molten salts**. As a result, they do not operate under high presume and cannot explode. Like all nuclear reactors, and indeed coal or gas furnaces, they produce high-grade heat, which is quantified as MWt (thermal), and when converted to electricity, the output is quoted as MWe. They have a lower power output (normally 100–300MWe), but some micro-reactors are much smaller.

While AMRs are small- or micro-sized, by coupling several small AMRs together, as with small LWRs, it is possible to increase a plant output to the gigawatt scale if appropriate. However, a key virtue of the physical smallness of each reactor which together create a giga-scale plant is that it enables an important aspect of SMR/AMR inherent safety. **This important safety feature of smaller reactors and from multiple smaller reactors linked in one location arises from the fact that their smaller core volume relative to their surface area means that unattended natural dissipation of decay heat to its surroundings during an upset condition is attainable. Thus, off-site emergency planning requirements can be scaled down to be proportionate to those reduced risks.** For example, the emergency planning zone (EPZ) can be much smaller than with conventional nuclear and can be within the plant site boundary fence rather than being 10–50 miles downwind of the reactor. In addition, AMRs can be sited underground with a blast resistant concrete cover, and many do not require human workforce to be in attendance when operating. They are autonomous, including for load-following. While human operating error has contributed to the small number of Gen II and III incidents, Gen IV has both inherent safety as well as self-regulating stability. Being located contiguously to the point of use avoids the cost and environmental disturbance of transmission inherent in larger plants or indeed large wind and solar farms. This, in turn, enables them to be located close their customers' points of use.

The higher heat output temperature of AMRs, compared to previous generations of reactor, also makes them suitable for many industrial processes currently using fossil fuels, while conventional water-cooled high-pressure reactors are not suitable for the provision of high temperature heat. Generation IV reactors, such as those using molten coolants, are also low pressure and thus do not require the expensive and difficult-to-manufacture pressure vessel of the previous generations of water-cooled reactors (Gen I, II, III, and III+ PWR plants).

When we compare the capital economics of building large and small reactors, we are comparing the **economies of scale** of Gen III+ with the **potential economies of mass production** of Gen IV. However, to enable Gen IV reactors to produce heat or power at an economic price, the manufacture of small reactors needs to employ a similar mass-production model to that used in modern automobile manufacture to counter the perceived economies of scale (large size) advantage of large reactors. Until now, it had always been assumed that the economies of scale as reactors increased in power output and physical size would reduce the cost per unit of power output. However, as reactors became larger, it was apparent that there are also substantial dis-economies of scale, which include the difficulty of constructing such complex behemoths and the time value of money lost during their construction. Recently, some gigawatt reactors have witnessed their cost doubling and the construction time more than doubling. The 'scaling effect', which was expected to hold as reactors grew ever larger and thereby reduce the unit cost of power, seems not to hold after a certain point. In addition, with the notable exception of France and Korea the fleets have been constructed only sporadically, so that each new one has had a somewhat different design, which has stopped costs falling by learning.

These additional capital costs of constructing giga-sized reactors and the longer construction period have greatly increased the total capital investment requirement and the time to start generating revenue. Both issues raise the final cost of power produced, particularly in an era of higher interest rates. Interest rates rose after 2021. In the case of the UK, for example, the Bank of England base rate rose from almost zero in 2021 to over 5% in 2023. As debt service (capital repayments plus interest) are the largest cost of nuclear (and of wind and solar), rising interest rates are a disincentive to investment. Large GW nuclear also has the problem that the time scale for the repayment of debt capital and a break-even on equity investment is much longer than for wind and solar. Large nuclear has a long build time and an operating life of up to 60 or even 80 years compared with 20 or 25 for renewables or for gas power plants.

Keeping individual units small also brings the advantage of being able to construct them in a factory and transport them to site by road or barge. It brings an additional advantage according to the US NRC in that the smaller amount of fuel in the core along with the inherent safety of Gen IV eliminates several potential emergency scenarios, resulting in much simpler systems. This reduces capital cost, commissioning time, maintenance, and operating complexity, including in some cases the elimination of all on-site staff. In contrast, regulatory proving and testing of active safety systems and their maintenance are an escalating cost of the legacy PWR family of conventional gigawatt reactors, which is one of several reasons why Gen IV reactors are currently being developed in most nuclear nations.

What is not yet known about any AMR is whether, in practice, their loss of the economies of scale that large reactors should in theory have can be more than compensated by the cost-reducing features of the new designs, including mass production. Thus, it is currently difficult to know which manufacturers' technologies will, in practice, enable their reactor to generate electricity and heat at a price which customers will be willing to pay while ensuring an appropriate return on capital for investors. AMRs cost-reducing features include less complex design, greatly simplified, and passive safety systems, greater speed of construction (hence less interest cost accruing during construction and faster revenue generation), higher operating efficiency resulting from the greater temperature difference between the coolant used for heat transfer and the ambient (outside air) temperature, and, importantly, economies of mass production.

The foregoing has predominantly dwelt on the capital and operating economics of AMRs for electricity. However, this misses three larger considerations for the investor:

Firstly, as stated in the previous section, the market for heat in a decarbonising world is equal or larger than that for electricity. The only tool currently known to address decarbonised heat is nuclear. More particularly, because heat is expensive to transmit, it must be generated near the point of use, which is only achievable from small reactors that match the quantity and dynamics of industrial, commercial, and domestic heat demand. This lack of competition except from the tail-end of fossil fuels gives small nuclear, especially AMRs, with their high-temperature heat output, huge market power as well as the possibility of being able to manufacture at scale.

Secondly, much discussion in decision-making circles so far has compared the output cost of electricity (or heat) *at the point of generation* (at the factory gate). However, this is not where the customer buys the product. The cost of electricity from large-scale wind or solar must also incur the system cost of storage systems to match the production variability of their output to the daily and annual use pattern of the customer. This is already proving problematic in Europe, where in addition, the scale of transmission reinforcement is proving to be considerable and transmission capacity only likely to be available many years hence. **This additional cost of renewables increases with the increasing renewable fraction in a region's total generation of power.**

Transmission, distribution, and backup are already over 20% of the cost to the end customer, while there should be no system surcharge for energy generated at the point of use and distributed without the use of a transmission system. Indeed, In the US, transmission and distribution costs are already 44% of the total cost of delivered electricity according to the World Economic Forum (WEF). In the UK, it was announced in December 2024 that the country's electricity transmission network owners plan to invest up to £77.4 billion per annum between 2026 and 2031. This is necessary to enable power to be transmitted from where it is produced to where it is in demand, e.g. from the North Sea to users 800 km (500 miles) or more distant in the south of England. The same issues arise in Germany, where industry is located in the south of the country, but clean power comes mainly from the North Sea. Such costs have to be recovered from customers and can be so recovered when the power generated is operating under a regulated asset base model (RAB). However, in most countries, it is currently also charged equally to all customers, including those consuming power close to the point of production. At some point, **location-based pricing** may be necessary to alter the seeming unfairness of this pricing mechanism. This would be in addition to 'time of use' tariffs to reflect imbalances in supply and demand when renewables have lower output.

Thirdly, large-scale nuclear, solar, and wind, along with their transmission requirements, are subject to political uncertainty and public perceptions, whereas a small nuclear plant inside a factory fence supplying all heat and power is, except in safety terms, entirely under the control of the industrial owner, just like the rest of the factory. Indeed, the outcome from universal standardisation of inherently safe small reactors for heat and power may logically be that their operation, except for transfers of fresh radioactive fuel once every seven or eight years, falls entirely outside of nuclear regulation if the containment can be designed to be unbreachable under any circumstances.

Cost competitiveness, whole life energy cost stability, and low political risk have been key drivers of all small nuclear propositions to industrial and artificial intelligence (AI) customers. Thus, just as technology has moved on from using a small number of large, centralised, fossil power stations in favour of distributed solar and wind power, so we are now seeing the possibility of distributed nuclear displacing single giga-sized reactors, which are also more exposed to enemy action in the event of war in Europe.

14.2.2 *Generation IV Small Advanced Modular Reactors Using Advanced Nuclear Technology*

14.2.2.1 *Generation IV International Forum*

The Generation IV International Forum (GIF) was initiated by the US Department of Energy in the year 2000 to help coordinate international work on advanced small modular reactors and to help realise their potential. It is an international collective representing governments of 13 countries where nuclear energy is significant now and also seen as vital for the future. The original charter members of GIF are Argentina, Brazil, Canada, France, Japan, South Korea, South Africa, the UK, and the US. They have been joined by Switzerland, China, Russia, Australia, and, through the Euratom research and training programme, the European Union. After some two years of deliberation and review of about 100 concepts, in 2002, GIF announced the selection of six reactor technologies (not companies) which they believe represent the future shape of nuclear energy. These were selected based on being clean, safe, and a cost-effective means of meeting increased energy demands on a sustainable basis while being resistant to diversion of materials for weapons proliferation and secure from terrorist attacks. They are the gas-cooled fast reactor (**GFR**), the lead-cooled fast reactor (**LFR**), the molten salt reactor (**MSR**), the sodium-cooled fast reactor (**SFR**), the supercritical-water-cooled reactor (**SCWR**), and the very high-temperature reactor (**VHTR**).

14.2.3 *Examples of the Technologies of Some Small ANT AMRs*

14.2.3.1 *TerraPower: Natrium (Natrium Is the Latin Name for Sodium)*

> *The beginning of construction in June 2024 on the Natrium demonstration project site in Wyoming marks the first ANT reactor project under construction in the Western Hemisphere!*

Bill Gates, the founder of Microsoft, provided the initial venture capital (Angel Financing) for this reactor company and is still a majority shareholder in the company as well as being its chairman. Since then, it has had institutional investor or corporate rounds from Series A to Series C. One such corporate investor is Arcelor Mittal, the steel manufacturing company, which is a large consumer of electricity and will require much more power if its use of (electric) arc furnaces increases, as is expected. It is thus a **'strategic'** investor rather than simply being a financial investor. Also, SK a Korean company and CRV, a US fund have contributed to financing. The US Department of Energy (DOE) has also provided a grant through prize money. Pillsbury, a specialised legal practice, has also facilitated the raising of $750 million from institutional investors.

TerraPower has already 'broken ground' on a site in the small town of Kemmerer, which will also be the demonstration project to showcase the abilities of TerraPower's Natrium energy system. When operational, the output of the plant will be provided to PacifiCorp, a state utility company, to supply Rocky Mountain Power customers in Wyoming and throughout the region. It is estimated that the project will cost $4 billion. Bill Gates will provide $1 billion with the DOE's Office of Clean Energy supplying half of the cost. In addition, TerraPower has a deal with Emirates Nuclear Energy Corporation to explore the export of TerraPower's Natrium reactors to the United Arab Emirates.

In March 2024, TerraPower was the first company to submit its application for a commercial advanced reactor construction permit to the US NRC for the Natrium reactor. Through the US DOE's Advanced Reactor Demonstration Program, which is a public–private partnership, this Natrium reactor demonstration project is intended to validate the design, construction, and operational features of the Natrium technology. This first AMR commercial project outside of China is being constructed near a retiring coal-fired power plant, and at the time the project started, it was the only *coal-to-nuclear project* under development in the world.

Natrium features a 345 MW ***sodium-cooled fast neutron reactor (no moderator required) with a molten salt-based energy storage system***. The storage technology can boost the system's output to 500 MW of power for more than five and a half hours when needed, which is equivalent to the energy required to power around 400,000 homes at times of peak demand. Unlike the statistics quoted about the ability of wind or solar to provide the power required for *x* number of homes nuclear, with its constancy of output (very high 'capacity factor') and low stoppages for maintenance (high 'availability'), Natrium can provide continuous (24/7/365) power and also flex its output up and down in accordance with the customer's varying needs, which many other generation systems cannot. The energy storage system in Natrium is simply to enable it to more economically balance supply and demand for power throughout the day (flex) without having to employ a larger, and hence, more expensive, reactor. Such a reactor might be less able to benefit from the higher surface to volume ratio (for incident cooling) that a smaller reactor provides. Other reactors may 'flex' output but constantly ramping a reactor up and down may reduce its life.

The heat energy storage system enables the reactor to **operate with an even load all day** (24/7), which leads to greater efficiency and less stress on the whole system. In the combined system, there are two sources of heat for the steam generator and the steam turbine. There is the heat energy provided by the evenly loaded nuclear reactor plus the heat energy from storage. Effectively, the steam turbine but not the reactor itself is being 'ramped up' to cover peak demand as well as to cover the times of day when renewable intermittency on a grid needs to be corrected. In Figure 14.1(a),[1] from midnight until 6 am when demand from customers is low, any excess heat from the reactor can be stored (this reactor produces heat 24 hours per day). Heat is conserved until 1 pm when customer demand rises and the combined system can then produce up to 500 MWe. After 7 pm, power demand starts falling, and by 9 pm, surplus heat can once again be stored. A visual of the Natrium facility is presented in Figure 14.1(b),[2] which shows the hot and cold molten salt tanks – a key feature of Natrium.

---

[1] https://www.revistanuclear.es/wp-content/uploads/2021/03/ART-Natrium.pdf.

[2] The Natrium heat-exchange process. Image courtesy of TerraPower.

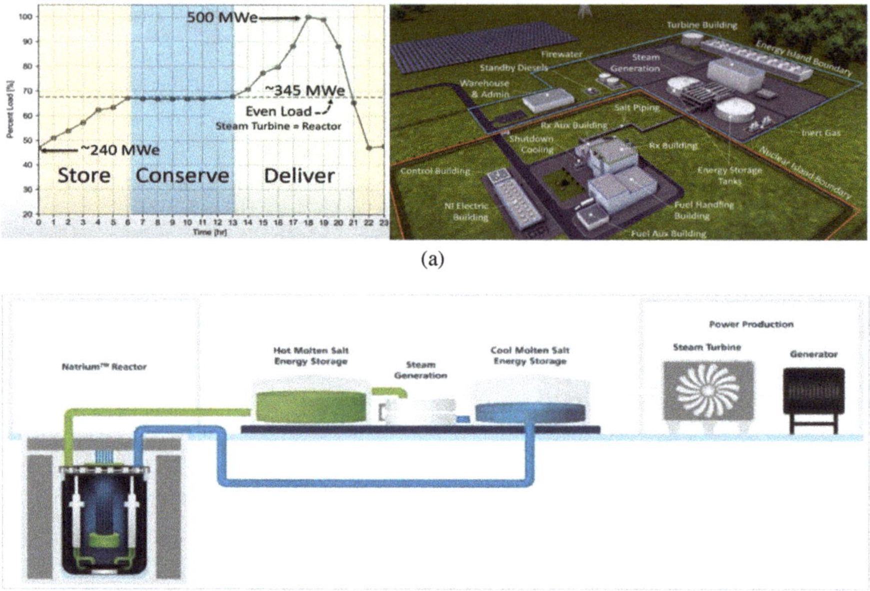

Figure 14.1. (a) How the TerraPower Natrium system functions. (b) The energy storage system.

14.2.3.2 *Terrestrial Energy Integral Molten Salt Reactor® and Its Techno-Economics*

This is a Canadian/US design, but for European, Middle East, and African (EMEA) markets, it could be manufactured in Europe. It would also almost certainly be available to generate power in the EMEA region within this decade, as it is at the licensing stage in Canada and the US. Speed is ever more of the essence since the European decision to phase out Russian gas and the warnings of the growing dangers of global warming in the third instalment of the IPCC's sixth assessment report published in early April 2022 has increased concern. Thus, if only for reasons of speed, it is now ever more critical for Europe to be willing to use non-European technology if this becomes available sooner than European-developed ANT reactors.

Terrestrial's Integral Molten Salt Reactor (IMSR) reactor uses molten salts (with the fuel dissolved in the molten salts), with the liquid salts also acting as the heat transfer fluid. The moderator is graphite. The reactor is designed to provide flexible dispatchable power, grid services to balance wind and solar, voltage and frequency stabilisation, etc., as well as industrial heat. It can also be used to produce hydrogen through high-temperature electrolysis and could be the ideal heat source for solid oxide electrolysers (for which heat is beneficial), which we discuss later. Unusually, it integrates the primary reactor components into a sealed, replaceable vessel. This unit, which contains the graphite moderator and pumps, is removed after seven years and replaced with a new core unit, thus simplifying the on-site infrastructure. During the seven-year operating life of each core unit, it can be topped up with uranium fuel. As the liquid fuel salt does not need to be kept under pressure, there is no need for the expensive and difficult to cast pressure vessel that traditional LWRs require.

The greater safety of this design is based more on physics than on engineered systems, which is a factor in its (and all AMRs) safe design. The IMSR approach is to rely on inherently safe design rather than complex active systems seen historically. For example, the cooling mechanism does not require backup electric power. Because the fuel is distributed through the coolant, the system avoids complex fuel fabrication processes, which are required for pebble bed reactors or rigid fuel elements required for some other Gen IV designs.

It is an axiom of public safety that, if a safer way of reasonably achieving the societal result is available, then statute law should require it to be used. This is the political version of the investor's correct and proper aversion to avoidable risk. This type of reactor has been designed with inherent safety. In October 2018, Terrestrial Energy started Phase 2 and completed the CNSC's Vendor Design Review process in mid-2023. It now has the 'green light' for commercial operation by large industrial users.

In terms of its **techno-economics**, the IMSR's higher operating temperature improves thermodynamic efficiency from around 33% on Gen III reactors to around 46% (as it does on all high-temperature reactors). According to the manufacturer, the IMSR should produce 40% more electricity than a comparably sized water-cooled SMR for the same or less fuel or waste quantity. The result is around 40% more revenue from the same reactor size. This can have a large impact on its economics. The intrinsic low-temperature waste heat from such a reactor could also be used for district heating, thus further improving its economics. Figure 14.2[3] shows the IMSR and its range of services.

[3] https://neutronbytes.com/2019/12/29/terrestrial-energys-imsr-to-get-joint-review-by-cnsc-and-nrc/ Courtesy of Terrestrial Energy.

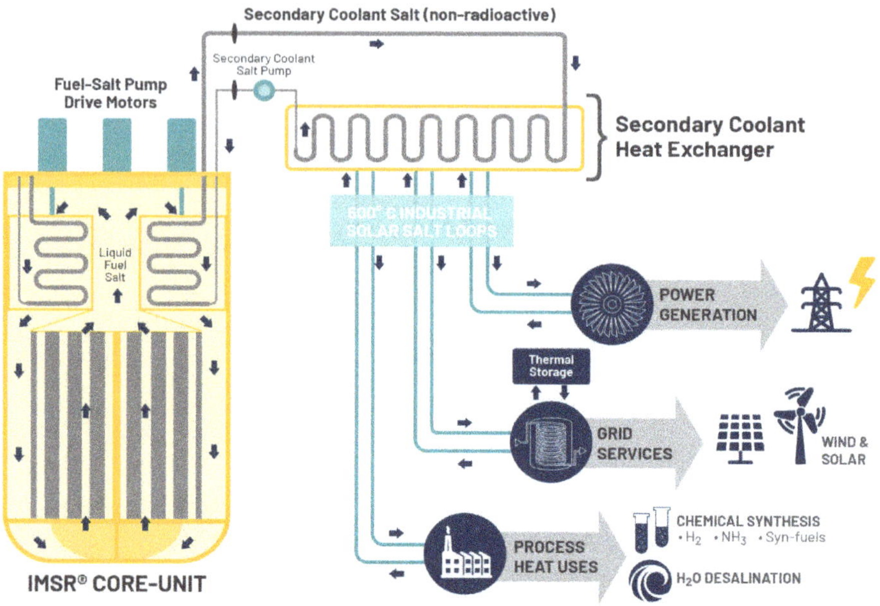

Figure 14.2. The Terrestrial reactor and its varied uses.

14.2.3.3 *Attractive Features of Both of the Above Reactors and Those Which Follow Below*

Both of the above AMR reactors have a small footprint and, due to their high level of safety, can be located (subject to properly developed regulatory permissions) in or beside existing boiler houses, which are usually near to built-up areas and communities. Such locations may also have suitable transmission and distribution capabilities already in place. Because they are small and manufactured in a factory and not on-site, the total capital cost is within the realms of conventional industrial and municipal projects (£200 million to £1 billion depending on size), and construction is much simpler resulting in a construction period expected to be only 3–4 years.

These types of reactor could be financed using, for example, the Finnish Mankala model (which we cover in Section 15.3.1.1 on Financing the Transition), where users of heat and power join up to own generating capacity and finance such a plant themselves. An alternative for utilities wishing to supply power to a national grid is to use the RAB model considered in Sections 4.4 and 15.3.2, possibly also with a government provided 'Contingent Financial Support Package'. Once the simplicity and performance of AMRs becomes known to investors, directly funded **merchant** plants (owners selling their power in the open market without a power purchase agreement or using the power for their own industrial use) should emerge and may become a standard part of a chemical, thermoplastic, iron and steel, and other industries, factory site hard infrastructure bought off the shelf in the same way as they buy standardised package boilers today. With reference to efficiency, if the different grades of steam (different temperatures) are utilised by means of an energy 'cascade' (including providing low-temperature heat to a district heating system), it is possible to utilise 80% of the heat, making it very highly efficient and a much more economic system.

14.2.3.4 *Newcleo*

Newcleo has large teams of researchers and engineers working on a 'fast' neutron 200 MWe AMR technology that uses molten lead as the cooling and heat transfer medium (an LFR). Among other potential benefits, it may eliminate the need for a separate heat exchanger between the reactor and the steam generating plant. It has already completed the preparatory phase of the French regulators' licensing process. Its head office was in London, but has now moved to Paris, with other offices in several countries in Europe. A key advantage of small reactors is that they can be installed next to the industrial sites or communities they supply. This is a great advantage in comparison to other clean energy technologies.

Investors include the Agnelli family's holding company Exor NV, Italian asset managers Azimut Holding SpA, as well as several wealthy Italian business people, such as ex-Goldman Sachs Group Inc. banker, Claudio Costamagna. It is said that the reactors would cost around £5 billion per gigawatt of generation capacity, compared to £15 billion per gigawatt cost of Hinkley Point C. The small reactors could also be built and installed within four years or less whereas Hinkley will require two decades, given all the problems it has encountered (though subsequent reactors of this type should not take nearly as long).

This $5 billion/MWe capex estimate is similar to that of the Offshore Wind Market Report: 2024 Edition actual capex for offshore wind, which of course operates for 40–60% of the year, whereas nuclear operates almost continuously for the same capex but does incur some fuel costs which are comparable to the additional costs of backup and transmission for intermittent wind and its much shorter operating life.

14.2.3.5 *Thorizon One: Can Recycle Waste Fuel as Well as Thorium*

Thorizon has evolved from a spin-off of NRG (Dutch Nuclear Research Institute) into a startup with offices in Amsterdam and Lyon. The company views small modular nuclear reactors as an ideal, stable, and clean contributor to an energy system, where wind and solar are the other important sources. The ambition is to develop a reactor based on molten salt as the coolant and heat transmission fluid that can be realised rapidly and smartly, is 'walk-away' safe, and *takes a first step towards circularity by using long-lived nuclear waste as a fuel source*. Thorium, uranium, and plutonium can all form stable fluoride salts.

Thorizon is working to further strengthen its financial position to be able to build a non-nuclear molten salt demonstrator in the short term and finalise the detailed design for a first-of-a-kind reactor by 2030. The output is intended to be 250 GWth and 100 MWe. Interestingly, like TerraPower, the 100 MWe can be converted to 50–300 MWe of flexible capacity by storing heat energy in liquid salt when demand is low and releasing it during peak hours.

In March 2025, the company reported that its latest funding included €16 million as the first tranche of its Series A round, led by the Dutch National Promotional Institution, Invest-NL, backed by an InvestEU guarantee from the European Commission for the research part, with strong backing from Positron Ventures, PDENH, and Impuls Zeeland. It said all of Thorizon's existing shareholders have reinforced their commitment in this investment round.

Thorizon has also secured an additional €4 million grant from the Dutch Province of Noord-Brabant in consortium with VDL Groep and Demcon. The investments follow an earlier €10 million grant from the France 2030 Innovative Reactor Programme of the French government in 2024. In total, including its first equity round, Thorizon has now raised €42.5 million to drive the commercialisation of its reactor technology.

With is latest funding, Thorizon is advancing the development of the Thorizon One, a next-generation reactor **that overcomes all traditional reservations against nuclear energy**. The Thorizon One is engineered to deliver carbon-free energy while repurposing long-lived nuclear waste as fuel. Its modular design and innovative cartridge-based fuel system provide a scalable pathway to a circular nuclear economy. By harnessing molten salt technology, Thorizon is developing reactors that are inherently safe, cost-efficient, and faster to deploy than conventional nuclear plants – offering a practical solution to Europe's clean energy transition.

Thorizon is also working closely with a range of partnerships, which increase the probability of success in the shortest timescale. It has forged key partnerships with Orano for fuel development, Tractebel for engineering, and VDL Groep for prototyping, while collaborating with EPZ for early operator input and with EDF on R&D. Dutch and French nuclear regulators have initiated a joint preparatory review of the Thorizon One design, and the company is conducting pre-feasibility studies at three nuclear-designated sites in France and the Benelux, targeting construction by 2030.

14.2.3.6 *Westinghouse Lead-Cooled Fast Reactor: The Next Generation of Nuclear Technology*

The LFR technology was selected by Westinghouse for their next-generation reactor after comparing all known nuclear technologies, based on criteria supporting global commercialisation and spanning safety, economics, market versatility, fuel cycle sustainability, technology readiness, and other features. It is designed to output 450 MWe when connected to a steam generator and turbine set. It requires only minimal pressurisation and, due to the use of liquid lead as coolant, has a smaller physical size than other types of Gen IV reactors. When used to generate steam to produce electric power, it is possible to use air cooling (driven by electric fans powered by its own electrical generation) to condense the steam emanating from the turbine. It uses 'fast neutron' technology and can use reprocessed fuel from LWRs. When reprocessed fuel is used, much more of the radioactive content in the fuel is used up and when it is removed from the reactor, the amount of spent fuel is smaller than in its first round of use.

14.2.3.7 *Ocean-Going Ship Propulsion & Floating Nuclear Power Plants (FNPPs)*

FNPPs, which can either be used for propulsion in ocean going ships or can be towed to a coastal location to provide power to the land, represent a game-changing approach to nuclear energy deployment. These plants can be centrally manufactured and easily transported where needed, combining ANT with shipyard efficiency. FNPPs can be deployed in ports and terminals, as well as in nearshore and offshore environments. The floating environment enhances safety by significantly reducing risks from earthquakes, tsunamis, and volcanoes compared to land-based systems. FNPPs can be readily moved on water, enabling fast implementation to meet urgent energy needs or decarbonisation goals. A fleet of FNPPs can be assembled and towed to customer locations.

14.2.3.8 *CORE POWER*

This company has focussed on the role of small advanced nuclear in marine locations. Core Power aims to extend marine use to floating nuclear generation to supply land-based needs and to civil nuclear ship propulsion and, in particular, container ships. Core Power's mission is to improve energy efficiency and security of ocean transport and heavy industry through innovative, sustainable advanced nuclear energy solutions that drive global progress towards a cleaner, more efficient future. Their reactor design is a molten salt fast reactor with the fuel mixed in with the salt. The company has offices in Washington and Tokyo, with its head office in London.

Key Benefits of FNPPs

- Serial construction of entire FNPPs in shipyards (which are experienced in this type of construction work) reduces costs, accelerates innovation/deployment speed, and ensures consistent quality, leveraging the efficiencies of modular construction and learning from previous iterations without site.
- Simplified design can reduce materials and construction costs as FNPPs eliminate the need for extensive, site-specific civil works required by traditional land-based reactors, significantly lowering both costs and deployment timelines.
- Decoupling of the plant construction and siting means that the construction of FNPPs can commence independently from the site licensing, reducing deployment time, particularly compared to terrestrial counterparts.
- Enhanced resilience to natural disasters, including tsunamis and earthquakes, if sited offshore.

- Flexibility in deployment and location: while terrestrial plants are immovable, FNPPs can be redeployed to new locations in response to medium-/long-term demand.

Key Benefits of Civil Maritime Nuclear Propulsion

The International Maritime Organization has mandated that the shipping industry eliminate 70% of its carbon emissions by 2040. The need for clean power is enormous in an industry that uses almost 300 million tonnes of emissions-intensive heavy fuel oil annually. Large nuclear-powered ships will be faster, have more cargo capacity, be more economical, and emit no greenhouse gases.

14.2.3.9 *SALTFOSS (Previously Known as Seaborg Technologies)*

Another company developing floating nuclear power is a Danish company set up in 2014 with the intention of producing an innovative AMR. The company is focused on the need to be able to 'fill the gaps' in the generation of wind and solar power and thus sees one of the uses of its technology as being to **complement renewable power**. An innovative idea that they have developed is to locate the reactor(s) on a barge, which can be towed to wherever power is required whether for grid backup, industries with a high energy requirement, desalination of seawater, manufacture of ammonia or e-methanol, or power to X.

The reactor modules each produce 100 MWe and are normally configured in groups of two modules providing up to 800 MWe output. The output temperature of the reactor is high enough to manufacture hydrogen and its derivatives efficiently. A number of AMR designs used liquid fuels in which the nuclear fuel is mixed with the fluoride salt, which is being used for cooling and heat transfer. Seaborg's design is one such design. The design also has the possibility of creating a 'closed fuel cycle', in which the waste from the reactor is recycled into fresh fuel salt with the addition of fresh uranium.

14.2.3.10 *Chinese ANT*

In China, the question as to whether nuclear should supply a substantial proportion of power in the country has already been answered with a long-term process of building SMRs as well as continuing to construct giga-sized reactors to speed up the power industry transition from coal. These high-temperature gas-cooled reactors (HTGRs) use graphite as a moderator and helium as a coolant, with uranium fuel in the form of 6 cm diameter 'pebbles', which can be sorted and switched in and out for on-load refuelling. Each pebble has an outer layer of graphite and contains some 12,000 four-layer ceramic-coated TRISO fuel particles dispersed in a matrix of graphite powder. The fuel has high inherent safety characteristics and has been shown to remain intact and to continue to contain radioactivity at temperatures up to 1620°C – far higher than the temperatures that would be encountered even in extreme accident situations, according to the China Nuclear Energy Association.

A pair of these pebble bed 100 MWe reactors have been built side-by-side to share a single steam turbine and are now running commercially. The original concept was to use three of these twins to repower China's very large fleet of 600 MWe coal power stations without disrupting their infrastructure. In addition, China has research programmes on small LWRs and in all of the other technologies summarised in this chapter.

14.2.3.11 *Copenhagen Atomics: Thorium Molten Salt Reactors*

One commonly raised objection to uranium-fuelled nuclear reactors is that they produce a large quantity of waste material after the fuel has been used. Much of the potential value in the fuel is in fact not used at all as only a relatively small proportion contributes to generating heat. In contrast, thorium produces much less waste material and its half-life is much shorter. Thorium is also much more available in the Earth's crust than uranium, although uranium reserves are nevertheless adequate for at least two generations.

The first thorium molten salt reactor was developed in the US in the 1960s and worked well. However, the US Navy opposed its commercial development because it produced only a small amount of plutonium compared with uranium 235-fuelled LWR. The military wanted large amounts of plutonium for the manufacture of atomic weapons, and as a result, further work on thorium reactors ceased.

Since then China has been first in issuing (in 2023) a permit for an experimental molten salt thorium reactor. It will undergo testing over the next few years. China also has large sources of thorium, as does India, where high-quality development has been continuing for decades. In Europe, Copenhagen Atomics is the most advanced company in this field and believes that taking advantage of the features of such reactors, it will be possible to scale up global nuclear output to double-digit terawatt (TW) levels in the coming decades. Japan, the UK, the US, and other countries have also demonstrated enthusiasm for research but do not yet have an operational thorium reactor.

Copenhagen Atomics is developing a thorium-based molten salt reactor with the same footprint as a 40 foot shipping container, which will deliver 100 MW thermal energy per module and is expected to reach an electricity price (LCoE) below $20/MWh in a mass manufacturing scenario. It expects to have an operational 1 MWth demonstration reactor ready by 2025. The project is being executed with Alfa Laval, Aalborg CSP, and Weel & Sandvig, with funding from the EU European Regional Development Fund.

14.2.3.12 *Korean AMR*

As well as developing the APR 1400 and Smart 100 reactors, KAERI is also developing a sodium-cooled fast reactor (SFR) known as SALUS. This will have an output of 100 MWe. It is expected to have flexible applications across various purposes, i.e. on and off-grid electricity generation, district heating, industrial heat supply, replacement of coal power plants, etc.

14.3 Generation IV (Gen IV) Micro Modular Reactor Using Advanced Nuclear Technology

Micro reactors are designed to be capable of providing a low-carbon, cost-effective, locally embedded, reliable source of power ***and*** heat for energy-intensive industries, hydrogen, and synthetic liquid hydrocarbon fuels and also to be used in remote locations, including space, while also being able to provide grid support. The smallest ones are sized mainly to provide electricity 'behind the meter', i.e. not necessarily connected to the grid but electricity to be supplied at an unregulated price and directly to customers whose individual demands are rarely more than 30 MWe. They can also be mobile to enable them to be moved to locations where they are suddenly necessary and linked together, if required, to provide the required total output of heat or electricity. They can be embedded in industry or in communities and are designed to be inherently safe and do not require external utility connections for safety or cooling.

14.3.1 *NANO Nuclear Energy*

NANO Nuclear Energy Inc. is an unusually diverse nuclear technology company which is vertically integrated across five business lines:

(i) portable and other microreactor technologies;
(ii) nuclear fuel fabrication – high-assay, low-enriched uranium (HALEU) fuel;
(iii) nuclear fuel transportation;
(iv) nuclear applications for space;
(v) nuclear industry consulting services.

NANO Nuclear is also the first portable nuclear microreactor company to be listed publicly in the US (on NASDAQ).

Its diverse set of microreactor technologies includes:

> 'ZEUS', a solid core battery reactor, 'ODIN', a low-pressure coolant reactor, and the space-focused, portable LOKI MMR™. The Zeus nuclear microreactor features a unique design with a completely sealed core that relies on a highly conductive moderator matrix to dissipate the fission energy. Heat is then removed from outside the vessel using a novel proprietary technology. The absence of an in-core fluid eliminates typical accident scenarios such as loss of coolant and loss of flow. Overall, Zeus is engineered for safe operation in remote locations. Its 'walk-away safe' feature ensures stability and safety, minimising risks in varied environments. This aspect is critical for ensuring consistent operation without the need for active intervention.[4]

14.3.1.1 *Acquisition of Ultra Safe Nuclear Corporation*

In December 2024, NANO signed an agreement to acquire select nuclear energy technology assets from Ultra Safe Nuclear Corporation. NANO Nuclear plans to highlight the high technology readiness level (TRL) of its newly acquired patented Kronos MMR® and also Pylon (now renamed LOKI) technologies.

Whereas 'ZEUS' and 'ODIN' (original NANO reactors) are being designed to be portable and produce 1–1.5 megawatts thermal ('MWth') of energy, KRONOS MMR™ (which was an Ultra Safe Nuclear

---

[4] From NANO website. https://nanonuclearenergy.com/zeus/.

Corporation design) is a stationary installation capable of producing power up to 45 MWth. The individual MMR units can be linked to provide the level of power required and be stopped and started to address the summer/winter heat demand ratio. The MMR can be located almost anywhere. It uses no external cooling water and does not require a grid connection or other infrastructure to operate. It requires no refuelling for its 20-year expected lifetime, which ensures security at the site, as the reactor is a sealed unit. It has a small footprint of less than five acres, is designed to operate flexibly at power levels ranging up to 45 MWt (15 MWe), and can utilise either low-enriched uranium (LEU) or HALEU.

Heat from the reactor is stored in molten salt tanks similar to concentrated solar power (CSP) plants. When electricity demand rises, hot molten salt is used to generate steam and power an electrical turbine. KRONOS MMR™ will typically store up to 10 hours of power plant thermal output and can be supplemented with hydrogen burners. The use of molten salt thermal storage allows for significant flexibility in the supply of both electricity and process heat. While the KRONOS MMR™ unit operates at constant power, electricity and heat are delivered on demand from the power plant.

While it uses a steam generator to drive the turbogenerator set, after passing through the turbine, the steam is condensed using air cooling and not water cooling. It can replace natural gas turbine generation and can match changing supply and demand if intermittent power sources (PV and wind) are also supplying power or if it operates as a heat and power island with no grid. Demonstration units are scheduled for 2026. The design is already in construction permit pre-application engagement with the US NRC in collaboration with the University of Illinois Urbana-Champaign (UIUC). Pending Canadian governmental approvals of the acquisition, further demonstrations of KRONOS MMR™ are expected to take place at the Canadian Nuclear Laboratories. KRONOS MMR™ is the first small modular reactor to enter the Canadian Nuclear Safety Commission's formal licensing review.

14.3.2 *Westinghouse e-Vinci Microreactor*

eVinci can produce 5 MWe with a 15 MWth core design. The reactor core is designed to run for eight or more full-power years before refuelling. It will use 19.75% enriched TRISO fuel. It will not use any fluid for cooling or for transferring the heat externally from the reactor. Instead, it uses the technology of 'heat pipes', avoiding the need for pumps. The heat output is at a temperature of 150°C as a heat source for a Brayton cycle for electricity generation. This reactor is thus using features that are new to reactors, unlike the AP300.

TRISO stands for TRi-structural ISOtropic 'particle' fuel, where each TRISO particle combines uranium in the centre, encapsulated by a triple layer of carbon/ceramic materials. This construction can avoid any release of radioactive fission products and has been tested at up to 1400°C, far above AMR decay heat temperatures. These particles are extremely small – said to be the size of a poppy seed – and can also be combined into much larger 'pebbles' around the size of a billiard ball, which are then used in so-called 'pebble bed reactors'. The heat transfer agent can be either helium or molten salt. The US Department of Energy has said that, 'simply put, TRISO particles cannot melt in a commercial high-temperature reactor and can withstand extreme temperatures that are well beyond the threshold of current nuclear fuels'.

14.3.3 *Rolls-Royce (RR) Micro-Reactor*

RR has not revealed much about the technology of its micro reactor (in contrast to its LWR SMR). What we do know is that it is likely to generate from 1–10 MW of power and is likely to use TRISO fuel. Its more compact size makes it a transportable source of power. It is likely to be of value in applications including defence, remote civil power, remote industrial power, and power in space, such as on the surface of the Moon.

14.3.4 *Blykalla LFR*

Blykalla AB was founded in 2013 by Janne Wallenius, Peter Szakalos, and Jesper Ejenstam as a spin-off from the Royal Institute of Technology in Stockholm, Sweden, where Janne Wallenius has carried out research on design and safety analysis on lead-cooled reactor systems since 1996. The goal of this company is to construct and deploy 1,000 SMRs. It is an LFR. The innovative fuel is uranium nitride (U_2N_3), which has 40% more uranium per volume unit than standard UO_2. This results in it being able to provide heat energy for 40% longer. It should also enable even better safety than some other SMR/AMR designs.

In the above sections, we have included only some of the many companies attempting to reach the stage of manufacturing small reactors (currently around 80). Which will succeed is not known at the present time. However, the industry (and society) will want to focus on those companies believed most likely to succeed to ensure economies of scale in mass production can be realised. Currently, we do not know the exact nature of their technologies, as companies are reluctant to disclose certain details.

Chapter 15

Financing the Global Deployment of Nuclear for Heat and Power

Financing growth in the provision of traditional (light water) nuclear power generation for heat and electricity has been difficult in many countries. Problems have frequently arisen in securing financing at an acceptable price for large light-water high-pressure reactors. An example of this is the 3.2 gigawatt electrical (GWe) Hinkley Point C reactor currently under construction by EDF in the UK. The problem in the case of this type of reactor, however, is not in the technology itself (though this is also a factor) but, as evidenced by build programmes in the past decade, that projected construction times have proven to be only a fraction of actual construction times and actual costs have exceeded forecasted costs by a substantial margin. In practice, state financing has almost always been the only way of financing.

Traditionally, the substantial financing need, coupled with long and often delayed construction timelines and potential liabilities in case of mishaps, has necessitated heavy government involvement in providing risk mitigation instruments and frequently financing through state-owned enterprises (SOEs). SOEs can often provide financing at highly favourable rates, as they can 'piggyback' on the strong credit rating of governments. According to the OECD Nuclear Energy Agency, in practice, a commercial lender may not achieve breakeven for up to 20 or more years, and the cost of financing can reach up to 67% of the levelised cost of electricity for a nuclear power plant. Minimising that cost is critical in keeping down the cost of nuclear power, not least by using lower-cost financing, such as that derived from government borrowing. We will not cover the financing of GW light-water reactor (LWR) generation facilities in further detail, as the more relevant issue is how to finance small and micro reactors.

Today's new reactor types comprise a wide range of Generation IV technologies, whose development and testing are progressing through what is known as the Generic Design Assessment (GDA) process, followed by a site-specific permitting process and then deployment. A reactor of this type represents an asset different from the GW-class reactor. Although this asset type is more likely to be financed privately upon development and many have been constructed, there is currently no well-suited private financing instrument that is well matched to the needs of nuclear. Project finance is not a solution at present (though with a state-provided regulated asset base (RAB) structure, it could be), but as with solar and wind, once Generation IV technologies have proved themselves and reduced the cost of electricity per megawatt-hour (MWh), project finance is likely to become available. However, a constraint that remains is the long lifetime of reactors, normally spanning around 60 years. Investors, as noted above, are not comfortable with projects that may not achieve breakeven within 20 years and are therefore unlikely to pay any dividends. If the construction cost is amortised over only 20 years, the unit cost of generation will be much higher than if amortised over 60 years, making the asset less competitive in the power market. A solution not rooted in the private sector is the RAB model, which we cover at the end of Section 15.3. Financing such long-lived assets through private means is a major innovation challenge we are faced with.

In most countries in the past (and even now in some), power generation (and the production of town gas prior to the use of natural gas) was implemented through municipal authorities or state-owned companies. An example is the Central Electricity Board (CEB) of the UK, created in 1926 and subsequently becoming the Central Electricity Generating Board (CEGB) in 1990, following the privatisation of electricity production and transmission in the UK. In the 1990s, following the privatisation of the UK electricity system, many other countries brought forward plans for their own privatisation. While the high absolute cost and risks associated with financing conventional nuclear generation have always been an issue, prior to privatisation, financing was obtained using public funds rather than through private investors.

Since the privatisation of electricity markets, the subsequent discovery of natural gas, such as that available from the North Sea and producible at a low cost, has resulted in very little nuclear construction during the private ownership period. For investors and utilities, the regulatory and political risks, a stricter operational regime, and a history of cost overruns, on top of large unit capital costs, meant that natural gas exploration and production was a far more lucrative proposition than nuclear, even though it was known as early as in the 1990s that CO_2 emissions lead to global warming.

One reason for privatisation was that some governments were no longer willing to be asset owners or to fund new power plants, which consequently impacted national accounts. While in most countries, nuclear fleet expansion took a backseat to the development of natural gas production, France took a different view. It specifically opted for nuclear for up to 80% of its electricity supply because it had little access to indigenous oil, gas, or coal (unlike, for example, the UK, which had extensive natural gas fields in the North Sea). Its intensive and standardised construction of a fleet of Gen II and III reactors serves as an instructive model.

It is likely that the world will continue to need GW-class reactors. Nonetheless, it now seems likely that distributed small and micro reactors and multiple small reactors linked to generate a GW or more of output could be the way forward for providing much of the continuously dispatchable power supply (baseload) that a country requires. Along with an appropriate mix of wind, solar, and geothermal, coupled with storage and demand interruption, we believe that nuclear will play a vital and significant role in the future design of the electricity system. Furthermore, non-fossil high-temperature industrial and other heat can be better supplied by small nuclear (both small modular reactors (SMRs) and advanced modular reactors (AMRs) depending on the required temperature) so that the number of such plants is likely to be quite high indeed. The advent of much smaller and safer AMRs, based on the physics of their design, means that they can be sited near users and even embedded in industrial facilities. The adoption of these new technologies has also opened the possibility of private financing to potentially dominate a new model for nuclear heat and power.

15.1 Two Types of Financing Required for Advanced Nuclear Technology

1. For AMR Development and Commercialisation
2. For Widespread Global Deployment of AMR/SMRs for the Provision of Heat and Power (this is covered in Section 15.3).

15.1.1 *Financing AMR Development and Commercialisation*

We refrain from covering the Technology Readiness Level (TRL) stages up to TRL 8 in any detail, as venture and other early-stage financing are addressed in Section 2.3. In addition, the companies we are focusing on are mostly already at, or near, TRL 8. If they have managed to reach TRL 8 (and many AMR project companies are unlikely to do so), then they must have avoided what we described in Section 2.3 as the 'Technological Valley of Death'. After this point, however, there is an equally risky stage – the 'Commercialisation Valley of Death', also known as 'The Missing Middle' – of financing, which we also covered at various points in an earlier chapter. This is the problem of scaling a viable product to mass production and global sales. This has been a significant issue for nuclear projects in the recent past, as well as for hydrogen projects. We also do not cover project and corporate financing in this section, as these were covered for all types of infrastructure in Sections 2.1 and 2.2.

The question we focus on in this section is how to finance companies that are **at the point where they can** enter the mass production stage (TRL 9) if only they could obtain the necessary funding. This resembles the point where, as Henry Ford demonstrated with his design of the production line system for making Model T Fords, huge economies of mass production were achieved through the use of production lines. In the case of modular reactors, this will require a large investment in automated production systems, involving the use of robots as in today's automobile factories.

Such mass production techniques for electric vehicles were clearly mastered by the upstart company, Tesla. The company's market value (MV) at the time of writing was just under US$1 trillion. At its peak, it surpassed the combined worth of the next 15 global carmakers, including Ford, GM, Stellantis, VW, and BYD. We need a similar approach to massively 'scale up' AMR production once companies reach TRL 8. Elon Musk managed to capture a large market share in electric vehicles despite Chinese competition, while many other development companies have fallen by the wayside. Another company that has mastered the art and science of low-cost manufacturing at scale is BYD.

Nuclear requires a similar strategy, although, with so many (at least 80 at the time of writing) SMR/AMR companies progressing through the TRL stages in their attempt to reach TRL 8, most will never even reach that level, let alone TRL 9, which represents the point when the scale of production becomes truly commercial. In the true spirit of capitalism, global competition and the survival of the fittest are the most likely mechanisms to enable the lowest-cost designs to survive and scale, thereby allowing for the production of competitive heat and electric power on a global scale. Investors have long held a belief in Musk's skills as an entrepreneur (at least prior to his involvement in his own and other countries' politics) that have so far enabled him to raise huge amounts of capital and expand the company's manufacturing capabilities. For investors, this has meant the prospect of substantial capital gains.

AMRs require a similar approach for mass production, involving both technical competence and capital. Energy utilities, conventional power generation equipment manufacturers, and energy customers are often highly leveraged, so only a few will have the balance sheet capacity or risk appetite to embark on such a development.

In 2024, an alternative source of capital emerged. Several 'Big Tech' entrepreneurs who do have such financing capacity decided to bring their advanced tech business skills and capital (along with a likelihood of signing power purchase agreements (PPAs)) to AMR developers. The reason is quite simply that the new AI models they are developing, specifically large language models (LLMs) such as OpenAI's ChatGPT and the pursuit of artificial general intelligence (AGI), require a substantial increase in power availability for their data centres. These companies are often referred to as 'hyperscalers'. In the current regulatory environment in most countries, they are unlikely to be able to obtain permission to construct a data centre unless they can provide 'additionality' to the country's power supply. Also, this additional power must be carbon-free (with EirGrid going one step further by issuing a moratorium on the development of new data centres specifically in Dublin until 2028). An example of such hyperscale datacentres is the proposal by Mark Zuckerberg's Meta to build a data centre somewhere in the US, which would cover an area nearly equivalent to that of Manhattan. This facility is projected to consume 'multiple' GWe, while their next data centre is expected to require 5 GWe (i.e. almost twice the power produced by Hinkley Point C, the latest large 3.2 GWe power station in the UK).

Data centres require 24/7/365 power, which is difficult and expensive for renewable power companies to supply even at a high price beyond normal market rates due to the extremely high investment cost of providing energy storage to cover the likelihood of a *dunkelflaute*, the German term for a period in which little or no energy can be generated from wind or solar because there is neither wind nor much sunlight. Periods without adequate wind can last for weeks, and the supply gaps are currently being met by switching on natural gas generators. Big Tech companies are therefore highly receptive to proposals for linking up with companies intending to offer electricity that is both carbon-free and dispatchable 24/7/365, such as that provided by nuclear power.

It remains to be seen whether these SMR and AMR development companies, working with Big Tech firms, can also bring in or acquire the skills required for establishing mass production lines. The production process is in some ways simpler than manufacturing an electric car, but it requires having to work under rigorous quality and safety regulations, areas where most developers and tech companies are unfamiliar (as they are more familiar with the 'move fast and break things' approach to development). But if they can, this will pave the way for greatly speeding up mass production and deployment of SMRs and AMRs worldwide, thereby making a substantial contribution to climate change mitigation.

15.2 'Big Tech' Facilitation of Nuclear Finance Through Strategic Investment and Power Offtake Contracts (Power Purchase Agreements)

Big Tech companies that need to be certain of 24/7/365 power availability on long-term contracts (PPA) at a stable and predictable price are now investing in AMR development. Some are also forming consortia of large power users to reap the diversification benefits that stem from drawing power from a range of carbon-free energy (CFE) electricity providers. For example, the US steel manufacturer Nucor, Google, and Microsoft are working together *to develop new business models* by, among other efforts, aggregating their demand for advanced nuclear clean electricity technologies. Consequently, the companies benefit from risk reduction through the diversification of their key input sources, namely electric power, and much greater security of supply if one generator were to go offline for whatever reason.

Big Tech firms, which comprise Alphabet (Google), Amazon, Apple, Meta, and Microsoft, are large users of electricity to power their data centres or chip production facilities. The advent of generative AI LLMs has dramatically increased their energy demand. As an example of this, according to the International Energy Agency, on average, a ChatGPT query needs nearly 10 times as much electricity to process as a Google search i.e. 2.9 Wh of electricity, compared to 0.3 Wh, respectively. Nvidia, although not operating a data centre, is at the heart of the AI boom. The demand for its chips from data centre providers could be greater if clean power were more readily available. For that reason, Nvidia may also decide to invest in a nuclear AMR company to encourage greater demand for its chips. In addition, such investment could prove to be profitable for them, regardless of its impact on chip demand. The only uncertainty concerning power demand by these US companies is whether they would be able to replicate the approach taken by China's DeepSeek, which is much less energy-intensive.

Given the huge increase in expected demand for power by their data centres, a potential solution for Big Tech would be the outright purchase of electricity utilities around the world. However, such a move would require permission from regulators, who are unlikely to agree to a prospect of scarce electricity supplies diverted away from existing users and becoming monopolised by Big Tech. This would come at the expense of other consumers or be handed over to enterprises with greater financial muscle and political influence than all but the largest nations. The alternative solution is for these companies themselves to facilitate an expansion of carbon-free electricity supply ('additionality', as per the European Commission's language).

Data centres have an absolute need for 24/7/365 power and therefore must be assured of consistency in power supply, which renewables currently cannot guarantee. In addition, renewable sources are almost always located at considerable distances from data centres, and in many places the installation of transmission/distribution lines to customers is an issue. These transmission lines themselves are also a source of unreliability, as storms can bring down power lines or cause ice accumulation on cables, causing them to fall to the ground (as in the film *Ice Storm*). The year 2019 also saw the **first climate change bankruptcy** when transmission lines owned by Pacific Gas and Electric (PG&E) fell onto parched land, causing an extensive wildfire. This resulted in claims by affected property owners and stakeholders seeking damages amounting to around US$28 billion, which led to the company becoming insolvent.

While transmission *costs* are clearly a concern, the time delays in obtaining planning permission for erecting electricity pylons across a country is a larger issue. In Scotland, the 'unsightliness' of pylons and transmission lines required for transporting renewable wind power from the North Sea to England has apparently become an issue driving some voters to switch to voting for the UK Reform Party. On the other hand, *advanced nuclear technology (ANT) reactors can be sited near many data centres, thus eliminating the need for the transmission system operator to provide new capacity.*

In the following, we look at some examples of Big Tech companies, AMR vendors, and utilities trying to overcome the problem of insufficient additional dispatchable carbon-free electricity.

15.2.1 *Strategic Users of Carbon-Free Energy in Datacentres as Investors in AMR Companies and AMR Construction Projects*

15.2.1.1 *Meta (Facebook, Instagram, and WhatsApp)*

On 3 December 2024, Meta, seeking help to meet its AI innovation and sustainability objectives, announced[1] it will release a request for proposals (RFP) to identify nuclear energy developers, targeting 1–4 GWe of new nuclear generation capacity in the US, as the company believes nuclear energy will play a pivotal role in the transition to a cleaner, more reliable, and diversified electric grid. Meta hopes that the chosen suppliers will assist in expediting the availability of new nuclear generators and creating sufficient scale to achieve material cost reductions by deploying multiple units, both to provide for Meta's future energy needs and to advance broader industry decarbonisation efforts. In its press release, Meta also reflects on a number of critical financing matters that we have considered earlier:

> Compared to renewable energy projects that we continue to invest in, such as solar and wind, nuclear energy projects are **more capital intensive, take longer to develop, are subject to more regulatory requirements, and have a longer expected operational life**. These differences mean we need to engage nuclear energy projects earlier in their development lifecycle and consider their operational requirements when designing a contract.

---

[1] https://sustainability.atmeta.com/blog/2024/12/03/accelerating-the-next-wave-of-nuclear-to-power-ai-innovation/.

15.2.1.2 *Amazon and X-Energy*

X-energy's 'flagship' reactor design, the Xe-100 small modular reactor, aims to serve either as baseload or load-following support for intermittent renewable sources such as solar and wind. Importantly, the company develops not only Generation IV high-temperature gas-cooled nuclear reactors but also the TRISO-X fuel to power them. TRISO fuel is a key component of many helium-cooled reactors. Xe-100 is an 80 MWe reactor that can be scaled into a 'four-pack' 320 MWe power plant, or it can be even larger if required. Each unit contains 220,000 graphite pebbles with TRISO particle fuel and can be refuelled on-line without interruption, giving an expected 95% plant availability. The plant has been designed for a 60-year operational lifespan.

In October 2023, Amazon, along with Energy Northwest (a public utility), said it would **finance the construction of ANT reactors in Washington State and invest in X-Energy**. Like other AMRs, the Xe-100 is expected to necessitate a mere 400 m safety perimeter, compared to the 10 miles required for conventional LWR reactors, which probably means that it can be sited within the data centre's normal security fence.

Also, in March 2024, Talen Energy sold a 960 MWe data centre campus to Amazon Web Services, a subsidiary of Amazon, for US$650 million. The data centre is directly powered by the adjacent Susquehanna Nuclear Power Station, which will provide a fixed-price nuclear power supply to Amazon.

15.2.1.3 *Google and Kairos Power*

Google has recently made a commitment to Kairos Power to support the construction of multiple AMRs. The power output of the Kairos plant will be 150 MWe generated by two 75 MWe reactors which function as baseload, as they will be available 24/7/365. The reactors will operate at near-atmospheric pressure and, like all ANT reactor designs, incorporate inherent design features for autonomous safe shutdown and decay heat dissipation in the event of power loss. It will also use TRISO fuel in the form of ceramic pebbles, similarly to X-Energy's Xe-100.

We noted in Section 11.3 while discussing energy storage that, on 25 July 2025, Energy Dome announced a global commercial partnership with Google, leveraging its CO_2 battery technology to enable CFE for the grids that power Google's operations. Alongside the commercial agreement, Google has made a strategic investment in Energy Dome. In contrast to Li-ion battery storage, which provides backup power for 4–6 h (or sometimes 8 h), Energy Dome's battery is capable of continuously dispatching energy for periods of 8–24 h, providing enough firm electricity to meet both the baseload and flexibility requirements of large energy users. This is likely to be critical for the use of nuclear or any other technology in supplying electricity to hyperscalers' data centres.

15.2.1.4 *Microsoft, Constellation Energy, and Helion*

Microsoft has entered into a 20-year PPA with Constellation Energy, the company which owns the Three Mile Island nuclear facility in Pennsylvania. This involves recommissioning one of the reactors to supply dedicated power to Microsoft.

Microsoft has also signed an agreement with fusion energy developer Helion Energy for the provision of electricity from its first fusion power plant. This is instructive because it shows that Microsoft's vision far exceeds that of any existing utility, manufacturer, or even developer, who cannot underwrite such an expensive and perhaps speculative venture, yet a tech entrepreneur can.

15.2.1.5 *OpenAI/Oklo*

Oklo is developing scalable nuclear power plants using liquid metal reactor technology. Utilities or others seeking to take power from Oklo's power stations **will not have to directly invest in the construction process, as is the case with a number of other ANT projects**. Instead, Oklo will use long-term PPAs. As the company has said, this takes the burden away from our customers of having to invest in their own plants to use the technology. It is hoping to have its first reactor generating electricity at the Idaho National Laboratory by 2030. It calls its reactor 'the Aurora Powerhouse'. Currently, the reactor is designed to produce 15 MWe but can be scaled up to 50 MWe and is expected to be able to operate for 10 years or more before requiring refuelling.

The CEO of OpenAI, Sam Altman, is also an investor in and chairman of Oklo. It should be expected that he will be signing a PPA with Oklo if the venture succeeds in reaching the point of generating electric power. In mid-December 2024, Oklo signed a 20-year power supply agreement with one of the largest data centre providers, Switch. Oklo will supply reactors and generating capacity of up to 12 GW (enough to power over 7 million homes 24/7/365, which wind or solar alone cannot offer). The deal is non-binding and the technology is not yet available, so the deal is an indication of intent rather than a true PPA. But Oklo is already hoping to have its first (test) reactor operational by 2027, though this test reactor will only have an output of 15 MW.

15.3 Financing Global Deployment of AMRs/SMRs for the Provision of Heat and Power and a Case Study of Nuclear RAB

The second financing issue is how to finance the **purchase of standardised small (including multiple reactors on one site) off-the-shelf reactors by potential customers** once the production lines are working satisfactorily. The facility may be used for electricity generation, for the provision of heat, or for both. In the OpenAI/Oklo example above, it seems that Oklo intends to finance the construction of the generating capacity itself and will enter into contracts with customers to sell them heat and power. We presume that financing would be available from banks and bond investors on the back of long-term PPAs that would guarantee the necessary revenue flow to satisfy lenders i.e. the project would have a very good chance of meeting its full debt service costs, thereby rendering it bankable. Such contracts are designed to ensure that, having financed the project, lenders have a high degree of certainty that the debt can be funded i.e. there is adequate 'guaranteed' revenue to service the debt (including the principal and interest payments). In addition, later institutional equity investors in the project, such as insurance companies and pension funds, will want to ensure that it generates sufficient revenue to provide an appropriate level of dividends to them.

More broadly, a nuclear generation facility (power station) or heat facility, in theory, can be owned and financed by a range of parties depending on, among other things, the final use(s) of the heat generated by the nuclear fission process, e.g. power generation, industrial heating, combined heat and power for space heating, desalination, hydrogen production, and powering data centres. Some of the different approaches to financing are summarised as follows:

The developer of the AMRs may manufacture them and also finance their deployment. This would result in the company operating almost like an electricity utility. The developer of the AMR may raise equity and debt and then build and operate the plant to provide heat or power to a customer at an agreed-upon price for an agreed-upon number of years, which avoids the need for the customer to source up-front funds for the project (see the case of Oklo above). This is analogous to the 'build, own, operate' (BOO) model, which is extensively used for conventional power stations. This approach involves the developer signing PPAs with prospective customers to provide sufficient certainty of revenues to satisfy lenders.

An existing power utility which could use its own balance sheet to raise the necessary financing (corporate financing, not project finance) and then sell the output to its existing or new customers. The revenue from this would contribute to servicing its general corporate debt and equity capital base. The smaller value of each small nuclear plant (which can accommodate additional reactors over time) compared to the GW-class reactors makes this model more digestible to a utility, its shareholders, and its lenders compared to a single large GW plant.

Large customers providing advance payments for future power to the developer and owner of the plant which provides the financing required or a proportion of it and thus minimises the need for the developer to find 100% external debt finance. Customers may include industrial firms, utilities, or a new utility set up by a local authority to supply power. Customers may opt for this when they are not happy to take on construction risk from the developer but are willing to help finance the project developer in return for an attractive PPA. Once the plant is operational, these risks reduce, allowing customers to scale their investments up to 100% of the constructed capex while either leaving the developer as the operator or taking that on as well. This is similar to the BOOT model – build, own, operate, *and transfer* – which is again common in the power industry.

Retail customers of an electric power utility, depending on the regulatory framework, may incur an additional charge, which would be used to make dividend payments to equity investors during the whole construction period (as in Thames Tideway, discussed in Section 4.4). The additional charge could also be imposed at a level sufficient to cover interest payments during the construction period when no revenue is being generated. Thus, interest would not accrue at the time before the plant starts to earn revenue and therefore increasing the capital sum outstanding to the bank. This approach would considerably reduce the cost of power when the facility begins its operation. It would be a particularly good approach when interest rates are 'high', as the cost of servicing debt constitutes the largest operating expense for a nuclear power company. This model has been the basis for GW-class nuclear in the US for several decades.

15.3.1 *User Ownership: Joint Ownership by a Group of Electricity Users*

15.3.1.1 *Mankala*

The Mankala model is not an uncommon ownership arrangement in the electricity production companies of the Nordic countries. For example, Teollisuuden Voima Oyj, owned by Fortum and Nordic Power in Finland, and nuclear power plants in Forsmark and Oskarshamn, cross-owned by Vattenfall, Sydkraft, and Fortum in Sweden are organised in this way.

The purpose of Mankala companies is to produce electricity for the joint owners at the lowest possible cost by producing energy themselves. The owners obtain electricity in proportion to their ownership at cost price, and hence the purpose of the model is to enable the sale of power to the owners but without generating profit for the captive operating entity.

The model has shown itself to be a way to execute new investments so that the costs and risks of a new plant investment are shared among participants who have an actual interest in the outcome beyond mere financial returns. In Finland, construction of unit 3 at the Olkiluoto Nuclear Power Plant began in 2005. Commercial operations, originally scheduled for May 2009, began only on 1 May 2023. This raises quite dramatically the risks associated with GW-class plant constructions using a first-of-a-kind (FOAK) design, even for a highly experienced nuclear owner/operator and plant contractor. In this case, the same design underlies the problems at Flammanville, Taishan, and Hinkley so that the rate of learning for such a behemoth plant, the largest ever built, is slower than what could be expected on simpler AMRs requiring a much reduced scope of the safety case.

Examples of customer financing and user ownership are as follows:

- **Customers – power**
  Both retail and wholesale customers could contribute to the financing of an AMR project. For example, long-term retail bonds are offered by an SPV but underwritten by the government and intended for pension savings. Holders of these bonds could also be offered price discounts for life from the local power retailer once power generation commences. This could be attractive to pensioners who may find the concept of lower future bills or even a whole-life 'energy annuity' attractive while still obtaining a higher yield from these bonds compared to their existing savings instruments.

  Wholesale purchasers of power with surplus balance sheet capacity may be offered attractive long-term energy offtake contracts in return for an upfront payment. An example of this approach is the Exeltium transaction under which EDF and the Exeltium (a consortium of electro-intensive users) signed a memorandum of understanding on a take-or-pay contract, as described in the following:

- **Exeltium – a consortium of large electricity users contributing to nuclear financing**
  France has the world's highest proportion of its power generated by nuclear. The fleet of reactors is state-owned through the state's ownership of Électricité de France (EDF). The Exeltium is a consortium of heavy power users, through which they have contributed to the up-front capital cost of a new nuclear power station by contracting with EDF to assist with its financing (in this case, Gen III+). In exchange, EDF has provided them with a long-term fixed-price supply contract which minimises price risk for the off-taking companies.

  The logic behind this transaction with customers lies in the fact that for electro-intensive industries (e.g. aluminium, chemicals, industrial gas, paper, and steel), electricity supply and its cost constitute a major competitiveness factor, as it represents between 15% and 50% of production costs. Industrial investments are long-term in nature (of at least a 10-year horizon), and therefore uncertainties regarding electricity price evolution threaten the sustainability of these industries, particularly in the face of competition from countries with much lower power costs. Since the liberalisation of the electricity market in 2000 in France, market prices have increased and become unpredictable, putting at risk companies' ability to invest in their core competencies which are not related to power.

In July 2005, the French government organised a roundtable with electricity producers and electro-intensive consumers to address the growing uncertainties surrounding long-term electricity prices. Following this, electro-intensive industrial companies founded the Exeltium consortium, with the objective of negotiating an optimal nuclear electricity supply contract on behalf of its shareholders. Note that nuclear is uniquely positioned to offer this kind of contract because its main costs are initial capex while its variable operating costs are only about 10% of the electricity cost. Arguably renewables have a similar high capex but a low variable operating cost structure, but large capital-intensive process industries cannot tolerate a potentially highly variable power supply, and the cost of ensuring a stable power supply through energy storage can be quite expensive. The shareholders of Exeltium included Air Liquide, Arcelor Mittal, Arkema, Rhodia, Rio Tinto Alcan, and Solvay. Despite difficulties in markets following the 2008 financial crisis, Exeltium successfully raised €2 billion in 2011.

- **Customers – heat**

  The economics of power stations improves quite substantially if used for combined heat and power. ANTs can be safely situated in built-up areas to supply industrial process heat or to new or existing towns for industrial processes, power generation, and, finally, low-temperature space heating, with each energy product (electricity and heat) having its own income stream.

  Individual space heating (gas boilers) for homes and offices, particularly in cooler latitudes, can be replaced by implementing a district hot water heating network. The addition of heat customers can alter the economics of small nuclear power, making it much more bankable. The technology of district heating is mature but is usually fuelled by fossil or biomass sources. Nuclear, combined heat and power or pure heat, offers a carbon-free heat alternative. At a conference on the proposed Oxford/Milton Keynes/Cambridge Arc project, which was intended to provide housing for more than one million people and was described as 'A long-term vision to secure sustainable growth and prosperity for all', the principal author of this book delivered a presentation on how decentralised power and heat for this growth could be provided by AMRs.

  In Finland, a simplified long-life small reactor designed exclusively for low-grade heat is being developed specifically to repower the country's existing investments in district heating networks. The UK government has chosen to promote individual electric heat pumps for domestic and commercial space heating. This is understandable since there are negligible existing district heat networks. However, the current electricity distribution network in most towns (through buried street cables) is inadequate for powering the deployment of both heat pumps and electric vehicle charging. Strengthening the distribution networks would be another capital-intensive project.

In July 2025, the European Commission proposed opening a part of its next long-term budget, worth €2 trillion, to include funding for nuclear energy. The draft budget opens up national EU funding streams to support 'new or additional fission energy capacity installed in GW'. Around €865 billion of EU funding would be accessible under these national spending plans. Conversely, the current EU budget explicitly excludes conventional nuclear power. While final approval will require consensus among all 27 member states, the draft suggests institutional acceptance of nuclear as a key part of Europe's energy mix.

Banks, institutional investors, and sponsors finance most infrastructure projects, including public–private partnerships. Generally, bank finance (project finance) is used for the construction phase, and the loan may be refinanced in the bond market at a later date when the early-stage risks have reduced. Banks are, however, unlikely to finance the FOAK of a new technology, as it is not their function to take on what would normally be equity risk unless, as in renewables, much of that risk is transferred to the government. Once a new reactor design has been successfully completed on time and without a cost overrun, it is more likely that banks would be willing to participate in funding.

In December 2023, the United Nations Climate Change Conference (COP28) in Dubai saw the 198 signatory countries to the UN Framework Convention on Climate Change call for an acceleration in the deployment of low-emission energy technologies, including nuclear power, to enable deep and rapid decarbonisation, particularly in hard-to-abate sectors such as energy-intensive industries requiring heat. In addition, **25 countries at COP28 pledged to work towards tripling global nuclear power capacity to reach net-zero by 2050**.

During New York Climate Week in September 2024, a group of **14 financial institutions expressed support for expanding nuclear power generation and the nuclear industry supply chain**. The institutions included: Abu Dhabi Commercial Bank, Ares Management,

Bank of America, Barclays, BNP Paribas, Brookfield, Citi, Credit Agricole CIB, Goldman Sachs, Guggenheim Securities LLC, Morgan Stanley, Rothschild & Co, Segra Capital Management, and Societe Generale.

An example of one of the banks with a strong focus on sustainability and nuclear financing is Crédit Agricole Group. At Euromoney's July 2025 Awards for Excellence, it was named World's Best Bank for Sustainable Finance, Europe's Best Bank for Sustainable Finance, and France's Best Bank for Sustainable Finance. The selection process followed a joint application by Crédit Agricole S.A., Crédit Agricole CIB, and Amundi. It also received the award for Asia's Best Bank for Sustainable Finance, recognising its leading role in sustainable finance. The bank is also known as *la banque verte*. Crédit Agricole CIB supported Korea Hydro and Nuclear Power in issuing Asia's first green nuclear bond.

Supranational agencies, such as the European Investment Bank (EIB), often provide some of the financing required by an infrastructure or power project. In principle, their financing is little different from private bank financing but may be at a lower interest rate or feature some different terms. Such agencies may also offer guarantees or other risk instruments. They may also be guided by the policy of their supranational region, such as the EU's recently more pro-nuclear energy taxonomy.

A public–private partnership (PPP) is a structure frequently used to provide a service to a government or local authority, for example, a hospital or school. In a PPP, private financing will almost always provide most of the financing to keep the debt off the public sector borrowing requirement. It also enables some or all of the risk to be kept within a special-purpose vehicle or with a construction contractor or a plant manufacturer. These latter may be more willing to come forward when the absolute sum of money per AMR plant is compatible with their balance sheets. This is common for process plants and gas power stations and may well be realistic for small nuclear but less so for GW scale.

Oil and gas majors are under pressure from investors first to reduce their carbon footprint and second to use their free cash flow to diversify away from fossil fuels. But they are equally under pressure to maximise returns in the short term, particularly when oil and gas prices are high. They have the potential to increase their investments in wind, solar, and nuclear, but in general, they do not seem to be doing so at a rate that could be achieved. Shell and British Gas were both early movers into thermal electricity and nuclear but withdrew when power generation shifted at a very large scale to gas, which was their core business. Regulators also 'punished' electricity technologies, except for renewables, which were subsidised. Gas and oil majors have plentiful international investment opportunities in their core businesses and, at the time of writing, have also cut back even their relatively modest funding for wind and solar assets because they currently perceive their customers and shareholders to be insufficiently green-sensitive.

The asset recycling (Australian) model takes advantage of a government's low cost of borrowing, which is used to finance a project up to the point when its operational revenue stream is relatively certain. The government therefore assumes construction risks for which it is rewarded. It will aim to gain a return on a portfolio of projects. It does so via a high debt coupon paid by the developer and by selling projects to institutional investors after completion. By seeking bids to buy an operational asset with known cash flows, the price received from the sale of such an asset should exceed the cost of constructing it.

If properly managed, the internalised risk of cost or time overruns or underperformance against expectations should, over a range of projects, be adequately covered by the final capex mark-up on sale. The buyer benefits from financing the project at the *sovereign borrowing rate, plus a risk premium to cover the risk in its total portfolio, rather than just a single project.* Once the sale has concluded, the government's proceeds can then be used/recycled to finance other projects.

The key advantage should be a reduction in financing costs due to the decreased cost of risk capital from the start of financing. However, the risks undertaken by the government may or may not be compensated for by the interest earned on the government loan, depending on the government's ability to accurately evaluate risk. This model resembles a 'build, operate, and transfer' (BOT) contract.

15.3.2 *Revenue Support Mechanisms: CfDs and Regulated Asset Base (RAB)*

A CfD is a contract between a power producer and a government. However, it is not an 'offtaker' contract for the sale of power (unlike a PPA) and thus does not guarantee the actual sale of power. It is only a mechanism to guarantee a minimum price for the sale of power generated by a project over a specified period. It is an agreement to pay or be paid the difference between the reference and strike prices while the generator is responsible for physical power sales. In the case of renewable power contracts, the generic CfD offers a 15-year contract term to provide equity investors with confidence that they will achieve their expected return, while for lenders, it should make the project bankable, meaning able to obtain debt financing. However, in the case of nuclear, where the design life of the project is likely to be 60 years, it would be desirable to have a longer contract period, although some projects may opt for a 'merchant' basis to generate revenue in which they sell into the market without any revenue protection in the hopes of generating higher revenue. In practice, few generators would consider 100% merchant sales (and, indeed, this would almost certainly make the project unbankable), but many would consider holding back a proportion of sales during the lifespan of a CfD as well as sales subsequent to the expiration of the CfD or PPA.

The Regulated Asset Base (RAB) model is the traditional, proven solution for financing capital-intensive infrastructure and is used in electricity, water, gas, and telecommunications worldwide. The RAB model, by providing regulated returns to investors, holds the potential to reduce the cost of raising private finance for new nuclear projects. As economic regulators are required by statute to ensure that investors obtain the agreed-upon rates of return, this financing mechanism has already proven successful in enabling private sector investment in other infrastructure sectors, such as the National Grid's UK investments in gas and electricity. In the US, it is used for nuclear, coal, and gas power stations.

Similar to the nuclear CfD model, the nuclear RAB model holds the potential to reduce the cost of private financing for new nuclear projects by distributing risk among investors and consumers (as in the **retail customers of an electric power utility model**, discussed in a section above). Unlike the CfD model, the UK RAB model also **allows the generating company to receive payments from customers during construction**. These can then be used to provide returns (dividends) to investors or/and interest payments to lenders (instead of interest being 'rolled-up'), thereby reducing the overall cost of the project.

Case Study
RAB Financing Model: Sizewell C in the UK

This is the first use of the RAB model for nuclear in the UK. The total cost estimate of £38 billion (at 2024 prices) for a power output of 3.2 GW represents a saving of around 20% compared with Hinkley Point C, of which Sizewell is a replica second-of-a-kind project (SOAK). The output of 3.2 GWe can provide the equivalent of sufficient electricity for 8 million homes. It uses Generation III+ light-water reactor (LWR) technology. The new Sizewell C owners include La Caisse with 20%, Centrica with 15%, and Amber Infrastructure with 7.6%. This comes alongside EDF taking a 12.5% stake in the project. as well as a proposed £5 billion debt guarantee from France's export credit agency, Bpifrance Assurance Export, to back the company's commercial bank loans.

The UK government is providing £3.8 billion in equity and £36 billion in debt. As the debt will be issued by the UK Treasury (gilts), the interest rate will be much lower than on bank or bond debt. Around £400 million in equity is being provided by the French Nuclear Liabilities Fund (owned by EDF) and others to pay for decommissioning costs, which should not arise for 60 years after the plant is commissioned.

Bringing in private investors requires offering them a package which transfers many of the risks, such as overspend and delayed completion, to other parties. Centrica's expected rate of return is 12%, which is lower than it would be without such risk transfers. However, it is considerably higher than many other infrastructure projects, which might return only 7% per year. The risk transfer involves removing the obligation from investors to provide funds beyond a project cost of £47 billion, but up to that point, they are responsible for cost overruns.

The above description of the project's financing needs to be followed by a consideration of the different risk management mechanisms used in previous nuclear projects involving a contract for difference (CfD). The key difference with RAB financing lies in the funding mechanism. The RAB starts paying dividends to investors and interest to lenders from the point when investors transfer funds to the project company. A surcharge on the electricity bills of all electricity consumers until the project is earning revenue, which might be 10 years away, should not result in more than £1 per month being added to consumer bills. But this will service debt and equity.

15.4 Project Developers, Independent Power and Heat Producers (IPHPs), and Community Projects: Kärnfull Energy, Kärnfull Next, and Community Nuclear Power Ltd

15.4.1 *Energy Communities*

Recognising the potential contribution of energy communities in achieving a more secure, affordable, and cleaner energy system for Europe, the REPowerEU Plan put forward the shared political objective of achieving one energy community per municipality with a population of 10,000 by 2025. The Commission has noted that:

> They can contribute to increasing public acceptance of renewable energy projects and make it easier to attract private investments in the clean energy transition. Energy communities can be an effective means of re-structuring our energy systems, by empowering citizens to drive the energy transition locally and directly benefit from better energy efficiency, lower bills, reduced energy poverty and more local green job opportunities. Energy communities allow local communities to join forces and invest in clean energy. Energy communities allow local communities to join forces and invest in clean energy.

What is not clear from the above is whether community **nuclear** would fall under the category 'clean energy', as used above.

15.4.2 *Kärnfull Energy and Kärnfull Next*

In 2021, Kärnfull Energy was started by two entrepreneurs, Christian Sjölander and co-founder John Ahlberg, in Denmark as a *supplier of electricity* which has been generated 100% from nuclear. It subsequently set up another separate but related business, Kärnfull Next, as the Nordic region's first *project development company for small-scale nuclear energy*. They have built up what is already a profitable and rapidly growing nuclear energy company group while also promoting the need for nuclear to accelerate the Nordic region's (and the world's) journey towards reaching the Paris Agreement goals. They consider the company to be a community and industry developer. The project has received backing from two investors, Climentum Capital and Granitor Growth

Management. Granitor has noted that they see great potential in Kärnfull Next's strategy to decarbonise energy-intensive 'difficult to decarbonise' sectors. The investment reaffirms their conviction that *nuclear energy is a key part of a sustainable future.*

In its role as an SMR project developer, Kärnfull Next is 'tech and vendor agnostic'. In practice, to have a generating project operational as soon as possible, they have chosen for now to use LWR reactors, as these are currently available. It is likely that, when AMR technology becomes commercially available and economically attractive, they will incorporate it into their projects, and if successful, they would drop the use of SMRs. The company has selected the GE Hitachi SMR BWRX-300 (300 MWe output) for its first project. It covers the full value chain of a new-build projects company, including everything from obtaining permits to licensing, siting, procuring land rights, designing, and engaging in local dialogue. The team assumes early-stage financial risk in projects and creates value through replicable processes across multiple projects which should generate economies of scale by lowering both the price of the reactors they purchase and the cost of their construction. This is in sharp contrast to the issues encountered in the construction of GW-class reactors, where new builds generally differ from previous ones. There also seems to have been little in the way of 'learnings' from previous builds being applied to new builds.

Their objective is to set up '**SMR Campuses**', as mentioned above, featuring the nuclear source of heat and power, electrolysers to produce hydrogen, and synthetic fuel plants which would combine captured CO_2 using CCUS and DAC with H_2 to manufacture synthetic liquid and gaseous fuels. Such a campus could also manufacture ammonia and fertilisers without relying on natural gas as the primary input. **The campus idea is in line with the idea of 'clusters' and a sector-coupled economy**, which we cover in Chapter 19.

Interestingly, Kärnfull Next is not the only new nuclear company to have been set up by entrepreneurs in the Nordic region. Denmark also has Seaborg Technologies and Copenhagen Atomics, both of which have developed highly innovative AMR designs which are moving closer to achieving certification. These new reactors are even safer than the current, already very safe, Generation III+ nuclear technology. In addition, they can operate as 'burners' and thus greatly reduce conventional nuclear waste by using it as fuel, i.e. circular nuclear power.

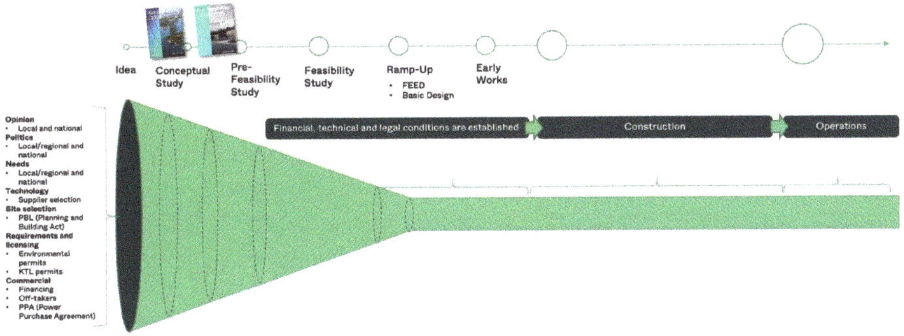

Figure 15.1. Kärnfull Next: From concept to generating electricity.

15.4.3 *Community Nuclear Power Ltd (CNP)*

The proposed business model of CNP has similarities with that of Kärnfull Next. They are not developing a new SMR or AMR themselves but have chosen, like Kärnfull, to use an LWR that has already been developed and is ready for construction. This enables them to get to market much more quickly than waiting for the appropriate AMR to become available on the market.

They view their project as a ‘regional development’ project, as well as an economic one, which involves deploying a low-risk nuclear heat, power, and hydrogen (and its derivatives) to enable a transformation programme in a region – initially in the northeast of England’s North Teeside industrial cluster. While clearly such a project must provide investors and lenders with an appropriate rate of return on their investment, they are also focused on what it can do socially for a region, as well as its economic returns. The community perspective arises in considerable part from the background of the founding entrepreneur, Paul Foster, the chief executive officer (CEO) of the company. He worked for 20 years at the Sellafield nuclear site, holding senior roles such as CEO. Sellafield, one of the world’s largest and most complex nuclear projects, has delivered some millions of pounds to community benefit projects.

As in the above graphic (Figure 15.1) from Kärnfull Next, the component parts and agreements needed to make this groundbreaking proposition happen, including the land, capability, technology, private capital funding, and community demand, are already in place and align with the Great British Nuclear (GBN) roadmap. GBN, a UK government body, was set up to facilitate the delivery of the government's long-term nuclear programme and support its ambition of delivering up to 24 GW of nuclear power in the UK by 2050.

The company is working with Westinghouse to deploy four AP300s, the design of which is based on the well-proven technology of the AP 1000, already in use around the world. The four reactors will provide around a gigawatt of energy, meeting the anticipated demand for the green energy and chemical hub planned on the North Tees Group Estate.

15.5 Changing Mindsets: How to Speed Up the Scaling-Up of AMR and SMR Mass Production for the Provision of Carbon-Free Electricity and Heat

Although AMRs and SMRs are expected, once approved by regulators, to be safe for installation in built-up areas, populations near intended nuclear sites will still need to be convinced that a nearby reactor poses no danger. It may also be necessary to prove to people that it brings them benefits such as reduced power prices within a local distribution network.

Legislators may also need to gain a better understanding of why nuclear is so necessary to achieve climate objectives, and this requires an increase in understanding of not only why these new technologies, both Gen III+ and – more particularly – Gen IV, are so much safer than even the existing global fleet of reactors but also how a nation's energy system as a whole will function with increasing proportions of variable renewable electricity (VRE) generation. There is a need for a 'systems' approach to determining the cost of generation and a regional approach to achieving a locally integrated energy system.

Increasingly, we believe that there will be regions where wind and solar are used along with unmitigated (or eventually mitigated) natural gas to 'fill in the gaps', while other regions will primarily use nuclear, i.e. for baseload. Baseload AMRs/SMRs will provide domestic heating and electricity, while micro reactors may be embedded in factories operating in industrial zones where surplus electricity and heat will be used to manufacture hydrogen and its derivatives, including synthetic liquid hydrocarbon fuels such as sustainable aircraft fuel (SAF), ammonia, and fertilisers, and waste heat will provide district heating. A good example in Europe where such a configuration could work would be the proposed Oxford/Milton Keynes/Cambridge 'Arc', where it is planned to build more than one million homes over several years and improve transport systems between these towns.

We outline in the following some of the 'changes in mindset' that will be necessary.

15.5.1 *Don't Think About Maximising the Proportion of Renewables in the Generation Mix; Think About Maximising CFE*

In 2023, Sweden's Tidö Agreement included a change in the energy policy goal. It was 'changed from 100% renewable to 100% fossil-free'. Other countries need such a shift in mindset to achieve the climate goals.

15.5.2 *Think Both Heat and Electricity*

Heat accounts for half of the world's total energy consumption, significantly more than electricity (20%) and transportation (30%), though proportions vary by country. Unsurprisingly, heat contributes to more than 40% of global energy-related CO_2 emissions, according to the World Economic Forum. Industrial heat accounts for half of all heating but requires innovation in 'hard to decarbonise' sectors of industry. Space heating in homes and offices constitutes most of the remainder. Carbon-free heating can be achieved through heat pumps and through district heating systems (DHSs). It can also be achieve by using waste heat from micro nuclear, installed either in new build or by conversion of natural gas DHS.

Therefore, the objective should not simply be attempting to maximise carbon-free *electricity* production. It should be trying to decarbonise the electricity system **and** the heat system (as well as, of course, the transport system) while minimising the unit cost of power and heat.

15.5.3 *Don't Think of Some Hundreds of Nuclear Heat and Power Generation Stations Worldwide: Think of Many Thousands and the Size of the Manufacturing Market as a Potential Boost to Economic Growth*

Nuclear can function either as baseload or as an aid to integrate renewables into the grid. If, however, the cost of nuclear power for a whole grid system falls substantially, it would make sense in certain locations to use small and micro nuclear to supply a major share of electricity with renewables providing the balance. This model for a power system network serves as an alternative to that where renewables provide the major share of electricity, with nuclear providing the balance.

15.5.4 *Think About Whole System Cost and Not Price of Power When the Sun Shines and the Wind Blows and Not at the 'Farm Gate'*

It is claimed that solar power is the lowest-cost power supply available; while this may be true in some geographical locations **'at the farm gate'**, there are other substantial costs that need to be considered. The first of these arises from the necessity to transmit electricity from where it is generated, i.e. on a solar or wind farm, to the customers' point of use. Solar is generated, typically, at rural solar farms, while consumers tend to be concentrated in cities far away from the source of generation. In the case of wind farms in Europe, most of the wind power is generated offshore, predominantly in the North Sea. The costs associated with transmitting power to the coast and then from the coast to the demand location can be very high. Often, objections to new pylon and cable routes are enough to stop such transmission projects or to delay them by many years, or their costs are raised by forcing the cables to be laid underground instead of overhead. This 'system cost' can substantially raise the true cost of solar and wind, as well as the time required to generate revenue from a project, due to the current long-duration 'pipeline' of applications for connections to the grid. As we noted earlier, in the US, transmission and distribution costs already make up 44% of the total cost of delivered electricity, according to the WEF. Small and micro nuclear can eliminate transmission costs and, on a closed network, greatly reduce the cost of power at the customer's premises.

An **equally important system cost** is that of converting the variable wind and solar output (which seldom provide their full so-called 'nameplate' power) into power that is dispatchable. Solar has a low annualised 'capacity factor', ranging from as low as 10% of the peak nameplate rating of its panels to perhaps 25%. The cost of short-term energy storage systems for wind or solar (which last for a few hours at most) remains considerable. Long-duration energy storage remains significantly expensive, as the assets involved are not called on frequently and must thus charge a very high price when called into service. During 'wind droughts', which can last for weeks, there remains a need for power, and this is currently being met by 'turning up the gas'.

A further consideration is the annual cost of **constraint payments to solar and wind farm owners** who are asked to stop production on very sunny or very windy days when supply exceeds demand. This cost can also be very high, and in the UK, for example, it is said to be of the order of magnitude of £1 billion per annum. Such payments, though obviously popular with plant owners, simply increase the price charged to consumers for each kWh of power they purchase.

As the percentage of renewables on the grid increases, the **system** costs of adding more increase non-linearly due to the need to compensate more for temporary overproduction, intermittency, transmission costs, and, in some cases, the rising price of suitable land or seabed.

Finally, for investors, it is well-known that an increasing share of renewables can **'self-cannibalise' the economics of renewable technologies**. This happens as a result of excessive wind or solar wishing to provide more power to the grid than the grid's customers require. As a result, electricity prices fall and can frequently turn negative. This leads to to lower revenue for wind or solar farms than was forecast when the project was at the point of Final Investment Decision (FID).

15.5.5 *AMRs and SMRs Are a Quite Different Reality from Our Feelings About Legacy Reactors*

The differences in technology, possible locations, construction time, safety, and required land area between legacy reactors, on the one hand, and SMRs/AMRs, on the other hand, are not widely understood. It's a shift from reactors occupying a huge ground area well away from cities to a decentralised generation model using relatively small plots of land, situated close to or, in the case of micro reactors, embedded into user facilities or residential and business districts.

It's a reversion to the old model of a utility – local electricity generation (and town gas production from coal) within a town or city or very close by. Bringing generation closer to users is important because we cannot quickly, or affordably, put in the extra transmission infrastructure to bring wind power from, for example, the North Sea or Baltic to the demand centres in southern Germany or even afford to do so.

Following this, the shift in mindset is the need to think about **community** power generation from nuclear, i.e. reverting to a model similar to the municipal electricity (and gas and water) model of the past, as mentioned in the paragraph above. Kärnfull Next in Denmark (see Section 15.4.2) is a good example of this approach. Another example is Community Nuclear Power Ltd in the UK (covered in Section 15.4.3). The latter has been set up by nuclear experts with the aim of making people understand that we can bring nuclear home with substantial benefits to local communities, both as offtakers and potentially as investors, allowing them to invest a portion of their savings and pensions in a real asset rather than outside their own community or even outside their own country.

15.5.6 *Remember That Decommissioned Coal Power Stations Still Have Useful Locations with Grid Connection*

SMRs and AMRs are well-suited for **repowering fossil fuel-fired plants**, allowing some employment in the area to be retained with well-paid, high-skilled job opportunities. Unless this is widely known among parliament members, legislators, and local government officials, this substitution is unlikely to occur. An example of a company attempting to achieve this objective is Terra Praxis, a UK company which has well-developed plans but is mainly active in the US and partially funded by a Microsoft founder.

15.5.7 *Small Distributed Nuclear Gives Much Greater Defence Against Foreign Threats, Including Missiles and 'Cable Cutting'*

Today, defence security in times of foreign military threats has come to the fore. The Nord Stream 2 explosions from sabotage, as well as the potential for unilaterally cutting gas supplies to Europe, also resulted in the worst release of methane gas in human history, with estimates ranging from 100,000 to 400,000 tonnes of methane emitted into the atmosphere.

Distributed small nuclear for heat and power in hundreds of locations is far less strategically vulnerable to concerted cyber or physical attacks (such as those Ukraine has experienced from Russian missiles) than a few large, concentrated installations and network hubs. More tactically, mobile microreactors, which we covered earlier, have the potential to provide critical defence and societal installations, as well as public services, with a continuous power and heat supply for several years, eliminating the need for refuelling. Because small and micro nuclear reactors can be sited close to energy consumption points, they are much less vulnerable to the easily inflicted damage (as already faced by Europe at the hands of 'hostile' powers) on submarine cables, international interconnectors, or overland transmission systems.

15.5.8 *Regulation of Nuclear Needs to Be Regional and International*

Regulation is an issue both within Europe and internationally. While EU countries can make their own decisions on the generation mix, we need to establish a Europe-wide (and we include the UK in this) regulatory framework for safety regulation requirements, such as the **Generic Design Assessment (GDA)** process for nuclear. Radiation hazard itself is of course transboundary, not national. A more prosaic reason is also that it is wasteful and slows down the energy transition when each country has to individually regulate these new technology assets.

It should be noted, however, that the **site-specific permitting** stage that follows the GDA stage holds less international applicability, as country-specific risks arising from seismic, meteorological, or geotechnical factors are unique to particular locations. A reconsideration of the types of risk that this entirely new design of reactor poses relative to our previous nuclear experience is needed. Indeed, without such a reassessment, mass production with its advantages of cost certainty, lower selling prices, and a faster rate of decarbonisation, is at risk.

Within Europe – but ideally worldwide – there is a need for a common regulatory framework applicable not only to reactors but, very importantly, also to the new fuels and methods of fuelling that have been developed. There are several new types of fuels and fuelling methods, with prominent ones being the so-called high-assay low-enriched uranium (HALEU), TRi-structural ISOtropic particle fuel (TRISO), and pebble bed fuelling, which have not been previously used in commercial reactors. It would be unfortunate if every country in Europe or worldwide were to conduct their own testing. It would certainly slow down, and raise the cost of, the energy transition.

15.5.9 *Scarce Land-Use Planning: Agriculture Versus Other Uses*

This issue is particularly relevant to countries with a high population density and with many people dependent on agricultural production. But it remains important for all countries in Europe as well. The issue lies in the significant amount of land taken up by solar or wind farms (or even legacy reactors) and therefore becoming unavailable for agriculture. While 'agrivoltaics', which is the joint use of land by solar farms and agriculture, has been suggested as a potential solution, it is not yet well established and may never become the hoped-for 'answer' to the land issue. We referred in Section 13.4 to the Italian government's disclosure that Italy is now planning to operationalise Generation IV SMRs within the next decade and that it has also imposed restrictions on new solar power projects, citing concerns that they may threaten food security by using up cultivatable land. India is another country which is seeing 'a revolt against solar' because it can destroy farmers' ability to earn a living as well as reduce agricultural output.

15.5.10 *Issues Specific to Europe*

15.5.10.1 *The French Nuclear Alliance*

Brussels to put nuclear and renewable energies on an equal footing when it comes to funding, research, and project development. The French-led nuclear alliance, which now has 12 EU member states, has noted that nuclear technologies do not benefit from all types of EU funding, such as the European Investment Bank (EIB) and innovation funds. They have pointed out the need for such funding to be applied equally to nuclear. We hope that readers of this section will take note of this and other points we have raised here and elsewhere in the book.

15.5.10.2 *EU Net Zero Act*

In the EU, the Net Zero Industry Act currently does not include nuclear as a key (strategic) technology. As it is clearly likely to be so, the act should be modified to include nuclear.

15.5.10.3 *EU Global Influence: Providing an Alternative to Russian and Chinese Nuclear for Developing Countries*

In our analysis (Section 14.1.2.8), we covered **Rosatom reactors and fuel supply** and referred to the contract that Rosatom recently signed with Uzbekistan, with plans to also sign agreements in Myanmar and India. It is crucial to ensure that European companies receive full support from Member State governments and the Commission as they try to compete in overseas markets, with the technologies being developed in Europe (or those developed in the US but made in Europe).

15.5.10.4 *Nucleareurope*

This is the Brussels-based trade association for the nuclear energy industry in Europe, with a membership of 15 national nuclear associations. Through these associations, nucleareurope represents **nearly 3,000 European companies working in the industry**. In November 2024, it issued a press release calling on the following actions by the Commission to speed up the decarbonisation of the European economy. Nucleareurope calls on the next EU Commission (now the current Commission) to:

- ensure a level playing field between net zero energy technologies (such as nuclear and renewables);
- provide equal support for nuclear under all EU funds, EIB financing, and the upcoming Clean Energy Investment Strategy and allocate more funding for fission activities within the Euratom budget;
- support the creation of an IPCEI for nuclear (including relevant state aid);
- encourage further development of the European nuclear supply chain (e.g. fuel cycle) to reduce dependence and strengthen Europe's autonomy;
- produce an updated Nuclear Illustrative Programme (PINC), which includes the latest forecasts and concrete recommendations to support European nuclear projects;
- enable coordinated actions with other sectors to tackle both basic and STEM skill needs.

15.5.10.5 *World Nuclear Association*

In January 2007, the World Nuclear Association established the Cooperation in Reactor Design Evaluation and Licensing (CORDEL) working group with the aim of stimulating a dialogue between the nuclear industry and nuclear regulators on the benefits and means of achieving a worldwide convergence of reactor safety standards for reactor designs. We hope this body will become much more active in furthering its objectives.

Although this organisation is not specific to Europe, we hope that European companies and the Commission will help ensure that this association can contribute to the objectives set by the European Industrial Alliance for SMRs.

15.6 Economies of Manufacturing Scale to Overcome Diseconomies of Small Size

As previously mentioned, there are currently some 80 modern nuclear designs at various points on the TRL scale. With so many companies trying to develop new designs, nuclear safety regulators are unlikely to have the necessary resources to evaluate all the designs submitted for certification.

The uncertainty surrounding which technology/technologies is/are likely to prevail could impact the time to market for every new design. Large-scale and global markets are essential to justify the investment in large-scale factory fabrication facilities, as fragmented regulations and excessive competition hinder the ability of a small number of manufacturers and technologies to achieve the necessary production scale, thereby establishing a selling price that will accelerate the energy transition. The scale of the required energy transition is such that, finally, an oligopoly comprising just a handful of companies may emerge (and perhaps should be encouraged), as has happened with similar multi-trillion-dollar endeavours. Therefore, a degree of well-informed regulatory steering would be helpful in reducing the time required to reach the oligopoly stage from the present **'let a thousand flowers bloom'** stage, which is both wasteful of talent and slow.

15.6.1 *European Industrial Alliance on Small Modular Reactors*

A potentially key initiative by the **European Commission** in this area has been its encouragement of the creation of the European Industrial Alliance on Small Modular Reactors (SMRs and AMRs). Under the heading 'Why Small Modular Reactors Matter', the Commission issued a statement in early 2024, expressing its view on the importance of SMRs/AMRs:

> Accelerating SMR development in the EU demands a robust and efficient nuclear supply chain. Collaboration among stakeholders, including project promoters, financial institutions, regulators, researchers, training centres, civil society organisations, and policymakers is essential. An alliance creates a European platform to enable and streamline such a collaboration.

The Commission document (https://ec.europa.eu/docsroom/documents/57934) on the Terms of Reference for the Alliance confirms that its work will focus on:

- reinforcing the European nuclear supply chain by identifying and addressing gaps and promoting close cooperation among stakeholders;
- supporting SMR project promoters to develop, demonstrate, and deploy their projects in the EU market and beyond by identifying and removing barriers and working together with financial investors, industry, and local authorities;
- establishing ways to inform and engage potential industrial users of SMRs, such as energy-intensive industries, hydrogen producers, and urban districts;
- facilitating and coordinating projects to address future research and innovation needs, particularly for AMRs under the Euratom Research and Training Programme, and by establishing a nuclear skills academy with the support of the Net-Zero Industry Platform;
- promoting public engagement on SMRs by working with relevant civil society organisations and NGOs, as well as cooperating with relevant international bodies to help European SMR projects reach international markets.

Later, in 2024, the Alliance started to elaborate a strategic action plan **to identify**, for example:

- **technically mature and commercially viable SMR technologies that could be supported under the Alliance**;
- potential gaps and solutions in the European supply chain for SMRs (including fuel and raw materials);
- investment barriers, funding opportunities, and new financial blending options to support SMR development;
- future needs for research on SMRs and AMRs and identify existing skills gaps along the supply chain to be addressed under the Euratom Research and Training Programme and at the national level.

In October 2024, the Alliance announced the first set of Project Working Groups (PWGs). These nine projects were selected from a competitive pool of 22 proposals, with another opportunity to apply for selection in Q2 2025. The nine project groups are as follows:

1. **EU-SMR-LFR** (Ansaldo Nucleare, SCK CEN, ENEA, RATEN)
2. **CityHeat** (Calogena, Steady Energy)
3. **Project Quantum** (Last Energy)
4. **European LFR AS** (Newcleo)
5. **NUWARD** (EDF)
6. **European BWRX-300 SMR** (ORLEN Synthos Green Energy)
7. **Rolls-Royce SMR** (Rolls-Royce SMR)
8. **NuScale VOYGR™ SMR** (RoPower Nuclear)
9. **Thorizon One** (THORIZON)

The first group on the list above is now managed under a consortium agreement and is seamlessly integrated into the European Union's energy strategy. The consortium, known as **Eagles 300** since June 2025, is working on a 350 MWe lead-cooled fast neutron reactor (LFR). This work is supported by the European Union through the Sustainable Nuclear Energy Technology Platform (SNETP) and the European Sustainable Nuclear Industrial Initiative (ESNII) for the development of Generation IV nuclear systems.

15.6.2 *Great British Nuclear (GBN)*

Great British Nuclear (GBN) is a state-owned company in the UK. It was established to boost UK energy security, reduce dependence on volatile fossil fuel imports, and deliver the government's priority of economic growth. Its aim is speed up the development and commissioning of SMRs and AMRs. Besides launching GBN, the government has launched a competition to award grant funding, totalling up to £157 million. Up to £77.1 million of funding is allocated for companies to enable the acceleration of advanced nuclear development in the UK and support the entry of advanced nuclear designs into UK regulation to maximise the likelihood of constructing small and advanced modular reactors in the next Parliament. Up to £58 million is allocated for the further development and design of AMRs and next-generation fuels. In November 2024, four companies – GE Hitachi, Holtec, Rolls-Royce SMR, and Westinghouse – were shortlisted for support following two rounds of assessment by GBN, the government's expert nuclear delivery body.

15.6.3 *International Bank for Nuclear Infrastructure (IBNI): The Geopolitical Angle*

Of the approximately 50 reactors currently under construction worldwide, the majority are funded by the Chinese or Russian governments. Russia, through Rosatom, is currently financing projects in Bangladesh, Egypt, and Turkey. Russian reactors are said to cost less than half of the selling price that American or European companies would have to charge and are around 60% of the price of South Korean reactors. Rosatom also offers a 'turnkey' service providing technology, construction, maintenance, and fuel, as well as financing. Therefore, unsurprisingly, most emerging countries choose Rosatom products, despite its business model of offering the reactor at a low price and recovering capital costs later through high-priced fuel sales (as with printers and cartridges) and through service and repair costs. If European countries, the US, South Korea, Japan, and other nations with an interest in nuclear power wish to compete in the global market against the current market leaders, they could benefit from a new multilateral institution similar in structure to the World Bank but exclusively dedicated to nuclear. Its role would be to offer risk reduction mechanisms and funding for the purchase of reactors. The IBNI is intended to be just such a bank. The IBNI Implementation Organisation has now been set up, and the targeted establishment date is 2026. Some of the types of financing and support tools they would intend to offer are as follows:

- emerging technologies supplemental grant and concessionary financing;
- committed equity financing of projects and companies through supply chains;

- committed senior and/or subordinate debt financing;
- hedging contracts: interest rates, inflation, and currency risks;
- completion risk insurance and committed contingent debt and/or equity financing (in case of delays and overruns);
- commercial guarantees and sovereign risk guarantees.

15.7 Nuclear Fusion: Iter, Tokamak Energy, and Helion

Although unrelated to Gen IV reactors or indeed with fission reactors, it would be negligent to conclude this chapter on nuclear without providing a brief discussion of nuclear fusion technology. There is currently no fusion reactor producing power on a commercial scale. Most forecasts indicate that commercial operation is unlikely to commence before 2035. This nuclear technology does not involve highly radioactive materials. Fusion is the process which is continually taking place in the sun, generating all the energy we harness in solar, wind, and concentrated solar power (CSP), and now humanity is trying to recreate this process on Earth. Fusion has the potential to produce four times more energy per kilogram of fuel than fission (used in current nuclear power plants) and nearly four million times more energy per kilogram than combustion of oil or coal.

Fusion, unlike fission, involves two lighter atomic nuclei combining to form a heavier nucleus. This liberates some of the binding energy in the atoms, thereby releasing energy in the form of heat. The first attempt to generate power from fusion was a UK experiment in 1957, known as the Zero Energy Thermonuclear Assembly (ZETA). The scientists working on the project believed that they had achieved nuclear fusion, but sadly, it was soon revealed that the measuring instruments were at fault and there had been, in fact, no energy output produced in excess of the input energy. **ITER** ('the way' in Latin) is one of the most ambitious energy projects of today. It involves a collaboration amongst 35 countries. The **tokamak** is a magnetic fusion device which is hoped to demonstrate the feasibility of generating power from the atom while avoiding the problem of nuclear waste.

Today, several companies are actively attempting to achieve commercial success with fusion. One of these is a joint UK–US venture, **Tokamak Energy**, which is a private sector spin-off from the UK Atomic Energy Authority (UKAEA). Another venture, **Helion**, is expected to begin operations by 2028, generating approximately 50 MW of electric power. The company claims that its **Trenta** fusion prototypes will be the first privately funded fusion devices to achieve commerciality. The company has already raised more than US$1 billion from private investors, including Sam Altman of OpenAI and the steel company Nucor. Its current valuation, based on its last fundraising, is over £5 billion. Helion already has a contract to supply power to Microsoft by 2028 (which we assume to be non-binding) and to supply a fusion reactor to Nucor (a steel company) at a future date.

Part VI

Innovation: Heat and Power from Concentrating Solar Power and Geothermal Heat from Decay of Natural Radioactivity

Chapter 16

Solar Heat from Concentrating Solar Power for Industrial Heat, Electricity, and Energy Storage: DEWA, Midelt, and NREL

16.1 Heat as an Energy Storage Medium: The Carnot Battery

While electric power itself cannot be stored, heat energy can be. Electric power has traditionally been generated from heat energy derived from the combustion of fossil fuels – oil, gas, and coal – and from the fission reaction in nuclear reactors. Thus, heat is a source of energy that can be converted to electric power by heating water in a steam generator, which then drives a turbogenerator set. This process can also operate in reverse: once electric power is generated, it can be passed through heating elements and thus converted back to heat.

If that heat can be stored, then it can be produced at one point in time and used to produce power at another point in time (representing a separation in time between the production and use of the heat to generate power). This is an extremely valuable attribute since it means that, while variable sources of direct power generation (such as solar and wind) are not producing power, if adequate heat storage is available, then that stored heat can be used later to generate power when those variable sources are not generating. It is thus one alternative to battery storage and long-duration energy storage, which we cover in Chapter 11. Heat storage is also a key feature of the TerraPower Natrium reactor, which we cover in Chapter 13.

Radiation from the Sun is the source of all natural forms of energy on Earth (excluding nuclear). Radiation striking the Earth's surface has three main components: visible light (around 45% of the Sun's radiation), infrared (around 50–55% of the radiation), and ultraviolet (which gives our skin a tan and accounts for about 5–10%). *Photovoltaic solar*, generally known as PV, converts visible light into electric power. This can subsequently indirectly generate heat by passing the electrical power generated through an electric resistance material, as in an electric fire or underfloor heating mat. The other type of solar is *solar thermal*, where longer-wavelength solar electromagnetic radiation (infrared) is converted directly into heat when it impacts a surface which absorbs it.

16.2 The DEWA Project in Dubai

A good example of a project using the heat from the Sun's infrared radiation is ACWA Power's 700 MW concentrating solar power (CSP) project for DEWA in Dubai, which was the world's first facility to offer a 35-year-long, day and night, round-the-clock solar contract. It also had a record-low price for solar with storage – of just 7.3 US cents per kWh. It is an example of an increasingly common request for solar combined with storage to enable the provision of power throughout the 24 hours of the day. This facility uses parabolic mirrors shining on a collector tube carrying a heat transfer fluid which transfers heat to a storage facility or directly to a generating unit. The ability to offer a long (35-year) contract at a fixed price arises from the fact that 'fuel' cost, i.e. sunshine, is zero compared with the highly variable cost of natural gas or coal, which, in the case of natural gas, rose by more than five times in 2021 and 2022 as a result of the Russian invasion of Ukraine.

Thus, a contract of this type can offer a longer power purchase agreement (PPA) compared with fossil fuel generation, as the key risk of natural gas generation is the volatility of gas prices. This enables the developer of the facility to amortise its costs over a longer period, which enables them to offer a lower cost of power in any PPA contract. ACWA has built a similar storage-equipped plant in Dubai, with one of the lowest tariffs in the industry.

An alternative design of CSP involves using a tower surrounded by mirrors focused on the top of the tower, where the molten salt storage medium passes through and is heated.

16.3 Midelt Project in Morocco – Hybrid Generation + Storage: CSP + PV + Molten Salts

Until recently, thermal storage tanks were connected only to CSP. However, in Figure 16.1, which shows the operation of the hybrid solar project at Midelt in Morocco, it can be seen that this plant is *also* connected to solar PV! So, while the CSP works by converting sunlight into heat energy (but at a temperature of only 400°C), PV works by converting sunlight into power. This power can then be passed through resistance heaters (similar to those in an electric kettle) to convert the power into heat, which then heats the molten salt to a higher temperature than achieved with CSP alone. Using PV in this way for generating energy for subsequent storage has only recently become economically viable, with the plunge in solar PV costs. It is less likely that CSP costs will fall to anything like the same extent. But by adding solar PV, it becomes possible to raise the storage temperature to 565°C, which means greater heat storage and hence greater power output and also higher thermodynamic efficiency. Figure 16.1[1] and the text that follows explain this system further.

MASEN, Morocco's clean energy agency, is executing a national renewable policy with an eye on how a future grid can operate reliably with dispatchable firm electricity from 100% renewables. Midelt in Morocco will be the first solar project in the world with heat storage from both PV and CSP. Midelt's first-of-a-kind hybrid solar and shared storage project will deliver dispatchable solar at 7 US cents per kWh. Morocco was paying up to 30 US cents per kWh for fuel oil electricity. The nation of 36 million needs five hours of power after dark and is on track to meet its target of having 52% renewable electricity by 2030.

To date, when PV solar projects have included storage, they have only been paired with batteries. But at Midelt, for the first time, the solar energy from not only the CSP plant but also the PV plant will be stored

---

[1] Kraemer (2020) https://www.solarpaces.org/morocco-pioneers-pv-to-thermal-storage-at-800-mw-midelt-csp-project/. Adapted from @DLR/Bauer: Experimental and Numerical Investigation of a 4 MWh High Temperature Molten Salt Thermocline Storage System with Filler.

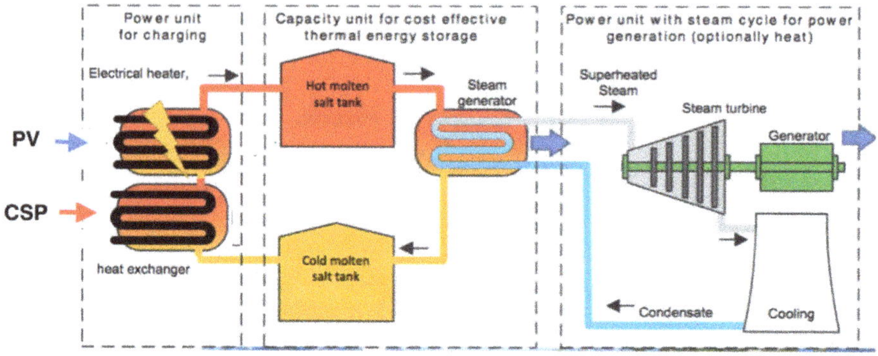

Figure 16.1. The Midelt project using both solar PV and CSP. How colocated PV and CSP would store solar energy in the molten salts thermal energy storage of a CSP plant.

in the thermal energy storage of the CSP portion of the project. CSP projects built today routinely include 10 or more hours of thermal energy storage in tanks of low-cost molten salts. Originally, the plan for Midelt was to include PV for daytime, with a small battery backup to cover transient daytime needs, and for the CSP to include five hours of thermal storage to cover the evening peak. But as PV prices dropped, the storage plan changed. ‘If we limit the number of storage hours, batteries can win. But in our studies we find we also need CSP to complete the mix, as we need five hours after dark’, Abderrahim Jamrani, MASEN technical director, said.

This idea of co-locating PV and CSP and sharing the CSP thermal storage is one that Schmitz (TSK Flagsol technical director) believes will be widely applicable as energy grids become more saturated with renewables, not just Morocco’s, and as a result, more regulators move from a lowest-cost to a ‘best-fit’ procurement approach. ‘If you are operating a grid which is still mostly based on carbon, then merely replacing as many kilowatt hours as you can with green energy is what you do; you just dump the PV in as much as you can. But at a certain point you will reach a level where there is just too much PV in the grid during sunny hours. So just feed it into the already existing tanks with the CSP’.[2]

[2]Kraemer (2020) https://www.solarpaces.org/morocco-pioneers-pv-to-thermal-storage-at-800-mw-midelt-csp-project/.

16.4 NREL

The US National Renewable Energy Laboratory (NREL) is in the late stages of prototype testing a new thermal energy storage technology that uses inexpensive silica sand as a storage medium. It is a low-cost, build-anywhere particle thermal energy storage technology in which silica particles are fed through an array of electric resistive heating elements to heat them to 1200°C. The heated particles are then gravity-fed into insulated concrete silos for thermal energy storage. The baseline system is designed for economical storage of up to 26,000 MWh of thermal energy! As it is a modular design, storage capacity can be scaled up or down relatively easily. This heating process takes place when power prices are low, normally when there is an excess of solar power available to the grid. Note that this system is drawing power from the grid, not from its own generating capacity. Thus, when the imbalance of total supply and demand (excess supply) would be overcome by selling solar to heat storage systems at a low cost rather than paying curtailment fees to solar producers.

When power is needed and therefore power prices are high, the hot particles are gravity-fed through a heat exchanger, heating and pressurising a working gas (which could be helium or steam) to drive the generators that produce electricity for the grid. The system discharges during periods of high electricity demand and when limited solar photovoltaic or wind power is available, such as early in the morning and evening, during dinner preparation, and when TVs are on. Once discharged, the spent, cold particles are once again fed into insulated silos for storage until conditions (and economics) are appropriate again for charging.

See also Chapter 19 on the use of CSP for the manufacture of synthetic liquid hydrocarbon fuels (or how to power aeroplanes on sunlight).

Chapter 17

Geothermal Heat: Earth's Heat as a Source of Carbon-Free Heat and Dispatchable CFE Power

The most comprehensive study of geological heat has been produced by the US Department of Energy in a report entitled 'Geovision – Harnessing the heat beneath our feet',[1] which concludes that:

> Geothermal is America's untapped energy giant. Geothermal energy is a renewable and diverse solution for the United States – providing reliable and flexible electricity generation and delivering unique technology solutions to America's heating and cooling demands. Geothermal resources can be found nationwide, are "always on," [firm power] and represent vast domestic energy potential. We could increase geothermal power generation nearly 26-fold from today, representing 60 gigawatts-electrical (GWe) of always-on, flexible electricity-generation capacity by 2050. This capacity would make up 3.7% of total U.S. installed capacity in 2050 and would produce 8.5% of all U.S. electricity generation.

The Earth is essentially a fireball with a central temperature of around 6,000 degrees (comparable to the Sun), which loses heat gradually at the surface but whose heat is replenished continually by natural nuclear

[1] http://www.osti.gov/scitech and https://www.energy.gov/sites/default/files/2019/06/f63/2-GeoVision-Chap2-opt.pdf.

processes (such as those occurring in nuclear reactors) in the core of the Earth. The Earth's temperature rises with increasing depth, and at its core it is similar to that of the Sun.

To evaluate opportunities for the success of geothermal energy, the US Department of Energy's Geothermal Technologies Office (GTO) initiated the GeoVision analysis. This technical analysis evaluated future geothermal deployment opportunities based on three core objectives:

- increased access to geothermal resources,
- reduced costs and improved economics for geothermal projects,
- improved education and outreach about geothermal energy.

It should be noted that for power generation, deep drilling is generally required, along with high-temperature water to generate steam, whereas district heating can be provided by shallower wells (but at sufficient depth to tap geothermal heat) and lower-temperature water. In addition, domestic heating can be provided by ground source heat pumps, which are typically no more than a metre underground. They are, in fact, operating mainly through incident solar heat on the ground (not geothermal heat), which penetrates to approximately this depth and maintains the ground at a relatively constant temperature.

The International Geothermal Association (IGA)

The role of the Association is to encourage, facilitate and promote the development of geothermal resources, as well as innovative research in geothermal technologies, through visible and integrated positioning and representation of geothermal power, heat, geo-exchange, cooling and other uses of geothermal resources.

The European Geothermal Energy Council (EGEC)

EGEC is a non-profit international organisation founded in 1998 to promote the European geothermal industry and enable its development both in Europe and worldwide by shaping policy, improving business conditions, and driving more research and development.

Based in Brussels, it provides a link between the industry and European institutions. More than 120 members from 28 countries, including developers, equipment manufacturers, electricity providers, national associations, consultants, research centres, geological surveys, and public authorities, make EGEC a unique network, uniting and representing the entire geothermal sector.

Innovative technologies in the field of geothermal energy include new drilling technologies, new exploration methods, closed-loop systems, smart systems, and 3D models.

Advanced plants and technologies include:

> **Ultra-deep geothermal systems, enhanced geothermal systems, and supercritical geothermal systems.** These enable access to resources, which would otherwise remain unexploited.

An example of a company that is advanced in using EGSs is the US company Fervo.

> **Fervo Energy** is commercialising a geothermal technology that could expand the regions of the world, tapping into underground heat by ‘fracking’, i.e. creating or widening cracks under the surface to allow water to more easily circulate underground.

EGS uses techniques not dissimilar to those used in fracking for oil and natural gas. There are therefore risks of setting off small ‘earth-quakes’. But this may be a price to pay for the benefits of dispatchable power and an increase in the portfolio of technologies available to achieve the energy transition. The temperature of the water coming to the surface will depend on geological conditions.

Part VII

Innovation: Other Technologies to Enable the Energy Transition

A number of the technologies covered in this part of the text require the use of innovative materials to enable their effective application. These include catalysts, which are materials capable of increasing the rate of a chemical reaction without undergoing any permanent chemical change, i.e. they are not used up in a process. They also include ceramic materials, which have a number of uses. Catalysts are provided by several companies, one of which is Johnson Matthey. It describes itself as a British multinational speciality chemicals and sustainable technologies company. Its strapline is 'Our purpose is to catalyse the net zero transition'. By doing this, the company is moving away from its previous key product, which was platinum as the catalyst in catalytic converters for internal combustion engines (ICEs), with battery and hydrogen energy sources set to take over in the production of vehicles.

Another company which specialises in products that enable several of the technologies discussed here is NGK, a Japanese company active in Europe (NGK Europe). Its strapline is 'Unique ceramic technology for a *future of coexistence between environment and humanity'. It is active in energy and climate tech, the H_2 economy, the Internet of Things, micro-electronics and semiconductors, and sustainability. Its 'New Value' business is transforming from an automotive industry business to a carbon-neutral and digital society business, as the production of ceramic components such as spark plugs for ICEs will diminish over time as the energy transition advances.*

One use of ceramic-based membranes is in carbon capture and storage, including for direct air capture. NGK is also developing materials that enable more efficient production and transportation of hydrogen. The energy transition also requires new processes enabling the manufacture of materials but with a lower need for the use of heat and electricity (to reduce costs). In Europe, the importance of this arises from the fact that many constituent countries have among the highest energy costs in the world, which can make a high energy input company uneconomic.

Chapter 18

Green Hydrogen (H_2): Manufacture and Applications

18.1 Green Hydrogen Value Chain

> *The World Economic Forum (WEF) identifies green hydrogen as one of the top technologies to have emerged in 2020 and as a technology with the greatest potential to transform industry and society. Green hydrogen is not something new, but its use is increasing and at an increasing rate, and hence it is regarded as a top emerging technology.*

Before we explore the importance of green hydrogen, we note that in the process of decarbonising the economy, direct electrification of industrial processes, transport, and heating rather than using hydrogen is the most technically efficient way to meet most energy needs. Interestingly, only 20% of total final energy demand is generated by electricity. However, this could reach 60–70% by 2050. For some sectors, direct electrification will likely remain impossible or uneconomic for many decades. Unfortunately, 2024 has not been kind to many hydrogen investors. The share prices of many electrolyser companies fell sharply in 2024 due to regulatory difficulties and a lack of eager offtakers. This is in contrast to a number of advanced nuclear technology reactor companies whose share prices have risen substantially throughout 2024.

It is in those uses and others where hydrogen may offer considerable advantages over current battery technology that hydrogen or its derivatives are likely to become sources of energy in a decarbonised economy.

As well as being a source of power generation from energy storage (see Chapter 5), it can also be used to provide heat for domestic and commercial space heating and for industrial processes. But in practice, it should be noted that given that carbon-free power to manufacture 'green' hydrogen will be in short supply for more than a decade, initially the most valuable uses of H_2 are likely to be in decarbonising fertiliser production (from natural gas) and also in decarbonising the H_2 used in fossil fuel manufacture.

Hydrogen, along with electricity, may be a solution in some 'hard-to-decarbonise' sectors, i.e. not easily electrified directly. For example, it can be used directly or in the form of derived fuels such as ammonia and synthetic liquid hydrocarbon fuels (also called synfuels, e-fuels, or drop-in fuels). In steel and long-distance shipping, for instance, hydrogen's (or ammonia's) potential new role is increasingly certain. For the manufacture of fertiliser, it is hoped that green hydrogen will displace hydrogen made from natural gas, with its consequential emissions of CO_2. In a number of other sectors, it is the leading decarbonisation option.

Today is not the first time we have heard that hydrogen is the fuel of the future, and it has certainly come 'into vogue' in the past few years as one of the elements of any successful decarbonisation of a country. But the idea behind the 'hydrogen economy' is, in fact, an old one, developed back in the 1970s, when it involved hydrogen being used mainly to power all vehicles in cities in order to reduce pollution, i.e. not for reasons of climate change mitigation. While in theory this was achievable, the energy majors and the major automobile companies, certainly in the US, were not in favour of such a change. And without their support, it was always unlikely the idea could succeed. But in practice, it was the low cost of carbon-based fuels along with the inertia of car manufacturers which ensured the continued use of the internal combustion engine (ICE).

While hydrogen is back 'in vogue' due to the need for decarbonisation of the economy, in fact it has been manufactured in large volumes for many decades and is continuing to be so. However, the starting point for such manufacture is fossil fuel, in particular natural gas (methane), though in China it is mostly coal, which can be converted to hydrogen. The hydrogen produced by this process is known as 'grey'. It uses a process called steam methane reforming (SMR). This results in a high level of CO_2 emissions because natural gas (methane) has the chemical formula CH_4 and therefore the carbon atom has to be broken away from the hydrogen atoms (creating CO_2), which is then released into the atmosphere.

Producing natural gas 'at the well head' can also create large 'fugitive' methane emissions.

Today, there are many so-called 'colours' of hydrogen. Grey hydrogen is hydrogen made from natural gas without capturing the CO_2 generated during that process. Green hydrogen is manufactured from water (H_2O) as the 'raw material', through which an electric current is passed to separate the two elements, although heat can also be used, as we noted later. Blue hydrogen is made from natural gas but with carbon capture and storage. While the cost of green hydrogen has, until very recently, been higher than grey, when the price of natural gas rose sharply due to Russia's invasion of Ukraine in some cases, it was lower. In addition, technological developments currently under way and the likelihood of carbon taxes being extended considerably in scope could make green hydrogen very economic.

The production of green hydrogen requires a large amount of electric power (or heat energy) to break the molecular bonds of water in order to separate the two gases – hydrogen and oxygen – which, in chemical combination, produce water (H_2O). Thus, when we are considering the cost of manufacturing green hydrogen and attempting to lower it substantially, the cost of power is a central factor.

Figure 18.1[1] provides a useful representation of the links in the value chain for H_2 from green power sources of energy to its applications. This value chain involves, firstly, the provision of the large amounts of green power required. The majority of operating costs (OPEX) for hydrogen production arise from the cost of power input. The capital investment (CAPEX) required for both power generation and electrolysis is also high. Reducing the cost of H_2 requires both lower OPEX and lower CAPEX.

The graphic shows solar, wind, and nuclear as the three sources of clean power. The price at which the power generated can be delivered to the hydrogen production facility depends, in part, on the distance between the power source and the electrolyser, as this gives rise to transmission charges. Also, whether it is 'dispatchable' or variable renewable power and if variable, whether there is energy storage to enable 24/7 operation of the electrolysers. These are much more efficient when operating 24/7.

---

[1]Zixuan Luo *et al.* Cost-Economic analysis of Hydrogen for China's Fuel Cell Transportation Field *Energy* December 2020 https://www.mdpi.com/1996-1073/13/24/6522.

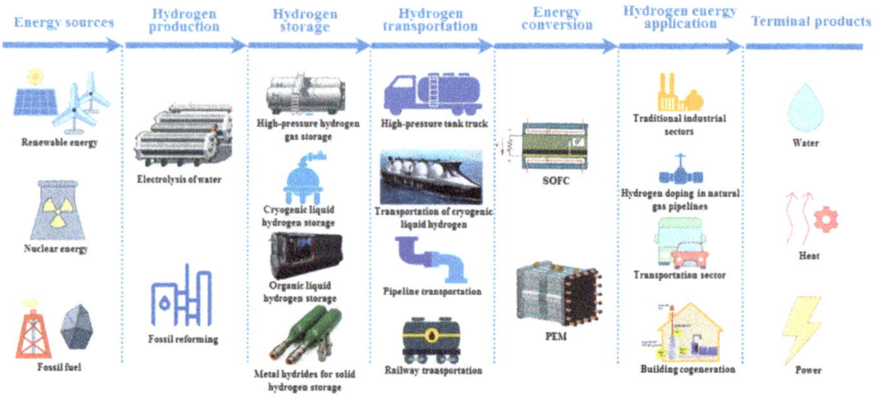

Figure 18.1. Links from power generation to applications for hydrogen.

Columns 1 and 2: Energy Sources (Generation of Power) and Connection to Hydrogen Production Facility

Regarding transmission costs, both wind and solar power have the drawback that good sunshine and strong winds are often found at a considerable distance from users, including those who might wish to install electrolysers. This is quite different from traditional power, which was once generated within each city and today is still often located near users. Thus, the 'delivered' price of VRE will depend on how the two facilities – power generation and electrolysis – are connected to each other. If power flows through the grid, there will be transmission charges, but if the two facilities are directly connected, it will not.

The construction of a wind or solar project capable of supplying the energy necessary for the manufacture of hydrogen is now relatively straightforward. Storage facilities are also relatively straightforward, but there is still not as much experience of constructing large-scale storage. Traditionally, investment in renewable power is undertaken with the intention of supplying power to the grid. This is a well-established business model. If, however, it is to finance hydrogen production, there is no such well-established business model, and therefore project financing might be more challenging.

The fact that there are few established business models for large-scale independents in this highly capital-intensive business (developing a power project alongside an electrolysis plant) explains why such projects are generally undertaken on the balance sheet of a large corporate and not as independent project financings. The logical type of corporate would be an energy or power company or utility (such as Ørsted) or a company which uses hydrogen in large quantities, such as oil refineries. This should greatly reduce the cost of capital compared with project financing, where lenders and investors would have a single, undiversified risk (including some technology risk), but instead places the risk on the corporate's larger and more diversified balance sheet.

Columns 2 and 5: Hydrogen Production from Power (Electrolysis) and Power Production from Energy (Fuel Cell)

The first key technology is the electrolyser. This produces hydrogen from water and electric power or heat. The original technology is the alkaline electrolyser. Somewhat more recently, the proton exchange membrane (PEM) technology has been developed and remains a highly successful alternative. The most recent technology, the high-temperature solid oxide electrolyser cell (SOEC), is examined in more detail in the section on 'electrofuels'. For those uses which require hydrogen to be converted to power, the other key technology is the fuel cell, which involves the opposite (reverse) process to that of electrolysis. It takes in hydrogen (or natural gas) and converts it to heat and power.

The technology of both the alkaline and PEM electrolysers is well established. The solid oxide cell (SOC) is a newer technology which forms the basis of both the SOEC electrolyser and the solid oxide fuel cell (SOFC). Although available on the market, these are relatively recent technologies and, like most technologies, are likely to be continually improved. They have higher efficiency in the conversion process (around 30% more efficient), and some models may be operated reversibly, that is, switching between operating as an electrolyser (SOEC) and operating as a fuel cell (SOFC). Such cells, known as reversible solid oxide fuel cells (RSOFCs), can perform a valuable function on a grid by arbitraging between use as an electrolyser when there is surplus (low-cost) power available, then switching to generation for the grid when power prices are high.

By alternating between SOFC and SOEC modes, RSOFCs can provide 'load shifting' services for the energy system. Solid oxide devices also operate at between 600°C and 850°C, a much higher temperature than other types of electrolyser, which operate at between 60°C and

160°C. This results in considerable efficiency gains compared with the other two main types of electrolysers. They can also be manufactured without the use of scarce and expensive chemical materials, which some other technologies require.

Columns 3 and 4: Hydrogen Storage and Transportation
An important part of the infrastructure for hydrogen, if it is to become a major contributor to a country's energy needs, is storage and transport. Companies in this sector of the value chain include SNAM, Lindt, and Air Products. While SNAM (originally part of ENI) remains a major natural gas distributor with the largest natural gas transport pipeline network in Europe, Air Products and Linde are industrial gas suppliers of oxygen, nitrogen, argon, helium, and hydrogen, as well as medical and speciality gases for welding and cutting.

18.1.1 *Financing Hydrogen Electrolyser and Fuel Cell Companies*

A key feature of many electrolyser and fuel cell companies is that they are relatively capital-light and seek to avoid debt, which is how larger companies obtain much of their financing. One way in which they achieve this is by licensing their technology to much larger companies. They are very different from utilities, which are the most highly indebted of all corporates, along with infrastructure companies (such as airports). Utilities are highly leveraged, with up to 80% debt financing. This high debt load is possible because such companies are expected to provide their investors with a steady, low-risk income stream from a low-volatility asset. They are not expected to be 'growth stocks'.

The types of investors in companies manufacturing electrolysers and fuel cells are quite different from those investing in utilities. Investors in electrolysers and fuel cells are not seeking regular, steady income such as utilities generally offer but rather large capital gains at the point of a dramatic improvement in technology and scale of operation, which will lead to much lower costs and selling prices, and consequently a dramatic increase in sales and, correspondingly, in profit and share price. In this characteristic, they are more like 'tech' companies such as Amazon and Google. However, this same investor expectation can lead to sharp price declines when market conditions change, as we note later. Also, because many of them are relatively new with relatively new technologies, many of which are still being developed and hence relatively risky, they cannot easily support debt financing. Hence, in their early stages at least, they will be almost 100% equity-financed.

18.1.2 *Sparc Technologies: Hydrogen and Photocatalytic Water Splitting (PWS)*[2]

This is an alternative method of producing renewable green hydrogen which avoids the need to convert solar energy into electricity to split water via electrolysis (direct solar-to-hydrogen technology). In a similar approach to that used in concentrating solar power (CSP) to produce heat, PWS employs a highly specialised photocatalyst which triggers the dissociation of water into hydrogen and oxygen. Sparc Hydrogen aims to commercialise PWS to produce green hydrogen at a large scale.

The potential advantage of Sparc technology is that it obviates the need to generate electrical power from sunlight and instead uses the whole sunlight spectrum to provide the energy required to break down the molecular bonds in water. The consequence is potentially lower capital investment requirements and higher efficiency by supplying the required energy via a CSP source rather than utilising electricity directly.

Sparc Hydrogen is a joint venture between Sparc Technologies, the University of Adelaide and Fortescue Limited. It has now commenced front-end engineering and design (FEED) for a global, first-of-its-kind pilot plant planned to be located at the University of Adelaide. The decision to commence FEED, obtain site approvals, and procure long-lead equipment is on the back of several key milestones achieved over recent months with respect to the pilot plant, including securing an in-principle agreement from the University of Adelaide, progressing pilot-scale reactor design and engineering, and the signing of a collaboration framework agreement with Shinshu University in Japan, a leading developer of photocatalyst materials. In parallel, work continues in the laboratory to test and optimise Sparc Hydrogen's photocatalytic water-splitting reactor

---

[2] See company website at https://www.sparchydrogen.com/blog.

under a range of conditions, using various high-efficiency photocatalyst materials.

Sparc Technologies is a good example of a company developing an idea which works both in theory and in practice. However, the first question is whether the technology and scale and then whether the unit cost and thus selling price of its hydrogen at the optimal scale of production will be below that of electrolyser H_2 production. The second question is whether a company of this type will suffer from the 'missing middle' phenomenon, leading to the 'valley of death', or whether it will be able to raise the necessary growth funding to enable it to scale. We covered this problem in the Preface and also cover it in Section 2.3.

18.1.3 *Repsol and Photo Electrocatalysis*

Repsol is a 'global multi-energy' company 'striving to become a net zero emissions' company. It is a Spanish based multinational. Repsol is the leading producer and main consumer of hydrogen in Spain, as hydrogen is a key component in the crude oil refining process. It is used in desulphurisation and hydrocracking treatments that improve the performance and environmental quality of the refined fuels.

Repsol is also developing photo-electrocatalysis to produce hydrogen directly from solar energy. The project also involves Enagás and several leading research centres, including the Catalan Institute for Energy Research, the University of Alicante, and the Aragon Hydrogen Foundation. In photo-electrocatalysis, 'the device receives solar radiation directly and with a photoactive material it generates the electrical charges that cause this separation' into hydrogen and oxygen.

18.2 Applications of Hydrogen: Energy Storage, Transport, and Hard to Decarbonise Sectors

There are four categories of use for H_2 – as a **fuel**, as a provider of **heat**, as an **energy storage medium** for time-shifting electricity production, and as a **feedstock** for the manufacturing industry. In the first two sets of uses, hydrogen is acting as an 'energy carrier', i.e. It needs to be converted to usable energy in some way, either by combustion or by passing through a fuel cell. On the other hand, when used as a feedstock, it is the hydrogen molecule itself that is used to create products, or, in the case of steel, to 'reduce' iron ore (iron oxide) by combining with the oxygen molecule in iron oxide to produce water (H_2O), in contrast to reducing iron ore using the carbon content of coal, where the carbon molecule combines with oxygen to produce CO_2, which is dissipated into the atmosphere.

Direct electrification is always likely to be a lower-cost means of achieving an objective than indirect electrification using hydrogen for the simple reason that conversion of power to hydrogen and back to power is a low-efficiency process. Hydrogen has a very low volumetric energy density, meaning that storing and transporting it is difficult. But it does have a very high energy gravimetric density (energy for a given weight of fuel). However, for long-distance aeroplanes, hydrogen would probably not be used directly but would be used to manufacture liquid fuels such as ammonia or e-fuels.

Batteries are the worst possible source of energy in terms of the weight of the 'fuel' to power a plane. Jet fuel has by far the highest energy content, at around 45 MJ/kg, whereas batteries are only around 0.6–0.8 MJ/kg. As a consequence, even a short-haul aeroplane would have to carry several hundred tonnes of batteries, which would be more than the weight of the plane itself.

The *current* major uses of hydrogen (mainly grey hydrogen) are in the manufacture of liquid fuels by the oil majors, e.g. petrol from crude oil (produced via hydrogenation to improve its quality) and the manufacture of nitrogen-based fertilisers. Thus, the initial markets for green hydrogen are likely to be in these two industries but only after the cost of manufacture falls to levels comparable to grey hydrogen or once governments legislate to prohibit grey hydrogen use (or its price rises dramatically as it has) or impose a carbon tax which makes it less economic in use than green hydrogen.

18.2.1 *Energy Storage for Dispatchable Power*

Wind power and photovoltaic (PV) solar are, as we noted earlier, variable power generators. Thus, provision of 'dispatchable power' is also needed in a power system with a high proportion of VRE (as we examined earlier) to 'fill in the gaps' and provide round-the-clock power.

One approach is to use hydrogen, which is in fact stored chemical energy used as an energy carrier. Hydrogen produced by electrolysis using green power can be converted back to power in two ways. The first is to convert it using fuel cells. The other option is to combust it in a conventional gas turbine or internal combustion engine, which turns a generator to produce electric power.

Once the cost of green hydrogen falls from present-day levels of around \$6 per kg to nearer \$2 – which some electrolyser manufacturers believe is achievable – it should become highly competitive for longer-term storage. However, it must still compete with alternative long-term storage solutions such as heat energy and the iron-based flow batteries we covered earlier.

18.2.2 *Transport*

In the European Union, transport is the largest emitter of GHGs (in part because other sectors have reduced their emissions). Electrifying transport, either directly or using H_2, is a key requirement for achieving the decarbonisation of the EU and many other countries.

Most electric vehicles (EVs) currently use batteries as the source of energy rather than hydrogen. Hydrogen vehicles also have quite substantial batteries in the electric drive trains. Power is not directly fed to the motors from the electrolyser but via the battery. The battery is also used to store power produced by regenerative braking. There are also electric vehicles (hybrids) where all the power is produced by a fossil fuel ICE and then passed to a battery from which it is delivered as required to the wheels. An example of this in passenger vehicles is the Toyota hybrid Prius (now also available as a plug-in hybrid, or PHEV). Daimler Trucks, the largest European truck manufacturer, has said that while diesel will dominate sales up to at least 2025, hydrogen could take off sometime between 2027 and 2030 and then rise steeply. In the UK, the government is planning to ban the sale of new diesel and petrol heavy goods vehicles (HGVs) by 2040. At this point, it is possible that heavy trucks will be an important source of demand for hydrogen and for H_2 refuelling stations across the country.

The benefits of using hydrogen as a source of energy for a fuel cell to power an electric vehicle (FCEV) rather than batteries to power an electric vehicle (BEV) come to the fore once large vehicles, heavy trucks and buses are considered or vehicles where a source of electric power for recharging is not readily available (such as earthmovers on building sites or forklift trucks in warehouses). When time is of the essence, hydrogen vehicles can be refuelled as quickly as traditional vehicles, whereas recharging a large vehicle's batteries is a lengthy process or, alternatively, requires very high rates of current flow if adequate power is available from the grid.

18.2.3 *Hard to Decarbonise Sectors: Heat from H_2 for Iron, Steel, Glass, etc.*

Some industrial sectors are much harder to decarbonise than others, yet decarbonisation is critical to achieving climate objectives. These include iron & steel, petrochemicals, aluminium, cement, glass, paper, and fertilizers, which in total are said to be responsible for around 22% of global CO_2 emissions.

18.2.4 *Iron and Steel Sector*

This has long been a sector where emission-free manufacture has been an aim of producers, but it has proven difficult to achieve. The Swedish company H2Green Steel (H2GS), now known as Stegra, is trying to make steel without emitting CO_2. It has stated that 'existing steel production technology leaves the steel industry as Sweden's single largest emitter of carbon dioxide. At the European level, the steel industry accounts for six percent of carbon dioxide emissions, and globally the figure is seven to nine percent'.

Stegra and Hybrit, both based in Sweden, are working towards producing iron and steel with virtually no emissions of greenhouse gases. In both cases, the technology involves 'reducing', i.e. removing the oxygen from iron ore (Fe_2O_3), whereby the oxygen in the ore combines with hydrogen to produce water (H_2O) as a by-product and produces pure iron (Fe). In traditional iron-making using coking coal, the ore is reduced by combining its oxygen with carbon from the coking coal, which produces CO_2 as a by-product. For both startups, the key to manufacturing 'green' steel at a reasonably economic price is to have access to 'green' power at the lowest possible cost. Sweden is fortunate in having a surplus of electricity produced mainly from hydropower. But this power is also wanted by datacentres (Meta currently has a datacentre in Sweden), and battery producers such as Northvolt (now insolvent) and fertiliser manufacturers such as Power2Earth. This latter company is set to establish Sweden's first fossil-free ammonia and mineral fertiliser plant in Luleå. It is hoped to commence production in 2028.

In France, in February 2022, the prime minister announced support from the French government for ArcelorMittal's decarbonisation programme in France, which involves a €1.7 billion investment in its Fos-sur-Mer and Dunkirk sites. This investment will enable a profound transformation of steelmaking in France and a total reduction of close to 40%, or 7.8 million tonnes per year in ArcelorMittal's CO_2 emissions in France by 2030. This transformation will represent a 10% reduction in greenhouse gas emissions from the manufacturing industry in France and put France's steelmaking industry on the path of the Paris Agreement.

In Dunkirk, ArcelorMittal will build a 2.5 million tonne direct reduction of iron (DRI) unit to transform iron ore using hydrogen instead of coal. This DRI will be coupled with an innovative technology electric furnace and completed by an additional electric arc furnace (EAF). Other investments are already underway to continue to increase the proportion of scrap steel used.

18.2.5 *Cement*

Cement manufacture is particularly hard to decarbonise since it is not only the fossil fuel currently used to manufacture it which emits CO_2 but also the inherent nature of the manufacturing process itself, which results in further CO_2 emissions from the chemical process. The cement industry alone is responsible for about at least 25% of all industrial CO_2 emissions. About two-thirds of these result from what is known as calcination, which is the chemical reaction that occurs when limestone is exposed to high temperatures. While the use of hydrogen would eliminate CO_2 emissions from the heating process, it would not eliminate the natural emissions from manufacture.

As cement production accounts for about 8% of all CO_2 emissions, it is a critically important industry to decarbonise. It has been said that if cement production were a country, it would be the world's third-largest emitter. While cement could be decarbonised if carbon capture and storage became available at a much larger scale and at a more economic price, other experiments are focused on absorbing carbon dioxide into cement at the time of making concrete. Carbon Cure is one such company. It has been backed by Microsoft, Amazon (Climate Pledge Fund), and Mitsubishi, but that does not guarantee its commercial success.

18.2.6 *Float Sheet Glass*

Sheet glass has been produced using hydrogen fuel at a plant near Liverpool in the UK, in an achievement that has been touted by developers as a world first and a major milestone in industrial manufacturing's transition towards net-zero emissions. Progressive Energy, one of a number of companies involved in the HyNet project to decarbonise industrial sites in the North-West, announced that hydrogen fuel had successfully replaced natural gas in the manufacture of float glass at a facility owned by Pilkington UK in the town of St Helen's in Merseyside. The achievement is believed to be the first full-scale, 100% hydrogen trial on a float glass line. In October 2021, Pilkington was in danger of having to shut to reduce losses as UK (and global) gas prices rose to unprecedentedly high levels. This shows an advantage of using green hydrogen, which can be supplied on very long-term contracts with no risk of price increases.

Chapter 19

Hydrogen for the Manufacture of Synthetic Liquid Hydrocarbon Fuels: Biofuels and Electrofuels

Transport is a large user of fossil fuels. To date, direct electrification (such as in electric railways or tramways) has been one way of avoiding direct use of fossil fuels, though clearly the power grid is currently not providing carbon-free electricity. But for ships and aircraft, direct electrification is not possible. Alternatives include compressed hydrogen, liquid hydrogen, methanol, ammonia, and batteries. Increasing attention is, however, now focusing on producing liquid synthetic hydrocarbon fuels.

The manufacture of synthetic liquid hydrocarbon fuels such as petrol and diesel for road vehicles and other uses, and sustainable aviation fuel (SAF) as a substitute for kerosene **may (or may not) become one of the major uses of hydrogen**. This is because the hydrogen molecule (H_2) is one of the two components of all hydrocarbon fuels, the other, obviously, being the carbon atom C. H_2 molecules and C atoms can be sourced in two different ways. The first is to use biological material, which comprises both, or, alternatively, to use what is known as power-to-liquids technology, involving the manufacture of hydrogen using electrolysis of water and the collection of CO_2 either from smokestacks or from the air.

Hydrogen-based synthetic fuels (e-fuels) are part of the EU's plans to decarbonise transport. In particular, for aviation and maritime transport, the EU has set minimum quotas for 'renewable fuels of non-biological origin' (RFNBOs) i.e. electrofuels (e-fuels), which include both hydrogen and e-fuels. Indeed, the Renewable Energy Directive (RED) requires 1.2% of all kerosene (aircraft fuel) demand to be met by 2030 through the use of renewable e-kerosene. This is expected to increase to 35% by 2050. EU countries are also required to set a minimum quota for RFNBOs of at least 1% of all fuels used across all transport modes, including road transport.

19.1 Biogenic Fuels: Hydrogen + Carbon from Biological Materials

Biofuel is a contentious area. First, there needs to be certainty that their use is at least carbon neutral. Second, if they come from agricultural crops, there needs to be an understanding of the extent to which they may damage food security by taking land out of agricultural production. The starting point for understanding the use of biological materials for the production of energy and the manufacture of gaseous and liquid fuels is to consider the process through which biomass is created (photosynthesis) and the different sources of biomass that are available from this natural process. Plants produce biomass through photosynthesis, which involves absorbing CO_2 from the air and taking in sunlight and water to produce biomass such as trees. Biomass can be burned directly for heat or converted to renewable liquid and gaseous fuels through various processes, as we consider in the following, but in either case, **the CO_2 taken in by the plants that have made the biomass is then released back into the atmosphere**. Figure 19.1[1] shows the process of photosynthesis and the sources of biomass available for conversion to fuels.

A key question is whether the use of biomass to generate carbon for manufacturing liquid fuels is compatible with the need for 'carbon neutrality'. The answer comes down to the 'length of the carbon cycle' and to 'what happens to the emissions of CO_2 in each case. Figure 19.2[2] tries to answer this question.

[1]US Energy Information Administration 'Biomass explained' https://www.eia.gov/energyexplained/biomass/. *Source* (figure on the left): Adapted from the National Energy Education Project (public domain).

[2]IEA 'Fossil vs biogenic CO_2 emissions' https://www.ieabioenergy.com/iea-publications/faq/woodybiomass/biogenic-co2/.

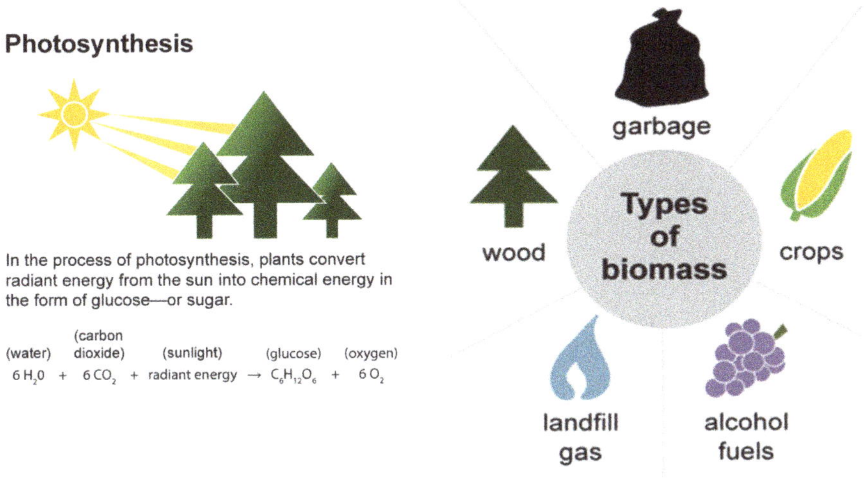

Figure 19.1. Photosynthesis and sources of biomass.

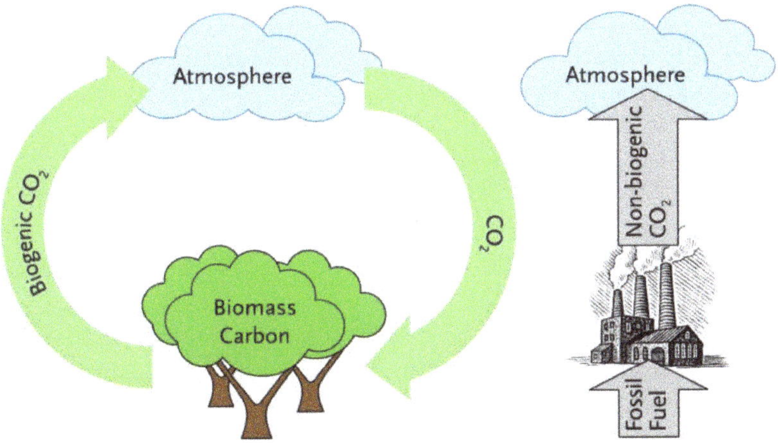

Figure 19.2. Understanding the biogenic carbon cycle.

On the left, the graphic shows the 'biogenic carbon cycle' – in this case, a tree absorbing energy from sunlight and using this energy to convert CO_2 and water into plant growth. If a tree takes 20 years to grow to a point where it becomes economically viable to cut it down and burn it for heat, then all the benefit it has provided by absorbing CO_2 from the atmosphere over that time is lost immediately and will not be recovered. Even if it is replaced one-for-one with a newly planted tree, it will take another 20 years to provide the same benefit. If the tree is not replaced, then burning it will be carbon additive in many people's remaining lifetime and hence will add to atmospheric concentration. Thus, CO_2 capture from trees may not be the answer to removing CO_2 permanently. Another important reason why trees are unlikely to be the answer is that forest fires covering huge acreages of land seem to be becoming more frequent. Once a tree is destroyed by the action of fire, the benefit of planting it is lost.

On the other hand, other types of biological materials such as sweetcorn, have a much shorter growth period (sweetcorn/ corn on the cob only three months), and the cob (core) of the corn can be used as biomass. If replanted each year, then the carbon in the cob when burned either by the farmer or when burned in the form of a liquid fuel is simply recycled over the space of a year, as shown in the first diagram. In the case of the liquid hydrocarbon fuel, the carbon and hydrogen atoms/molecules it contains can be converted into synthetic hydrocarbon fuels (biogenic fuels), as these two elements are the only ones required. When a biogenic fuel is burned in an internal combustion engine (ICE) or jet engine, the carbon is returned to the atmosphere, making the biogenic cycle carbon neutral over a period of less than a year. Carbon in the form of CO_2 is then reabsorbed by the replacement crop over the crop-growing season.

The graphic on the right (above) shows something quite different. In this case, CO_2 which has been locked-up underground for millennia is being released straight into the atmosphere. Even if this were captured and used as a component of synthetic fuel, when that fuel is burned, there is no mechanism to recycle the emissions from combustion, as there is no room on Earth to plant the number of trees that would be required to absorb all the carbon released by vehicles, and the time taken to reabsorb the carbon would, anyway, be very many years. Thus, this process is creating non-biogenic CO_2 emissions, as it allows continued use of fossil fuels in factories and elsewhere.

Such fossil fuel combustion releases carbon that has been locked up in the ground for millions of years. Thus, at the point of combustion, atmospheric CO_2 concentration is increased. Burning biomass, on the other hand, emits carbon that is part of the biogenic carbon cycle, which is much shorter and thus not increasing carbon concentration over the biogenic carbon cycle period, though it is not reducing it either.

Factory chimney emissions as a source of carbon for liquid hydrocarbon e-fuels rather than direct air capture (DAC) is controversial. But it may not necessarily be bad in the short run. Iron and steel works are an example of industrial assets that cannot be converted overnight to using hydrogen as their source of energy and thus are likely to be emitting CO_2 for many years into the future. Thus, when considering any possible benefit to decarbonisation in the case of an e-fuel manufacturer purchasing carbon atoms from a steel company's emissions (so that the cost to the steel company of CCS is either zero or relatively small), we need to consider the alternative – namely that the emissions are simply released straight into the atmosphere, where they clearly add to GHG concentration.

When factory emissions (from the use of recently produced fossil fuels) are used as the source of carbon atoms in the manufacture of an e-fuel, the vehicle using that fuel (for example, an aircraft) is not carbon neutral – it is still adding to atmospheric GHG concentrations. However, the steelworks (if all the CO_2 were removed from its processes and the carbon content reused) would in fact be carbon neutral. Thus, all that is happening is that the original CO_2 from the steelworks is being recycled by being used twice. So, which process is carbon neutral? The steel factory or the vehicle, such as an aircraft which uses synthetic fuel sourced from the steel factory's chimney? The answer is that only one (it hardly matters which) can be considered carbon neutral but certainly not both, as is sometimes claimed.

By using such an e-fuel, atmospheric CO_2 concentration does increase, though in a slightly delayed timeframe from its first round of use in the factory. What has happened, however, is that this roundabout process has saved the additional fossil fuel emissions that the plane would have produced had it not been converted to an e-fuel. Therefore, although the whole process is not carbon neutral, it reduces what would have been a greater amount of emissions if this reuse of carbon had not taken place. However, this only applies in the short run.

In the longer run, to make the whole process carbon neutral in this and other industries, steel producers and others must eliminate their emissions by transitioning to non-fossil fuels, particularly hydrogen. E-fuel manufacturers would then be forced to source their carbon from the air (DAC). At this point, we have overall carbon neutrality, assuming the electric power producing the hydrogen now used by steelworks and the hydrogen used to manufacture the synthetic fuel were both 100% green.

19.2 Electrofuels: Synthetic Liquid Hydrocarbon Fuels from Power-to-Liquids

CO_2 should be captured from the atmosphere in order to reduce atmospheric concentration of this greenhouse gas. However, the current markets for CO_2 (such as in sparkling drinks) are not very large. But the other reason for wishing to capture CO_2 is to obtain carbon atoms to use in combination with hydrogen in the manufacture of liquid hydrocarbon fuels.

The first thing we can say about e-fuels is that hydrogen itself is an e-fuel but in gaseous form. The key input in its manufacture is electric power. However, making hydrogen, which is a gas, from water (hydrogen was traditionally known as water-gas) involves 'forcing' considerable energy in the form of electricity or heat or both into the water. To break the atomic bonds between the hydrogen and the oxygen, which originally converted these two gases into water (H_2O), a huge amount of energy is required. After electrolysis, a proportion of the energy used in the electrolysis of water is then available as a source of energy in the form of hydrogen. Unfortunately, the hydrogen produced has considerably less energy content than the energy used to create it, resulting in an efficiency loss.

The processes by which we extract the CO_2 for manufacturing e-fuels from chimney stacks and even more so from the air to produce carbon atoms for an e-fuel is also highly energy intensive. There are substantial losses in each link in the chain of manufacture – both in the electrolysis of H_2 and CO_2 capture. It is for this reason that it is always said that if an activity can be electrified directly (such as by using overhead power lines for railways), then it should be. In the case of passenger vehicles, it means using battery power, even though this is slightly less efficient than a direct supply of power (such as electric trains using overhead power cables). If neither means of direct electrification is possible, then the activity might use hydrogen or ammonia, and if they are not suitable, then, as a last resort, the activity or process should use e-fuels.

19.2.1 *Haldor Topsoe (Now Simply 'Topsoe')*

Topsoe is a Danish chemicals company founded in 1940. It specialises in carbon reduction technologies. Topsoe is focused on carbon emission reduction technologies for the chemical and refining industries. It therefore has a particular focus on liquid e-fuels.

Topsoe has been selected by Zhejiang Jianglan Bio-Energy Technology Co. Ltd (Zhejiang Jianglan) to deliver technology and services for SAF production. Once fully operational, the plant is set to produce 300,000 tonnes of SAF annually, which will initially be exported to European markets. Construction is expected to start in September 2025, with operations expected to commence in December 2026.

19.2.2 *Porsche/Volkswagen Group*

One vehicle manufacturer which has invested substantially in the manufacture and use of e-fuels is Porsche. This is initially for the Porsche racing team. The e-fuel is manufactured in Chile by HIF. It will use wind power (wind in this part of Chile is strong, relatively constant, and from the same direction) to make hydrogen and will extract CO_2 from the air using DAC. The project will integrate a wind-powered electrolyser to produce green hydrogen, which will then be combined with CO_2 captured from the atmosphere to produce synthetic methanol. The methanol will be converted into e-fuel via a methanol-to-gasoline technology. Porsche has been testing a 911 running on synthetic liquid hydrocarbon fuel made by HIF from electricity, air, and water rather than crude oil. This highlights one of the many possible directions of the energy transition. On 5 May 2025, HIF Global signed an agreement with Porsche AG and Shell for the delivery of e-fuels produced by its Haru Oni e-Fuels facility in Punta Arenas, Chile. HIF is delivering e-gasoline produced at the HIF Haru Oni e-Fuels facility in Chile to Shell's Technology Centre in Hamburg, Germany. The e-gasoline will be blended by Shell and then used by Porsche at different events to showcase its compatibility with existing combustion engines.

19.2.3 *Norsk e-Fuel: Aviation*

A good European example of the production of e-fuel for aviation (SAF) is provided by Norsk e-fuel. This uses CO_2 captured from the air, which is converted to what is called 'syngas' (also called synthesis gas), a mixture of hydrogen and carbon monoxide (CO). This syngas can then be converted into liquid fuels in a 'reactor'. The production of this type of fuel involves a closed carbon cycle in which carbon is first extracted from the air (DAC) and then returned to the air as an aircraft burns it in a jet engine. The process is, therefore, carbon neutral.

19.3 Thermochemical Processes for Manufacture of H_2 and Synthetic Liquid Hydrocarbon Fuels – Case Studies: Synhelion and Heliogen

19.3.1 *Synhelion (or How to Enable Jets to Fly on Sunlight)*[3]

Synhelion (now incorporating Heliokon) is focused on synthetic liquid hydrocarbon fuels produced by converting sunlight into fuels but without the need for an electrolyser and producing the heat required for manufacturing cement and energy storage to enable RTC manufacture. The CO_2 is sourced from chimney stacks, through DAC, or from methane. The H_2 comes from water, which, as well as being available from a water main, may also be obtained via DAC.

It is the first company able to produce syngas on an industrial scale using only solar heat as an energy source.

It has developed a solar thermal process to produce synthetic fuels that does not require electricity by instead using high-temperature solar heat to produce syngas, which can then be used in standard industrial processes to synthesise liquid fuels, such as kerosene, gasoline, or diesel, that are compatible with conventional jet engines and ICEs. The 'formula' is $CO_2 + H_2O$ + sunlight energy = fuel (in the form of a hydrocarbon, with the H coming from water and the C from carbon dioxide). As with other synthetic hydrocarbon fuels, the process is carbon neutral (defined in Section 19.1), as the process releases an equal amount of CO_2 during its combustion as was previously used for its production.

Making liquid fuels requires, in the first instance, heliostats (which are the mirrors in Figure 19.3), a solar receiver (in the tower), a thermochemical reactor (to enable the production of syngas from CO_2 and water), and thermal energy storage (to enable RTC operation of the 'refinery' or industrial process such as cement manufacture). A great advantage of the technology is that it can be sited anywhere. The company already works with international partners such as Lufthansa Group, Wood, Eni, CEMEX, and Zurich Airport.

DAC could also be a source of water (for making hydrogen) as well as CO_2 as the air is full of moisture. Differences in the chemistry of CO_2 and water adsorption may mean that it will be possible to simultaneously remove water from the atmosphere alongside CO_2.

---

[3]This section based on information from company website at https://synhelion.com/.

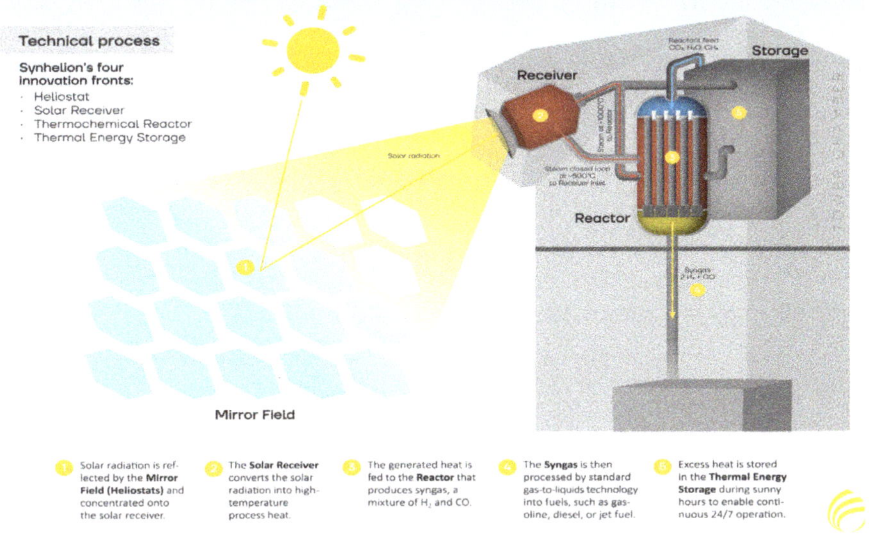

Figure 19.3. Synhelion's technical process.

19.3.2 *Heliogen*[4]

This company was founded in 2013. It was trying to integrate related aspects of energy – heat, power, and hydrogen (which is one of what they call HelioFuels). It developed a technology using heat from concentrating solar power (CSP) to manufacture clean hydrogen, which can then be used for the manufacture of liquid fuels. The company's technology produces primary energy from the sun in the form of heat rather than electric power, utilising CSP to produce heat rather than using photovoltaic (PV) panels, which use sunlight to produce power. They call it a 'Sunlight Refinery'. The technical process is shown in Figure 19.3.

Heliogen is focused on industrial decarbonisation. As we have noted previously, electrolysers and synthetic fuel plants operate more efficiently and hence at a lower unit cost of production if RTC clean power is available. As the energy produced by CSP is heat, not power, it can be stored economically in the form of heat (heat storage is covered in Chapter 16), allowing a manufacturing plant to operate 24/7. Clearly, however, the addition of storage raises the total cost of RTC clean power above that of

[4]This section based on company website, which also provides a video explanation of this particular CSP design; see https://heliogen.com/.

intermittent heat or power that the CSP system would produce without heat storage.

The first thing Heliogen emphasises on their website is that their CSP technology is best employed where solar 'insolation' is high. Although they say that sunlight is 'one of the most evenly distributed natural resources', in fact the economic value of that energy is determined by the 'insolation' of sunlight at any particular location. Most parts of Europe, apart from Spain and southern Italy, are not obvious places to use CSP or even PV, whereas California, the northern part of South America, Africa, parts of Asia, and Australasia are all good locations for solar, either CSP or PV or both. Indeed, Hydrogen Europe in its 'Compass for the Hydrogen World' publication noted that a photovoltaic plant in Chile, Namibia, or Morocco produces around **double the yield of a plant in central Europe**. In North Africa, a one kWp (p = peak nameplate output) plant generates 2,190 kilowatt-hours, whereas in European Union countries, the average is only 1,051 kilowatt-hours.

In the solar industry, the capacity factor and dispatchability are key. If they are low, as in Europe, then it is hard to compete with North Africa. Figure 19.4, also from Hydrogen Europe, shows that with the same solar panel 'nameplate' power, the actual power generated is so much less in Europe (apart from Spain and southern Italy). As a result, even though currently the efficiency of fuel cell electric vehicles (FCEVs) is much lower than that of batteries, the result 'at the wheel' is the same.

Heliogen sees a competitive advantage in the ease with which it can provide 'dispatchable' power by means of heat storage, despite using variable sunlight. In this way, local power in parts of the US may become nearly as economically efficient as distant sources of energy that have to be transported over long distances. Such a combined heat and storage system for power generation is ideal for those operating power distribution networks as well as for those manufacturing hydrogen or synfuels.

Figure 19.4 shows a CSP system without storage (in yellow), which can produce 600 MW of power output when the sun is at its peak. However, that power is wasted or can only be sold at a low price compared with the demand at other times of day. By storing that heat, the system can make truly dispatchable power available for industries that require 24-hour availability.

The other key competitive advantage is the higher temperatures, up to 1000°C at some point, which is necessary for many industrial processes (Heliogen's main target market).

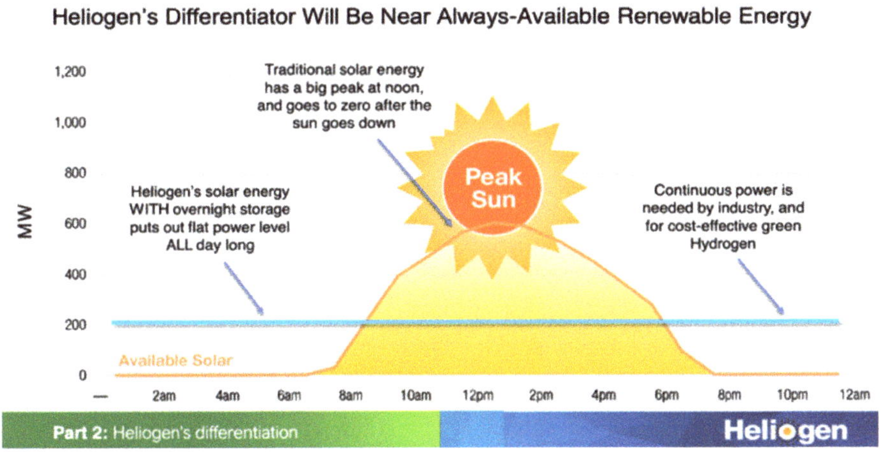

Figure 19.4. Smoothing out peaks and troughs of electricity production.

What they have achieved is a system using cameras that focus on the mirrors continuously and which provide a signal to a micromotor which can make continuous, slight adjustments to the positioning of each mirror to ensure it is working at full efficiency. This is in contrast to previous technologies, which used 'tables' of sun position at every time of day in all locations but which is inevitably not so accurate as Heliogen's positioning.

Heliogen's storage system enables all peak power to be easily stored as heat, which can then be used to provide continuous power supply to industry, hydrogen production, or for the production of synthetic liquid fuels. While solar PV installations can construct special storage facilities using batteries, hydrogen, or heat storage with CSP, the energy starts as heat and does not need to be converted to heat. Thus, adding a storage facility to a CSP heat generator should be much simpler (and hopefully lower cost) than adding it to a PV facility. By this means, Heliogen can provide RTC heat directly to a steam generator set for the production of power. Alternatively, heat and power can be supplied to electrolysers to make hydrogen 24 hours a day. In this latter use case, the added value stems from the much higher 'capacity factor' than with only a PV power supply. Also, of course, it can provide heat for industrial processes, replacing natural gas.

Heliogen makes claims for several innovations that will sharply increase the efficiency of their system. First, in the size of the mirrors (the heliostats). They are considerably smaller units compared with earlier generations of CSP technology and therefore easier to install. Because they are so much lighter, they are also easier to continually adjust to ensure they always point directly at the sun.

Heliogen's so-called closed-loop system (rather than simply using tables of sun positions linked by time of day and day of the year) uses computer vision (four cameras that focus on each heliostat to identify exactly how to move it to maintain perfect alignment with the sun) to identify the precise position of every single mirror and make micro-adjustments in real time to achieve perfect focus. Closed-loop AI ensures that the mirrors are pointed more accurately using optical feedback.

They have also developed an 'Installation & Cleaning Autonomous Robot & Utility Solution' (ICARUS). By bringing the same advanced technologies used by its AI-enabled CSP systems to the task of installing and maintaining its systems, the company's latest innovation is expected to significantly reduce the time to deploy its concentrated solar facilities, as well as the costs associated with construction and ongoing maintenance, such as heliostat cleaning. Their 20-module 100 MW plant is forecast to produce power at a cost of 5 US cents per kWh and use much less ground space than comparable PV or other CSP.

Plant estimated to produce:

- >85% capacity factor renewable power,
- <5 cents per kWh power cost,
- <1/6th square mile footprint (650 m × 650 m),
- >850,000 kg of hydrogen/year.

In July 2021, it was announced that Heliogen would merge with Athena Technology Acquisition Corp, a so-called Special Purpose Acquisition Company (SPAC), which would enable Heliogen to raise funds and to become a public company on the New York Stock Exchange, with an expected value of approximately $2 billion. The SPAC is a company led by two women with extensive experience in venture capital and who have also served as board members of public and private companies. Sadly, it was delisted from the New York Stock Exchange in November 2023. It is still operating, however. But now the business model is 'commodities-as-a-service'. It provides steam, power, or green hydrogen to municipalities and heavy industry, with charging based on quantity supplied (e.g. steam-as-a-service).

19.3.3 *Price of SAF, eSAF, and Kerosene*

The European Union Aviation Safety Agency (EASA) has calculated 2024 average prices for SAF from biofuels, advanced biofuels, and recycled carbon biofuels at €2,085/t ($2,169/t). The reference price for synthetic aviation fuels (eSAF) is estimated at €7,695/t ($8,007/t), while conventional aviation fuel is €734/t ($763.8/t). This creates price differentials of €1,351/t ($1,406/t) for SAF and €6,961/t ($7,244/t) for eSAF compared with conventional fuel (kerosene).

Under Article 12 of ReFuelEU Aviation, penalties for non-compliance must be at least twice the price difference between SAF and fossil kerosene. Based on 2024 average prices, SAF non-compliance penalties would be around €2,700/t ($2,809/t), while eSAF penalties would be significantly higher at €13,992/t ($14,560/t), though the eSAF sub-mandate does not begin until 2030.

Significantly, jet fuel suppliers who fail to meet EU mandates must pay these penalties but are not exempted from meeting quotas. Unfulfilled quotas carry over to subsequent years, making non-compliance increasingly costly over time.

Chapter 20

Carbon Credits, Carbon Capture, Usage and Storage, Direct Air Capture of CO_2[1]

Until recently, it was common practice for companies which were large emitters of CO_2 to claim that they were carbon neutral or carbon negative through the use of voluntary carbon markets. However, the real challenge for those buying such credits is knowing how to separate the genuine, impactful credits available on the market from those that won't deliver the promised emission reductions, removals, or avoidance. These latter are a form of greenwashing.

Reducing, Removing, or Avoiding Emissions

A carbon credit is a mechanism by which one party compensates another for the action of reducing, avoiding, or removing carbon. Based on their net emissions impact, there are three types of carbon credits: carbon reduction, carbon avoidance, and carbon removal. Carbon avoidance is an action that prevents a carbon-emitting activity from occurring. Carbon reduction is an action that decreases the amount of greenhouse gas

[1]This introduction is based on an abbreviated and edited version of information on the website of Carbon Direct (https://www.carbon-direct.com/), which offers services to measure, reduce, and remove carbon emissions with Carbon Direct science-backed carbon management solutions.

emissions compared to prior practices. Carbon removal is the process of removing carbon dioxide from the atmosphere and locking it away for decades, centuries, or millennia. For all three types, credits are measured and issued by assessing or estimating the climate impact from carbon credit purchases – specifically, how much carbon is avoided, reduced, or removed based on the purchase of the credit.

Avoidance, reduction, and removal are three very different actions. But without accurate baselines, the climate impact cannot be determined. According to Carbon Direct:

> **Carbon Avoidance Credits**
> Examples of carbon avoidance include avoiding deforestation that would result in the release of carbon dioxide into the atmosphere. **Roughly 75% of all certified credits available are avoidance credits.**
>
> **Carbon Reduction Credits**
> While carbon reduction must drive most of the push to net zero, the transmission mechanism of carbon reduction to a purchase of an external credit is challenging. Examples of carbon reduction credits include reducing fossil-fuel use by improving fuel efficiency, or programs that reduce the methane that is generated from farms or municipal waste processing. **Reduction credits make up roughly 22% of certified credits on the voluntary carbon market.**
>
> **Carbon Removal Credits**
> Projects that remove carbon come from a diverse set of solutions, from nature-based solutions like reforestation, to engineered solutions such as direct air capture and storage. **Roughly 3% of credits on the carbon market today are classified as removals.**

While currently only 3% of credits originate from carbon removal activities, we focus on the technologies for achieving this, as removal provides greater certainty regarding the actual impact on global warming. However, to date, there is little sign of actual removals taking place on more than a tiny fraction of emissions. A key reason is the high cost of removals and the absence of revenue for most removals. One market, however, is that for synthetic hydrocarbon fuels, which combines green hydrogen with carbon atoms extracted from CO_2 to create fuels for planes and road vehicles. Today, however, the cost of such fuels is substantially higher than that of natural hydrocarbon fuels.

In February 2024, the European Commission welcomed the provisional agreement between the European Parliament and the Council on the first EU-wide voluntary framework for the certification of high-quality carbon removals. The agreed criteria will ensure that carbon removals are correctly quantified, store carbon for an agreed long-term period (a minimum of 35 years for carbon stored in products), go beyond existing practices and do not simply reward the status quo, and contribute to broader sustainability goals, for instance by providing positive impacts on biodiversity.

20.1 Capturing CO_2

When CO_2 is removed from 'smokestacks' or from the air, it can either be 'stored' or 'used'. It may be removed for two reasons. The first is to reduce the concentration of CO_2 in the atmosphere and hence limit global warming; the second, as we noted above, is to provide the carbon atoms to enable the production of synthetic hydrocarbon fuels by combining them with hydrogen molecules.

Reducing the concentration of greenhouse gas emissions in the global atmosphere is a critically important need. In 2022, emissions amounted to 53.8 billion tonnes (CO_2 equivalent). According to the IPCC, any pathway that could limit global warming to 1.5°C requires carbon removal of between 100 billion and 1 trillion tonnes over the remainder of the century. In effect, the IPCC is saying that we will not achieve the Paris Agreement temperature target without a massive investment in carbon removal and storage while, at the same time, eliminating most new carbon emissions by moving towards the rapid phasing out of most fossil fuel use.

CO_2 can be 'captured' in two ways:

> The first is from industrial processes that emit a high tonnage of CO_2 from their 'chimney stacks/flues'. It thus comes directly from hydrocarbon fuels being burnt shortly after they have been extracted from the

> Earth. If not captured, these emissions would immediately be increasing the concentration of CO_2 in the atmosphere. When this CO_2 is stored or sequestered, the process is known as carbon capture and storage (CCS).
>
> The second approach is to capture or extract it from the air, known as direct air carbon capture (DACC). Given that the concentration of CO_2 in the air is only 422 parts per million (ppm), this is very much lower than that from flue emissions. It has thus been considerably more costly, at least up to now, to extract directly from the air. When this CO_2 is stored or sequestered, the process is known as direct air carbon capture and storage (DACCS).

Capturing flue CO_2 or atmospheric CO_2 to obtain carbon atoms for use as a component in the manufacture of synthetic liquid hydrocarbon fuels, such as aircraft kerosene (see Section 19.3 on synthetic fuels), only enables carbon from the air or flue to be 'reused'. If recycled in this way, then CO_2 extracted from flue gases will still increase atmospheric concentrations of CO_2, though with a slight delay, as carbon is still being combusted and released into the atmosphere in its second time of use, for example as aircraft fuel. **Thus, this process is carbon additive, as it still involves combusting newly extracted fossil fuel. In the case of DACC capture, the CO_2 concentration in the air will initially be reduced, but when the synthetic fuel is used, the concentration returns to what it was before. This process is carbon neutral.** We cover these concepts in more detail as follows, using aircraft synthetic kerosene as an example.

20.1.1 *Carbon Neutral*

The use of carbon captured via DACC is carbon neutral because it recycles existing CO_2 from the atmosphere and therefore reuses CO_2 that has already been emitted into the atmosphere from the combustion of fossil fuel sometime since the start of the industrial revolution. When combusted a second time, it is simply returning that CO_2 to the atmosphere. There is thus no net impact from the two processes – DAC (CO_2 removal) and an aircraft flight (CO_2 addition) – provided the capture process uses carbon-free energy. But one company, Tree Energy Solutions (TES), has developed a system whereby carbon removed from smokestacks and collected to manufacture methane (CH_4) is, in fact, carbon neutral.

20.1.2 *Carbon Additive*

If the carbon atoms that are being used for the manufacture of electrofuel are sourced from a factory or power station exhaust, the combined process from the two linked manufacturing processes (for example, producing aircraft fuel from the emissions of a factory making steel) is carbon additive.

The double use of CO_2 can be considered as:

> carbon-neutral steel-manufacture and carbon-additive flying, *or*
> carbon-neutral flying and carbon-additive steel making, *or*
> the two activities each being only half as carbon additive as the alternative.

But certainly both are not carbon neutral, as is often claimed.

In the longer run, however, there should be almost no industrial processes (apart perhaps from the manufacture of cement) which emit CO_2. As we noted above, at that point, the manufacture of synfuels would have to rely entirely on DAC. These fuels are always likely to be expensive.

20.1.3 *Carbon Negative*

Rather than reusing the carbon obtained from DACC which returns it to the atmosphere, if we 'sequester' it, i.e. put it somewhere for very long-term storage, normally underground in caverns, then the concentration of CO_2 in the atmosphere actually declines (or rises less than it otherwise would).

Microsoft recently announced that it aims to become carbon negative by 2030 and by 2050 to remove from the atmosphere all the carbon that it has emitted since its founding in 1975. This is a significant commitment by the company, but it underscores the potential for approaches using negative emissions to play an important role in meeting international climate goals.

We discuss later some of the companies which already provide or are developing methods of carbon capture. We then consider how carbon can be sequestered and also cover companies with different methods of sequestration or (re)use. The first DAC company we cover (Climeworks) is one which has teamed up with another company (Carbfix), providing an example of an interesting method of carbon sequestration.

Other methods of sequestration involve converting CO_2 to solid carbon. This could be black carbon (with the oxygen being released from the carbon bond). An example of this technology in use is provided as follows.

20.2 Direct Air Carbon Capture Companies – Some Examples: Climeworks and Phlair

20.2.1 *Climeworks*[2]

Rather than supplying carbon dioxide to electrofuel manufacturers (which remains carbon additive), Climeworks main business is using DACC to permanently remove CO_2 from the atmosphere and mineralise it underground. As they note, 'air-captured carbon dioxide can either be recycled and used as a raw material or completely removed from the air by safely storing it. Direct air capture, therefore, is one of the key ways to stop global warming by reducing the amount of potentially harmful CO_2 in the Earth's atmosphere'.

One of the ways they generate revenue is by 'selling' subscriptions to individuals and companies to help finance an expansion of their DAC facilities:

> Why should I subscribe or what should I use the subscription for?
>
> This is of course up to you! Do you want to enable the removal of some carbon dioxide to reduce your personal footprint? Good! Are you planning a long flight and want to build up some carbon dioxide "allowance" first? Even better! Do you believe that the Climeworks technology is

[2] https://climeworks.com/news/climeworks-builds-first-commercial-scale-direct-air-capture-plant.

> surprisingly cool and deserves some more publicity? Great! Join us on the climate positive journey.
>
> Those who contribute are called Pioneers and these are the people who finance Climeworks (it's a bit like crowdfunding). There are now over 20,000 Pioneers.

It would not be feasible to scale up to the CO_2 removal capacity required to make a significant difference to the current CO_2 concentration in the atmosphere (around 430 ppm) using only crowdfunding. On 6 April 2022, Climeworks announced that it had signed an equity round of CHF 600 million ($650 million), which marks a major milestone in the company's history and an even bigger step for the carbon removal industry.

The financing is co-led by Partners Group (acting on behalf of its clients) and GIC, with further participation from (in alphabetical order) Baillie Gifford, Carbon Removal Partners, Global Founders Capital, John Doerr, M&G, and Swiss Re, as well as other new and existing shareholders, including long-term investor and anchor shareholder BigPoint Holding AG. Moreover, J.P. Morgan Securities LLC served as the sole placement agent for Climeworks in connection with the transaction.

Since its foundation in 2009, Climeworks has pioneered direct air capture technology and is the most advanced industry player, with the world's largest direct air capture and storage plant launched in September 2021. The new investor base includes several of the most renowned and largest institutional technology and infrastructure investors globally. The funding will unlock the next phase of Climeworks' growth, scaling direct air capture to multi-million-tonne capacity and implementing large-scale facilities as carbon removal becomes a trillion-dollar market.

In terms of what has happened on the ground (and below ground), in September 2021, Climeworks launched its flagship facility, Orca, in Iceland. It is the world's largest direct air capture and storage plant. As carbon capture is only the first stage of CCS, they joined forces with the Icelandic company Carbfix (which we cover later), which has a strong scientific backbone and is one of the world's leading experts in the rapid underground mineralisation of CO_2. Orca removes 4,000 tonnes of CO_2 per year, while Carbfix mixes the CO_2 with water and pumps it deep underground. Through natural mineralisation, the CO_2 reacts with basalt rock and turns into stone within a few years.

It is probably fair to say that CCS has not yet taken off at a sufficiently large scale to make more than a minimal contribution to reducing CO_2 concentration in the atmosphere. However, over time the contribution will increase, particularly if carbon taxes become more widespread.

Its latest (Generation 3) technology is said to use novel structured sorbent materials, replacing the packed filter beds used in previous technology generations. The new structures increase surface contact with CO_2, reducing the time to capture and release it by at least a factor of two, thus capturing more than twice as much CO_2 as previous filters. The new filter materials consume half the energy and are designed to last three times longer than prior materials. The company's cost reduction strategy aims to achieve costs of $250–350 per tonne captured and total costs of $400–600 per tonne of net removal by 2030. This represents an overall cost reduction of up to 50% compared to today.

20.2.2 *Phlair*[3]

This is a relatively new company which has developed a carbon removal process from the air, which is less energy intensive than other methods. It uses what is known as a hydrolyser. It claims that its system has three times lower energy capture costs than thermal DAC approaches, yielding lower removal costs. It can also operate intermittently to take advantage of periods when electricity prices are low or negative. Phlair already has contracts with companies that will benefit from being able to use this new DAC technology.

20.3 Sequestration, Mineralisation, and Reduction of CO_2 by Removing the O_2: Storegga, Carbfix, and CarbonBuilt, Converting CO_2 to Solid Carbon

Most CCS projects inject CO_2 into sedimentary basins and require an impermeable cap rock to prevent the CO_2 from migrating to the surface. A good example of a company using this technique is Storegga.

[3]Company website https://phlair.com/technology.

20.3.1 *Storegga*[4]

Storegga uses proven geological stores many kilometres below the Earth's surface to safely and permanently lock away harmful CO_2 which has been captured from industrial emissions processes or drawn from the atmosphere.

CCS has been proven in industrial-scale projects globally, with the world's first industrial CCS project operating continuously since 1996. Now, CCS projects are being developed in industrial regions across the globe because it is the most effective way to prevent or remove CO_2 emissions from our atmosphere in volumes significant enough to slow the pace of climate change. These emissions have been generated from modern society through activities such as power generation and industrial processes, including manufacturing.

CO_2 storage in deep geological formations is also known as 'geo-sequestration'. It has a significant role to play in a sustainable future.

Storegga is also developing both electrolytic and CCS-enabled hydrogen projects in Scotland.

20.3.2 *Carbfix*

Captured carbon can, alternatively, be stored through injection into reactive rocks, provoking CO_2 mineralisation and thereby permanently fixing carbon.

Mineral carbonation extends the applicability of CCS by enabling storage in areas previously not considered feasible. The rapid mineralisation of CO_2 through injection into reactive rock formations also increases storage security. Carbon mineralisation in basaltic rocks offers a global storage potential that exceeds anthropogenic emissions.

20.3.3 *CarbonBuilt Inc*

CarbonBuilt's Reversa™ process includes innovations to both the concrete mix design and its curing process. On the formulation side, it

[4]Company website https://storegga.earth/co2-storage.

introduced portlandite (a widely available commodity chemical, also known as calcium hydroxide or hydrated lime), which reduces the usage of traditional Portland cement and increases the use of low-cost and low-carbon supplementary cementitious and filler materials. Concrete products are then formed using the same processes and equipment as used today.

After forming, the concrete products are cured with CO_2 using a process that does not require expensive capture, compression, or purification of the CO_2, high temperatures, or elevated pressures. This flexibility enables the use of a wide range of CO_2-containing streams, ranging from point-source or distributed emissions that can be used directly to merchant gas sourced from such sources and emerging direct air capture technologies.[5]

[5] From https://www.carbonbuilt.com/.

Chapter 21

Connecting Up the Energy Value Chain – The Sector-Coupled Economy and Clusters: Hydrogen, Electric Power, Synthetic Fuels, and Fertilisers with Case Studies of Iberdrola and HyDeal España

Hydrogen will be the driving force of our energy system of the future, but the bloc will need to rely on imports. Europe is never going to be capable of producing its own hydrogen in sufficient quantities.

Frans Timmermans, European Commissioner for Climate Action

The statement above is one of hope. But that hope will not be realised unless it becomes possible to provide hydrogen to the market at a much lower cost than today, whether manufactured in the EU or abroad in locations with higher capacity factors for wind and solar. Such locations would be in parts of Morocco or Saudi Arabia. Alternatively, legislation could be introduced to require manufacturers of ammonia-based fertilisers and oil refiners (who are currently the largest users of hydrogen) and others to switch to green hydrogen.

21.1 Sector Coupling

Synergies exist on a wider scale in the energy sector than those explored above and, if exploited, can result in a considerable reduction in unit costs. The European Parliament Committee, which covers the energy sector, has produced a report entitled 'Sector coupling: how can it be enhanced in the EU to foster grid stability and decarbonise', where the Committee made it clear that the energy system will have to function without coal, oil, or gas.

The question then is what will replace these carbon-based *primary energy sources* in the future provision of power (currently ~20%), heat (currently ~40%, including industrial heat), and transport (currently ~40%). The fact that electricity represents a lower proportion of energy use than either of the other uses points up the illogic of considering power decarbonisation on its own. Hydrogen can provide energy for power, heat, and transport and is therefore one integrating element of the equation for all three main energy uses. There are, however, technology issues in transporting hydrogen over large distances.

While there remain some differences in usage across sectors, sector coupling essentially means interconnecting and integrating energy-consuming sectors – such as buildings (heating and cooling), transport and industrial heat – with a diverse energy sector encompassing electricity, hydrogen, and the associated transport, storage and grid systems for these energy carriers. It may also involve the use of two-way fast charging points for electric transport vehicles to provide power storage (vehicle-to-grid) and hydrogen refuelling points, alongside increased contractual automatic disconnection of some power users at times of high power prices (demand management) through the use of interruptible contracts.

Sector coupling includes the conversion of power to hydrogen, ammonia and fertiliser and to liquid fuels, and the use of nuclear heat and power to produce hydrogen. Moreover, for the nuclear power station owner, when spot prices of power are high, arbitrage should occur between the revenue generated from hydrogen production and the revenue generated from selling power to the grid at times of such high prices (normally when wind and solar are making little or no contribution). In addition, some nuclear installations will be selling both industrial heat and heat for district heating systems, which will provide them with an important source of revenue.

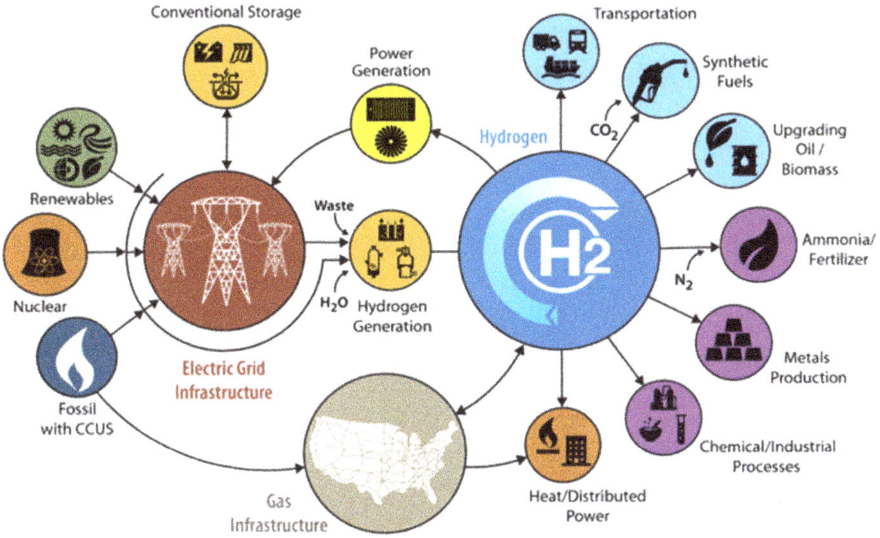

Figure 21.1. Sector coupling.

Figure 21.1[1] indicates examples of sector coupling and integration based around:

- the conversion of power to hydrogen (through electrolysis);
- the reverse conversion of hydrogen to power (though fuel-cell regeneration);
- the conversion of hydrogen to liquid fuels and ammonia for use in decarbonising shipping and aviation;
- hydrogen in use for space heating and combined heat and power;
- hydrogen in the transport sector;

hydrogen as a source of heat for industrial processes, such as steel manufacture.

For investors, such integration potentially provides investment opportunities. In this vision of the very near future, hydrogen is a pivotal technology in sector integration.

---

[1]US Department of Energy (DOE), National Renewable Energy Lab (NREL), 'Hydrogen in future integrated energy system.' https://www.nrel.gov/news/program/2020/study-shows-abundant-opportunities-for-hydrogen-in-a-future-integrated-energy-system.html.

21.1.1 Clusters

We have already noted that there are considerable advantages in siting renewable or clean power (such as small nuclear, wind, and solar) in close proximity to hydrogen manufacture and other users of hydrogen. This is designed to give economies of scale and scope, and such agglomerations are known as industrial clusters. They are characterised as geographic areas that co-locate companies across multiple related industries. Concentrating manufacturers and users within a relatively small geographical area enables the scaling of low-carbon technologies. By sharing risk and resources among multiple companies working in partnership, industrial clusters enable economies of scale, scope and learning. A good example of a cluster is the WESTKÜSTE100 project in Germany.

WESTKÜSTE100 is a project which aims to achieve sector coupling in a region of Germany. It is trying to achieve both decarbonisation and the use of green hydrogen across a range of industries, such as air transport, construction and space and industrial heating. It has the benefit of strong winds in the region, as well as geological storage possibilities for CO_2. It also plans to use the CO_2 emissions from cement manufacture to provide the required CO_2 for the production of synthetic hydrocarbons. What it is doing is in line with the EU's 'Fit for 55' Hydrogen Strategy and, in particular, the EU's plans in its ReFuelEU Aviation initiative, which is targeting an increase in the use of sustainable aviation fuel in Europe from 0.05% in 2020 to 5% by 2030. The 2050 target is 63%, which may be achievable, as investors are encouraged by the fact that synthetic liquid can use existing infrastructure for storage, delivery, etc.

It is also aimed at obtaining the efficiencies gained from a cluster and an integrated energy/industrial system. The objective in this case is to build regional infrastructure to enable sustainable production of steel, ammonia, ethylene, and fuels for the Netherlands and Belgium, accelerating carbon reductions in a world where wind is the new North Sea oil and gas.

21.1.2 Case Study: Iberdrola

Iberdrola, a Spanish utility, provides a good example of cluster or sector coupling. It is a global energy corporation which describes itself as the 'Utility of the Future' and is the number one producer of wind power in the world, as well as one of the world's largest electricity utilities in terms of market capitalisation. Of the group's long-term investment plan, 90% is aligned with the green taxonomy of the EU. It has a global reach with assets in the UK (Scottish Power), across Europe, Australia, and Japan, and in Mexico, Brazil, and the US.

Iberdrola SA and fertiliser maker Fertiberia SA have proposed a €1.8 billion ($2.1 billion) plan to transform the country into an exporter of ammonia produced using renewable energy. Ammonia is a key component in fertilisers.

21.1.3 *Integrating the Value Chain: HyDeal España as an EU Exemplar*

It is all very well building large electrolyser manufacturing plants or operating a large green hydrogen plant, **but without end customers guaranteeing offtake of the product (as in any project financing) to mitigate volume and price risks and a means of transporting the end product to customers, a project will not be bankable**. To overcome the issues involved in successfully scaling up the hydrogen market, 30 energy players in Europe announced in February 2021 that they had developed a plan for an integrated value chain, known as HyDeal, to deliver green hydrogen across Europe at the price of fossil fuels. The idea is to create an industry platform covering the complete green hydrogen value chain, from upstream to midstream, downstream, and finance, including solar power generation, electrolyser manufacturing, project engineering, gas transmission and storage, and industrial applications in steel, chemicals, and electric power.

HyDeal España is the first industrial implementation of the agreement between the parties. It will manage the development, funding, and construction of renewable hydrogen production and transmission infrastructures. It should deliver competitively priced renewable hydrogen to an industrial complex in Asturias from facilities based in northern Spain. Anchor sponsors include ArcelorMittal, Enagás, Grupo Fertiberia, and DH2 Energy. Production is planned to start in 2025, with a total installed capacity expected to reach 9.5 GW of solar power and 7.4 GW of electrolysers by 2030.

HyDeal España has developed an industrial and financial model based on value chain integration, captive solar power, electrolysers operating at an industrial scale, hydrogen pipelines, and demand aggregation to give it a cost advantage. Mass-scale renewable hydrogen will sustainably replace fossil fuels across industry, energy, and mobility and will provide Spain with a domestically produced, secure, competitive, and zero-carbon energy, steel, and food supply alternative. HyDeal España aims to provide the equivalent of 5% of Spain's natural gas imports, contributing to the country's energy independence. ArcelorMittal and Grupo Fertiberia – together with other large hydrogen clients set to join the project – plan to purchase 6.6 million tonnes of renewable hydrogen over 20 years (avoiding annually the equivalent of 4% of current Spanish CO_2 emissions) to produce green steel, green ammonia, green fertilisers, and other low-carbon industrial and energy products, positioning the companies as European leaders on their respective markets. The direct connection of mass-scale renewable hydrogen production and long-term bankable consumption creates enormous value for the system.

21.2 Potential Future Energy Systems

The US Idaho National Laboratory (INL) has set its objective to advance integrated energy generation, storage, and delivery technologies needed for a net-zero future (Figure 21.2).[2] The integrated systems approach represents a marked change from traditional energy system designs typically focused on single generation sources to support a single energy demand. The footnote provides useful information and graphics on how the future might look.

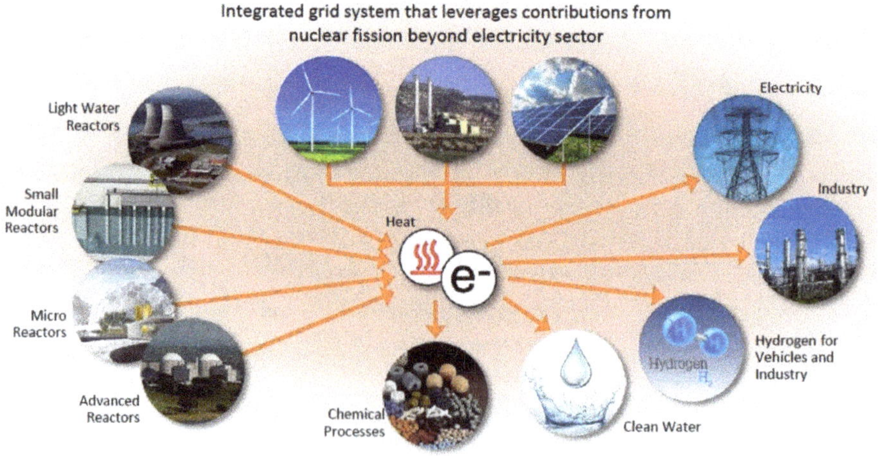

Figure 21.2 Potential future energy system.

[2]Power Electronics News, 'The role of technology in the energy transition,' January 2024. https://www.powerelectronicsnews.com/the-role-of-technology-in-the-energy-transition-2/.

Appendix 1

The Science of Climate Change and Its Relevance to Lenders, Investors, and Insurers: Climate Tipping Points and Global Cascades

A1.1 The Earth's Carbon Cycle

The science starts with the fact that:

> *The sun delivers more energy to Earth in an hour than we use in a year from fossil, nuclear and all renewable sources combined.*

Fortunately, we can now convert *a little* of that sunlight (photons) into electricity (electrons) at low cost using photovoltaic (PV) cells (at least when the sun is shining). Other electromagnetic energy frequencies from the sun (infrared) can convert directly into heat, which happens when infrared rays strike matter, such as the air (or, indeed, you or me). The energy in the air can then generate powerful air currents (winds), which can be converted into electric power through wind turbines. Infrared rays can also be captured by what is known as concentrating solar power (CSP) mirror collectors and then focused on a collector, normally in the form of molten salts, which then reach a high temperature suitable for industrial heat use, for the manufacture of hydrogen by breaking down water molecules into oxygen and hydrogen or for raising steam to generate power.

Some of the sunlight not used by our solar or wind farms or CSP is used by plants for photosynthesis, as plants absorb CO_2 from the air, which 'locks in' carbon in all plant life, at least until a plant is used for animal feedstock or burned or simply decays.

Clearly, after these uses of sunlight, there is still a huge amount of energy from the sun incident on the Earth. It is this energy that keeps the Earth at an average global temperature (across the whole globe and averaged over day and night) of around 15–17°C, which is the equilibrium temperature of the Earth. But there is a problem! Basic thermodynamics tells us that if all the solar energy that arrives on Earth while the sun is out is not dissipated back into space at night, Earth will heat up day-by-day as it tries to restore a temperature equilibrium, which will be higher than before.

The problem we are now facing (and its manifestations in higher temperatures, more violent and more frequent storms and hurricanes, and rising sea levels) arises from the fact that the Earth's energy systems are *not* in equilibrium. The only way in which Earth can achieve a new equilibrium rather than suffering ongoing rising temperatures is for us to cut greenhouse gas (GHG) emissions to close to zero. The deadline for doing so to have a good chance of keeping the global temperature rise to less than 1.5°C or, alternatively 2°C, is 2050. However, equally critical is the pathway to 2050. Cutting emissions only from, say, 2040, would not achieve the same objective as halving emissions between 2022 and 2030 and continuing the reduction at that rate until 2050, something which has now become necessary.

All life forms on Earth, including humans, are carbon life forms. Without carbon, humans and animals, as well as plants, would not exist. Carbon is a key element for life (as we know it) to exist. However, CO_2 generated by coal, oil, and gas is emitted into the atmosphere, where, unless stored in some way, it would cause global warming. It is partly stored in the atmosphere, partly in the oceans. Much is also stored in calcium carbonate ($CaCO_3$) deposits, which are a product of the remains of sea creatures from millennia past. This CO_2 is released in large quantities when calcium carbonate (limestone) is converted to cement for building purposes.

The CO_2 emissions absorbed by the ocean (around 25% of fossil fuel emissions) result in acidification of the ocean, which is now 30% more acidic than before the industrial revolution. This has dramatic effects on ocean life, including damage to corals and to creatures with shells such as oysters. But it also affects fish, and while the full impact of acidification (a change in the pH measure of the water) is not yet fully researched, it is one of the major issues arising from fossil fuel emissions. The rising temperature of the ocean also results in the extinction of some forms of life, including corals. Already, the Great Barrier Reef off Australia has suffered considerable destruction as a consequence.

A1.1.1 *Earth's Energy Budget,*[1] *the Natural Greenhouse Effect, and the Anthropogenic (Human-Related) Contribution*

The climate's heat engine (an engine in the sense that heat provides the energy that causes weather such as storms and tornadoes) must not only redistribute solar heat from the equator towards the poles but also from the Earth's surface and lower atmosphere back to space. Otherwise, Earth would heat up endlessly, as it has since the start of the industrial revolution in 1750 when anthropogenic emissions started to grow.

Fortunately, when sunlight strikes the Earth, land and ocean surfaces and the atmosphere are simultaneously radiating heat back into space. The heat is generated from the infrared component of sunlight incident on the Earth. Earth does not suffer the temperature extremes of the Moon (−173° at night to 127° during the day) because the blanket of atmosphere around it creates the natural greenhouse effect. This is the process by which radiation from the Earth's atmosphere warms the planet's surface to a temperature above what it would be without this natural 'blanket'. This is due to the greater transparency of the atmosphere to visible radiation from the sun (light) than to infrared radiation (heat) emitted from the Earth's surface. This net flow of energy into and out of the Earth system during the day and night is Earth's energy budget, as explained in the following section.

As is shown at the top of Figure A1.1,[2] the incoming solar energy on average across the Earth's surface is 340 watts per square metre, or one-third of a kilowatt – a very substantial amount of power when measured over the whole area of Earth! Some is reflected back into space (around

[1]The material following is based on a NASA publication: https://earthobservatory.nasa.gov/features/EnergyBalance/page1.php.

[2]NASA Earth Observatory https://earthobservatory.nasa.gov/features/EnergyBalance/page4.php.

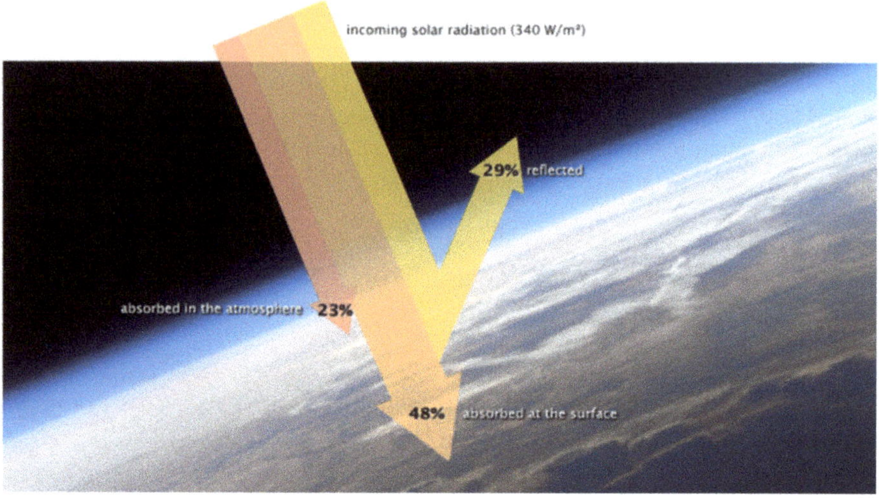

Figure A1.1. Incoming solar radiation must balance outgoing solar radiation to keep Earth's temperature constant.

29%), leaving around 240 watts of solar power per square metre. About 23% of incoming solar energy is absorbed in the *atmosphere* by water vapour, dust, and ozone, and 48% passes through the atmosphere and is absorbed by the surface.

On the surface, absorption of this energy drives photosynthesis (plant life), photovoltaic (PV) power generation, wind turbine-generated power, and CSP. It also provides the energy that causes evaporation of water from the oceans and rivers (and when this falls back to earth as rain, enables hydropower), melts snow and ice, and warms the Earth, which on average provides an equable temperature which enabled the development of animal life (including human life). This is in contrast to everywhere else in the Universe (at least as far as we know). The natural greenhouse effect from which Earth benefits has enabled humans to evolve on Earth and to survive and generally to prosper. However, humankind's footprint on Earth is now giving rise to concern for future generations.

Earth's natural greenhouse effect was critical in enabling life on Earth to come into existence and thrive. It is the atmosphere above Earth that enables it to retain sufficient heat rather than being a frozen planet like the moon. So, in an (imperfect) analogy to a garden greenhouse, the atmosphere enables Earth to have a much more equable temperature range than otherwise. During the day, light from the sun falls on (is incident on) the Earth and, as a result of striking the solid body of the Earth, changes from light energy (shorter electromagnetic wavelengths) to heat energy (longer wavelength infrared radiation), which warms up the land and the sea. The land and sea then radiate some of this heat (infrared radiation) back to the atmosphere, and in turn the atmosphere absorbs some of it and radiates some of it back to Earth.

This balance between Earth's heat being radiated back into space and or being radiated back to Earth is particularly evident at night in the desert. During the day, a cloudless desert region is extremely hot. At night, it can cool down dramatically not only because the sun is no longer incident on that part of the Earth but also because the absence of cloud cover enables more infrared energy to be radiated back out to space. In contrast, in a country with frequent cloud cover, it will be warmer at night when it is cloudy compared with when the sky is clear when it may be frosty.

The key question is whether or not the *balance* between incoming and outgoing energy, i.e. an equilibrium, is achieved without a change in average Earth temperature. This energy balance is most meaningfully evaluated at the top of the atmosphere, where the net outgoing energy encapsulates all the complexity of the greenhouse effect and climate feedbacks. If more heat is trapped within the atmosphere, equilibrium can only be achieved through the Earth (land and oceans and atmosphere) warming in order to radiate more heat out into space. It is the factors that lead to more or less heat being trapped which determine whether the Earth's temperature remains constant, rises, or falls.

Of course, the Earth has had temperature cycles throughout time, including a number of ice ages. We have also had temperatures very much higher than today. Thus climate change has been a fact of life even before life existed on Earth. Temperature changes can be caused by changes in the Sun's energy emission, which can be related to 'sunspots' or to changes in the amount of snow cover or cloud cover, resulting in a higher albedo (whiteness or reflective coefficient) (Figure A1.2).[3] This can lead to feedback effects. As the brightness of the Earth increases for whatever reason, the Earth will cool more, which leads to snow covering greater areas, which then leads to an even higher albedo. In contrast, as the Earth heats up, for example, as a result of human-caused (anthropogenic) GHG emissions, snow melts, the albedo decreases, and the feedback effect then leads to the earth heating up more and the snow receding further, leading to even higher temperatures depending on the so-called 'climate sensitivity', i.e. an estimate of how much Earth's surface will warm.

---

[3]Data source: NASA's Goddard Institute for Space Studies (GISS). Credit: NASA/GISS.

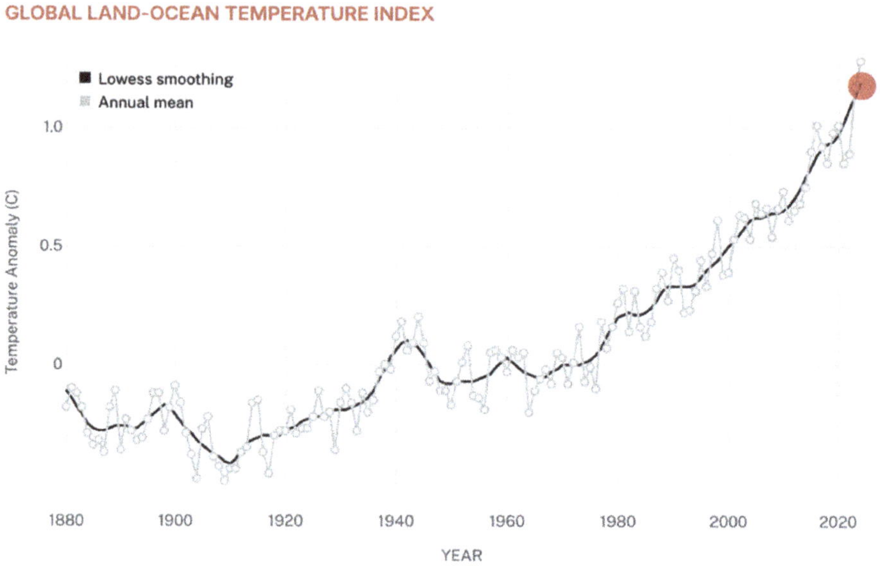

Figure A1.2. Change in Earth's temperature over time.

A good up-to-date source of information on the impact of rising temperatures leading to climate change is summarised by Carbon Brief on 10 January 2025. It can be accessed at https://www.carbonbrief.org/state-of-the-climate-2024-sets-a-new-record-as-the-first-year-above-1-5c/.

A1.1.2 *Why CO_2 Causes Global Warming*

In the atmosphere, the major atmospheric gases (oxygen and nitrogen) are transparent to incoming sunlight (solar light energy) and also to infrared radiation (heat). Thus, infrared radiation (heat) would be able to pass in or out through the atmosphere if these were the only constituents of the atmosphere. The key to the greenhouse effect is that water vapour, carbon dioxide, and methane are opaque to thermal infrared energy.

While water vapour comprises the major part of greenhouse gases, it is not the direct cause of warming. It is carbon dioxide (CO_2) and methane (CH_4) which cause what is known by scientists as 'radiative forcing' or climate forcing. This is the difference between sunlight absorbed by the Earth and energy radiated back to space. Changes to Earth's radiative equilibrium that cause temperatures to rise or fall over decadal periods are called *climate forcings*. Positive radiative forcing means Earth receives more incoming energy from sunlight than it radiates to space. This net gain of energy will cause warming. Conversely, negative radiative forcing means that Earth loses more energy to space than it receives from the sun, which produces cooling.

> Natural climate forcings include changes in the Sun's brightness, Milankovitch cycles (small variations in the shape of Earth's orbit and its axis of rotation that occur over thousands of years), and large volcanic eruptions that can inject light-reflecting particles into the stratosphere. Human-made forcings include particle pollution (aerosols), which absorb and reflect incoming sunlight; deforestation, which changes how the surface reflects and absorbs sunlight; and the rising concentration of atmospheric carbon dioxide and other greenhouse gases, which decreases heat radiated back into space. A forcing can trigger feedbacks that then intensify or weaken the original forcing. The loss of ice at the poles, which makes them less reflective, is an example of a positive feedback. A system in thermal equilibrium has zero radiative forcing.[4]

This is the stability that needs to be restored to Earth quickly before excessive damage to life on Earth is the outcome.

GHG molecules absorb thermal (infrared) energy, resulting in a rise in their temperature. Because GHG molecules radiate heat in all directions, some of it spreads downward and ultimately comes back into contact with the Earth's surface, where it is absorbed. The temperature of the surface becomes higher than it would be if it were heated only by direct solar heating (sunshine). This supplemental heating of the Earth's surface by the atmosphere is the natural greenhouse effect. So, it is important to note that while much of the ability of Earth to support life depends on the water vapour in the atmosphere providing a 'blanket' around the Earth – the greenhouse effect – it is the additional molecules in the atmosphere added by humankind (principally CO_2 and CH_4) which strengthen this greenhouse effect and lead the Earth's energy system's attempt to return to equilibrium.

---

[4] Wikipedia https://en.wikipedia.org/wiki/Radiative_forcing and NASA Earth Observatory, *Op. Cit.* https://earthobservatory.nasa.gov/features/EnergyBalance.

Table A1.1. Radiative forcing over time.

Year	**Radiative Forcing (RF) Relative to 1750 (W m^{-2})**
1750	0.0
1950	0.57
1980	1.25
2011	2.29

Note: Prior to 1750, before the Industrial Revolution, Earth's average RF remained relatively stable. To document how the atmosphere has changed since then, scientists calculate current RF levels as if it were zero in 1750.

However, as the concentration of GHGs in the atmosphere increases, this equilibrium can only be achieved at an increasingly higher temperature. If the proportion of GHGs in the atmosphere is continually increasing as a result of the annual release of CO_2 and CH_4 into the atmosphere, then over time the Earth's temperature will rise and continue to rise and remain high, possibly for centuries. However, estimates of climate sensitivity are hugely important to climate policy and have been a key focus of the IPCC. However, the exact number has been difficult to determine due to the complexity of the processes involved and the relatively short historical records. The AR5 reported a likely range of 1.5C–4.7C and AR6 reported 2.5C–4C per doubling of CO_2.

A1.1.3 *Radiative Forcing and Its Impact on Earth*

In accordance with the basic laws of thermodynamics, as Earth absorbs energy from the sun, it must eventually emit an equal amount of energy to space if its temperature is to remain constant. The difference between incoming and outgoing radiation is a planet's radiative forcing (RF). Thus, Incoming Energy – Outgoing Energy = Radiative Forcing. So, if more of the energy that reaches Earth stays on Earth, this will force a change in the Earth's temperature. Table A1.1[5] shows how much radiative forcing there has been up to 2011. Thus, while the energy beaming onto the Earth is, on average, 340 W/m^2 by 2011, an additional 2.29 W/m^2 of Earth's surface was being absorbed by land and ocean rather than being sent back out into space, thus raising Earth's temperature.

[5]NOAA Climate.gov https://www.climate.gov/maps-data/primer/climate-forcing. *Source*: IPCC AR5 WG1 Figure SPM.5.

A1.2 Human Influence Causes Heating of the Climate System

In Figure A1.3,[6] Earth's energy budget compares the flows of incoming and outgoing energy that are relevant for the climate system. Since at least the 1970, less energy is glowing out than is flowing in, which leads to excess energy being absorbed by the ocean, land, ice, and atmosphere, with the ocean absorbing 91%.

Climate scientists[7] have defined four possible *scenarios for our future* based on possible amounts of radiative forcing. Which one will come to pass depends on the actions that are taken now to limit the causes of such forcing. Each scenario is based on a plausible future *pathway* regarding global emissions of GHGs. The scenarios, known as *Representative Concentration Pathways*, or RCPs, in Table A1.2,[8] specify the amount of

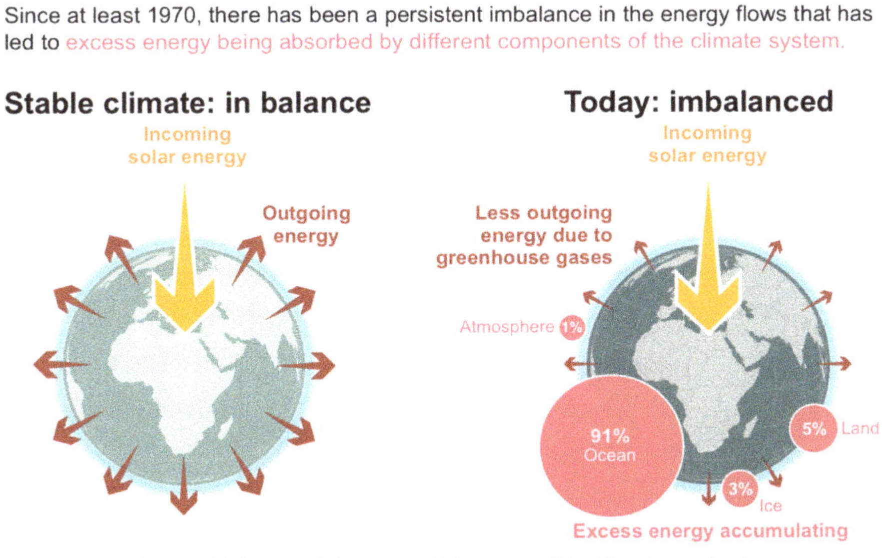

Figure A1.3. Explaining a stable or unstable climate on Earth.

[6]From IPCC AR6.

[7]IPCC Fifth Assessment Report Summary for Policymakers: IPCC, 2014: *Climate Change 2014: Synthesis Report. Contribution of Working Groups I, II and III to the Fifth Assessment Report of the Intergovernmental Panel on Climate Change* [Core Writing Team, R.K. Pachauri and L.A. Meyer (eds.)]. IPCC, Geneva, Switzerland, 151 pp. https://www.ipcc.ch/site/assets/uploads/2018/02/SYR_AR5_FINAL_full.pdf.

[8]IPCC Fifth Assessment Report Summary for Policymakers, Figure SPM.5 and Box SPM.1.

Table A1.2. Four scenarios for our future.

Representative Concentration Pathway (RCP)	Forcing Compared to 1750 ($W\ m^{-2}$)	Climate Policy Associated with Scenario	CO_2 Equivalent (ppm)	Projected Global Average Temperature Increase Over 1986–2005 (°C)
2.6	2.6	Mitigation	475	1.0
4.5	4.5	Stabilisation	630	1.8
6.0	6.0	Stabilisation	800	2.2
8.5	8.5	None	1313	3.7

radiative forcing (RF) in 2100 relative to 1750 and are determined by the parts per million of CO_2 in the atmosphere. Thus, the first physical connection is between how much CO_2 (and other GHGs) we put into the atmosphere and the extent to which this disturbs the balance between the energy landing on Earth in the form of sunlight and the heat leaving the Earth as infrared radiation – hence radiative forcing. The final column shows the impact of each pathway in terms of temperature rise.

The RCPs shown above were developed in 2014 by the Intergovernmental Panel on Climate Change (IPCC) and are a key part of their Fifth Assessment Report (AR5). The pathways show four possible scenarios for CO_2 output leading to a higher concentration in the atmosphere and hence the resultant climate forcing in watts per square metre. The Bank of England uses the following explanation in its Report on the Impact of Climate Change on the UK Insurance Sector:

> The scenarios of human influence underlying the AR5 projections are known as RCPs, because they are expressed in terms of greenhouse gas *concentrations* (the result of emissions) rather than emission levels. Each RCP implies a different amount of human-driven climate change (i.e. each RCP results in a different amount of extra heat energy being stored in the Earth system because of greenhouse gas emissions). The scenarios are developed using assumptions concerning economic growth, choices of technology and land-use and reflect a wide range of possible mitigation actions.[9]

[9]University of Cambridge (2013), 'Climate Change: Action, trends and implications for business'.

There are clear linkages between the factors we have been discussing, from radiative forcing through surface temperature change and economic and other impacts and ultimately to policy actions, such as the likelihood of the so-called 'inevitable policy response', including a high carbon tax. CO_2 emissions in the atmosphere are already changing the pH value of the oceans, making them more acidic, which gives ocean life survival problems. That is a direct effect of CO_2 emissions. But it is the indirect effect from CO_2 leading to higher Earth surface temperatures that is critical in relation to life on land.

Land-use change (LUC) is an important source of GHGs. LUC emissions account for around 11% of the global total GHG emissions. The British Ecological Society has recently focused on an area of LUC often not considered and that is changes in emissions from peat bogs (which, like coal, comes from decayed vegetation laid down over centuries). Peat is a fuel in many countries such as India, and in developed countries it is used to improve the soil in gardens and in horticulture. Globally, according to Wikipedia, peat stores up to 550 gigatons of carbon, representing 42% of all soil carbon, which exceeds the carbon stored in all other vegetation types, including the world's forests. This is more than 50 times global GHG emissions in 2019.

The problem that peat presents is that if peat bogs are drained as they are when peat is being harvested or a peat bog catches fire as they often do, then the CO_2 is released. As growing human populations expand, peat harvesting might be expected to increase and peatlands may be stripped. As global warming raises fire risk, peat bogs may dimmish substantially or disappear as a result of fire, thereby putting large volumes of stored carbon into the atmosphere.

A1.2.1 *Impact of Climate Change on Life on Earth*

The degree of temperature is linked to impacts on humans, animals, and plants on Earth. But the impact of climate change does not scale linearly with temperature increase (some impacts scale much harder than others; every half a degree counts). A major result of the post-Paris 1.5°C special report from the IPCC is an evaluation of the impacts of climate change on life on Earth at different warming levels. The following text is from the Carbon Brief website[10]:

> At 1.5°C, risks will increase for "health, livelihoods, food security, water supply, human security and economic growth", the report says. At this level of global warming, many low-elevation and small glaciers around the world would lose most of their mass or disappear, the report says with high confidence. Coral reefs are expected to decline by a further 70–90%, it adds with high confidence.
>
> At 2°C, risks associated with extreme weather events will transition to "very high", the report says with medium confidence. At this level of warming, changes in food availability and diet quality could increase nutrition-related diseases and undernourishment for up to "hundreds of millions of people", particularly among low-income households in sub-Saharan Africa, south Asia and central America, the report says with high confidence.
>
> At 3°C, "risks in many sectors and regions reach high or very high levels, implying widespread systemic impacts", the report says. The number of endemic species in biodiversity hotspots at a very high risk of extinction is expected to be 10 times higher than at 1.5°C, it says with medium confidence.
>
> At 4°C and above, around half of tropical marine species could face local extinction, the report says with medium confidence. Around four billion people could face water scarcity, it says with medium confidence. It adds that the global area burned by wildfires could increase by 50–70% (medium confidence).

A1.2.2 *Climate Tipping Points (Too Risky to Bet Against) and Global Cascades*

'Tipping point' refers to a threshold at which a tiny disturbance (sometimes jokingly described as a butterfly flapping its wings) can alter the state of, or the development of, a system. In the case of the Earth system, a 'tipping element' is something that causes a component of the Earth's

[10] https://www.carbonbrief.org/qa-ipcc-wraps-up-its-most-in-depth-assessment-of-climate-change/#warming.

climate system to pass a tipping point. This is explained in an article in *Nature*[11]:

> If current national pledges to reduce greenhouse-gas emissions are implemented – and that's a big 'if' – they are likely to result in at least 3°C of global warming. Some economists, assuming that climate tipping points are of very low probability (even if they would be catastrophic), have suggested that 3°C warming is optimal from a cost–benefit perspective. However, if tipping points are looking more likely, then the 'optimal policy' recommendation of simple cost–benefit climate-economy models align with those of the recent IPCC report. In other words, warming must be limited to 1.5°C. This requires an emergency response.
>
> In our view, the clearest emergency would be if we were approaching a global cascade of tipping points that led to a new, less habitable, 'hothouse' climate state. Interactions could happen through ocean and atmospheric circulation or through feedback that increase greenhouse-gas levels and global temperature. Alternatively, strong cloud feedback could cause a global tipping point. We argue that cascading effects might be common.

Cascading effects (multiple tipping points triggering one after another) suggest the need to protect the world from such possible climate 'outliers'. The normal mechanism used by society for such risks is insurance. But in this case, our societal response needs to be through regulation and other mechanisms. Such mechanisms would include relative price changes between fossil fuels and clean fuels by, for example, ceasing to subsidise fossil fuels and putting a tax (carbon tax) on fossil fuels. Another would be subsidising clean energy technologies until they reached the scale of manufacture at which they were highly economic in use. Although Integrated Assessment Models (IAMs), which we discuss in the following, are widely used as a basis for policymaking, they are unable to account for the *tail risks*, i.e. the risks resulting from tipping points and cascades, which lead to much more dramatic climate change than is envisaged in IAMs.

[11] https://www.nature.com/articles/d41586-019-03595-0#ref-CR4.

A1.2.3 *Integrated Assessment Models and IPPC AR6*

General Circulation Models (GCMs) that we looked at above are driven by the fundamental physics of the atmosphere, the oceans, and the Earth's land surface. They were later enhanced in the so-called Earth System Models (ESMs), which considered in greater detail the complex interactions between the atmosphere, the oceans, and the land surface. Later still, so-called Integrated Assessment Models (IAMs) were developed.

IAMs are an interdisciplinary research platform which combines a climate science module in which GHG emissions are linked to temperature increases with an economic module which links such increases in temperature to economic and policy outcomes.

The interaction of human and Earth systems requires that these models analyse not only physical interactions between atmosphere, oceans, and land surface but also the human aspects, which, in this case, means incorporating economic and social data. This should enable these models to project future GHG emissions and climate impacts. Most models use the Representative Concentration Pathways (RCPs), which we discussed earlier, as the foundation for various temperature pathways that the world may take. As decision-making is what we are concerned with, IAMs enable decision-makers to move towards the necessary requirement of understanding the potential economic impact of climate change on industries, countries, regions, and corporate real assets. This then allows those in the finance industry to estimate the likely impact on the financial assets in their portfolios.

Carbon Brief have produced an excellent 'Explainer'[12] on IAMs.

Depending on the IAM, researchers must answer myriad questions before they can start their work. The answers to these questions are the assumptions that regulate how the model behaves. For example:

> What is the relationship between income and energy use?
>
> How does demand for private transport change over time, as incomes increase and as populations move away from rural areas towards urban centres?

---

[12] Carbon Brief 'Q&A: How integrated assessment models are used to study climate change 2018' https://www.carbonbrief.org/qa-how-integrated-assessment-models-are-used-to-study-climate-change.

> What can we say about average diets in relation to wealth and in particular demand for beef and other ruminant protein sources relative to previous diet and the likelihood of the populations of developed countries shifting dramatically from animal protein to other protein sources including plant and insect sources?

Researchers have worked on modelling possible changes in socioeconomic factors over time. These changes include population growth, economic growth, education levels, the rate of urbanisation, and the rate of technological change. These groups of researchers have agreed on what are known as 'Shared Socioeconomic Pathways' (SSPs).

Although IAMs are used by the IPCC to explore some of the relationships between society and the natural world, their modelling limitations must also be acknowledged. They use what are known as 'general equilibrium models' based on 'neoclassical' economics assumptions. Many would consider these assumptions about how capital markets function as unrealistic and in particular the use of the rational expectations hypothesis. Newer models also allow for 'cascade effects'.

Models which allow for nonlinearities may in future be able to provide better predictions. Such models may also, of course, demonstrate the high probability of much more severe outcomes than general equilibrium models. In addition, because of different assumptions built into different models, their outputs differ, sometimes substantially. Carbon Brief also points out that those models that suggest there is less need for urgent action on reducing carbon emissions are likely to be those that rely on a *deus ex machina* being developed and available on a large scale, sometime in the future that will enable carbon capture and storage and direct air capture of CO_2, to 'solve' the problem at some point.[13] They may also use a high discount rate, which weighs near-term costs much more than future costs. This means there is a tendency in such models to delay action today in favour of hoped-for technology in the future, particularly given the very high cost of carbon capture, transport, and storage.

Although they could cause governments and regulators to limit their climate intervention, they do provide a guide and a platform to enable rational debate on how to move ahead with the additional actions urgently

[13] Translated as 'God from a Machine'. *Deus ex Machina* is a phrase used to describe any situation where something unexpected or **implausible** is brought into a storyline to resolve a situation, in this case an unwillingness to do anything about climate change.

required to give the world a chance of minimising adverse outcomes for Earth and its human, animal, and plant life. The fact of uncertainty about the severity of outcomes from GHG emissions, rather than being a reason to slow down action, is a reason to speed up inter-governmental action given the high probability of 'tail' outcomes resulting from, for example, further loss of Amazon forest or Antarctic ice sheet.

It must also be accepted that whilst climate models are based on physical laws which don't change, IAMs involve human behaviour which does change in response to incentives such as taxes, subsidies, new technologies like plant-based foods, and regulation. Socio-economic models cannot easily predict human behaviour in response to price changes, diets, need for transportation, etc. Indeed, some would argue that we are in a situation of 'Radical Uncertainty' (see Section 5.3.1), where it is not possible to 'scientifically' calculate probabilities of damage from physical or transition risk. In such a world, they would argue that the only answer is what is known as 'Precautionary Policy', where the focus must be on avoiding *tipping points* and building resilience into systems.

Recently, there has been greater concern in the United States about the implications of climate change – in particular on the implications of migrations from south of the globe to the north of the globe as a result of starvation in the global south. There are also concerns about a newly resurgent Russia on the back of its increasingly cultivatable land area as global warming leads to the tundra melting and increasing crop yields over larger areas of land at the same time as US crop yields are declining due to the same warming climate. In addition, Russia would benefit from an ice-free northeast passage for ships between Europe and Asia along the north coast of Russia. More recently, the high temperatures and forest fires in the northwest US and in western Canada in July 2021 and even more in the years since have made people ever more aware of the likely impact of climate change.

A1.2.4 *Limitations and Assumptions of Commonly Used Climate-Change Scenarios in Financial Services*[14]

History is replete with incidents in which we have placed too much faith in models, including the global financial crisis, the collapse of long-term

[14]The text provides excerpts from a paper (July 2023) by the Institute and Faculty of Actuaries and the University of Exeter written by Sandy Trust, Sanjay Joshi, Tim Lenton and Jack Oliver entitled 'Limitations and assumptions of commonly used climate-change

capital management, and, more recently, the UK pension scheme LDI crisis. A common theme is that models underestimate the level of risk because it is mathematically tractable to do so, in some cases due to underpinning assumptions that do not hold in the real world.

Consequently, many climate models are severely underestimating the economic impact of climate change because of the following reasons:

1. Modelled scenarios do not incorporate our experience on climate change and may not be realistic when compared to experience. There is also limited consideration of higher warming scenarios.
2. There is considerable uncertainty in key climate-system modelling assumptions, including how rapidly the climate will warm for a given level of emissions, which are likely to be prudent, meaning the carbon budget for limiting global warming to 1.5°C may now be exhausted. However, no margins are currently included to represent this uncertainty.
3. Damage functions that are used to estimate the economic impacts of climate change exclude many of the risks we expect to face, such as those impacts from tipping points or societal consequences such as involuntary mass migration.
4. General equilibrium models that are widely used to estimate economic impacts contain a number of simplifying assumptions that do not hold in the real world.

We conclude that it is essential for model users to be climate literate in order to understand the context in which models are used and the limitations that apply to climate models. They should be used alongside narrative scenarios to provide direction and rank on risks – but users must not interpret results as accurate. In particular, the observed benign results for the hot-house world are deeply flawed and underestimate the impact of the risks we expect to face.

---

scenarios in financial services' https://actuaries.org.uk/media/qeydewmk/the-emperor-s-new-climate-scenarios.pdf.

Appendix 2

Frameworks: The EU and the European Commission as Exemplars in the Creation of Sustainable Assets and Achieving the Paris Agreement

A2.1 The EU Regulatory Framework: Shaping Markets, Not Just Fixing Market Failures

It is certainly the case that the European Union is well ahead of other countries and regions in taking necessary measures to mitigate climate change risk. But it is possible that much more is needed in terms of phased-in regulation that prohibits more activities rather than simply trying to 'nudge' investors, industrialists, and the public to act 'responsibly'.

In conditions of Deep Uncertainty where there are threats of serious or irreversible damage from climate change, governments and regulators need to employ the Precautionary Principle, which in practice means radical regulatory action rather than market pricing alone. We strongly believe that governments (including the EU) and regulators should *shape* markets, not just think in terms of fixing market *failures*. It has to be active government policy rather than simply relying on market forces through, for example, small changes such as to capital ratios for 'brown' or 'green' assets. The *Inevitable Policy Response* is an example of the possible likely radical action. Such a large sudden crackdown on emitting industries, including farming, would result in large falls in values of affected assets.

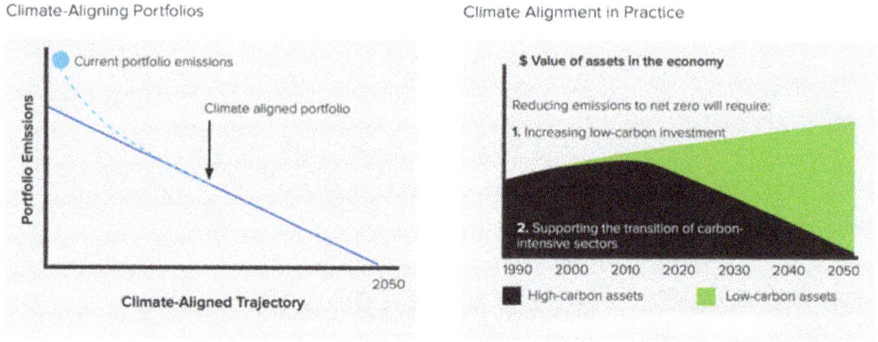

Figure A2.1. How to achieve climate aligned portfolios.

The European approach to sustainable finance and 'Paris compatibility' is still dominated by what old (and many young) bankers and others were taught in university finance classes. As a result, policy intervention has essentially been devoted to stimulating market efficiency based on what many consider to be the erroneous efficient market hypothesis (EMH) model. But market forces simply work too slowly.

If these policies are to be successful, they must help ensure that carbon emissions fall as shown in the first graphic in Figure A2.1 as a result of a transition shown in the second graphic (both from the Rocky Mountain Institute).[1] This shows the path that emissions must take for investors to achieve 'climate-aligned portfolios', i.e. what needs to be done to align investor portfolios with Paris.

The second graphic in Figure A2.1 shows that the transition involves the phasing out of carbon-intensive activities and the investment required to replace this through investment in low-carbon-producing assets. Thus, we are looking here at how to achieve climate-aligned finance. At present, world CO_2 emissions continue to increase annually; record temperatures have been felt across the world in summer 2022 and wildfires and other manifestations of climate change continue to cause huge losses to society.

We may now have to wait for the impact of a sudden climate crisis even greater than the heat and drought in Europe and China and other

---

[1] https://rmi.org/insight/navigating-five-barriers-to-climate-aligned-finance/. *Source* (figure on the right): Climate Finance Leadership Initiative, *Financing the Low-Carbon Future*, 2019, https://www.bloomberg.com/cfli.

countries in 2022 to drive governments to meet their Paris commitments. It must be remembered, however, that it is not currently possible to 'reverse' what has already happened, as climate impact depends on atmospheric concentrations of GHGs, not annual emissions (which just add to concentrations). It can take 100 years or very much longer (depending on which GHGs we are considering) for the impact of GHGs emitted today to dissipate, and thus we cannot reverse what has been happening over past decades, as it is unlikely that carbon removal technology (DACC) will be able to remove more than a small proportion of the existing concentration. All we can do with today's and tomorrow's technologies is to stop emitting GHGs as soon as possible – much sooner than 2050 if we are to have any chance of meeting 1.5°C or even 2°C.

A2.1.1 *The EU Action Plan and the European Green Deal for Financing Sustainable Growth*

The Action Plan[2] outlines ten reforms in three areas:

Reorient Capital Flows Towards Sustainable Investment in Order to Achieve Sustainable and Inclusive Growth

(1) establishing an EU classification system for sustainability activities,
(2) creating standards and labels for green financial products,
(3) fostering investment in sustainable projects,
(4) incorporating sustainability when providing investment advice,
(5) developing sustainability benchmarks.

Mainstreaming Sustainability into Risk Management

(6) better integrating sustainability in ratings and research,
(7) clarifying institutional investors and asset managers' duties,
(8) incorporating sustainability in prudential requirements.

Foster Transparency and Long-Termism in Financial and Economic Activity

(9) strengthening sustainability disclosure and accounting rulemaking,
(10) fostering sustainable corporate governance and attenuating short-termism in capital markets.

[2] https://www.unpri.org/download?ac=5173.

The four legislative proposals adopted in May 2018 address several areas of the Action Plan:

1. Taxonomy: A proposal for a regulation to establish a framework to facilitate sustainable investment. The proposed regulation establishes the conditions and the framework to create, over time, a unified classification system (or taxonomy) on what can be considered environmentally sustainable economic activities. This is widely seen as a first and essential enabling step in the overall effort to channel investments into sustainable activities.

2. Disclosure and Duties: A proposal for a regulation on disclosures relating to sustainable investment and sustainability risks, amending Directive (EU) 2016/2341(IORP 2 - Pensions). This regulation will introduce obligations on institutional investors and asset managers to disclose how they integrate ESG factors in their risk processes. Requirements to integrate ESG factors in investment decision-making processes, as part of their duties towards investors and beneficiaries, will be further specified through delegated acts.

3. Benchmarks: A proposal for a regulation amending the benchmark regulation. The proposed amendment will create a new category of benchmarks, comprising low-carbon and positive carbon impact benchmarks, to help investors better understand the relative carbon impact of their investments.

4. Sustainability Preferences (Consultation): In addition, the Commission has solicited feedback on amendments to delegated acts under the Markets in Financial Instruments Directive (MiFID II) and the Insurance Distribution Directive to include ESG considerations into the advice that investment firms and insurance distributors offer to their clients.

The European Green Deal, which was approved in 2020, is a roadmap with a much wider plan than the Action Plan for Financing for making the EU's economy sustainable by turning climate and environmental challenges into opportunities across all policy areas and making the transition just and inclusive for all. The European Green Deal is the EU's new growth strategy. It shows how to transform ways of living and working and of producing and consuming so that people live more healthily and make businesses innovative.

The Green Deal provides a roadmap with actions to boost the efficient use of resources by moving to a clean, circular economy and to stop climate change, revert biodiversity loss, and cut pollution. It outlines investments needed and financing tools available and explains *how to ensure a just and inclusive transition.* The Green Deal covers all sectors of the economy, notably transport, energy, agriculture, buildings, and industries such as steel, cement, ICT, textiles, and chemicals.

Achieving the 2030 climate and energy targets is estimated to require €260 billion of additional annual investment, representing about 1.5% of 2018 GDP. This will need the mobilisation of both the public and private sectors. The Green Deal Investment Plan will mobilise EU funding and create an enabling framework to facilitate and stimulate the public and private investments needed for the transition.

Complementing other initiatives announced under the Green Deal, the Plan is based on three dimensions: financing, enabling, and practical support. It will mobilise at least €1 trillion of sustainable investments over the next decade, provide incentives to unlock and redirect public and private investment, and provide support to public authorities and project promoters in planning, designing, and executing sustainable projects. The Commission has committed to at least 25% of the EU's long-term budget being for climate action, and beyond this, the European Investment Bank, Europe's climate bank, will also provide further support.

A2.2 The Role of the European Investment Bank

At the end of 2020, EU Member States approved the EIB Group Climate Bank Roadmap 2021–2025. In summary, the key points in the roadmap are as follows:

- **Support €1 trillion in investment for climate action and environmental sustainability in the decade to 2030.**
- **EIB Group is now aligning all financing activities from end of 2020 with the goals of the Paris Agreement.**
- **More than 50% of annual financing is dedicated to green investment by 2025.**
- **More green advisory services and financing of innovative low carbon technologies.**
- **Support for green capital markets, climate change adaptation, and Just Transition projects.**

Announcing the Roadmap, EIB Vice-President Ambroise Fayolle, who is responsible for climate action and environment, said in a speech:

> 2021–2030 is the critical decade to mobilise the trillions of investment that is required to limit global warming to 1.5 degrees and protect the environment and biodiversity. As requested by the European Council and the EU member states, I am proud to announce that the EIB Group has achieved an important milestone in its transformation to become Europe's fully fledged climate bank.

A2.2.1 *The Technical Expert Group (TEG) Report[3] and the Green Taxonomy*

The EU has been active in a wide range of ways to help Europe achieve sustainability. 'Sustainable finance' in the view of the EU refers to the process of *taking due account of environmental and social considerations when making investment decisions, leading to increased investment in longer-term, sustainable activities.* In July 2018, the European Commission set up a Technical Expert Group (TEG) on sustainable finance to assist it in four key areas of the Action Plan through the development of the following:

Mission

The group's tasks shall be to assist the Commission, notably in the development of:

- an EU *taxonomy* or classification system of climate change mitigation, climate change adaptation and other environmental activities;
- an EU *Green Bond Standard*;
- *benchmarks* for low-carbon investment strategies; and
- *metrics* for climate-related disclosures.

---

[3]TEG (2020) *European Commission Report on Climate-related Disclosures* https://ec.europa.eu/info/sites/info/files/business_economy_euro/banking_and_finance/documents/200309-sustainable-finance-teg-final-report-taxonomy_en.pdf.

The *Taxonomy* is the key output of the TEG and is the foundation for the other aspects of the Action Plan – the Green Bond Standard (GBS), benchmarks, and metrics. We cover in the following the Taxonomy, the Green Bond Standard, the EU Paris-Aligned Benchmarks, and the EU Climate Transition Benchmarks. Metrics have not yet been tackled by the TEG.

The EU Green Taxonomy Regulation (which resulted from the TEG Report) notes that the Taxonomy is 'the most important and urgent action envisaged by the EU Action Plan' and an important aspect of the EU Green Deal package.

Until now, there has been no common classification system at EU or global level that provides a definition of 'environmentally sustainable' economic activities. The green list[4] aims to:

- reduce fragmentation resulting from market-based initiatives and national practices;
- reduce 'greenwashing', i.e. the practice of marketing financial products as 'green' or 'sustainable', when in fact they do not meet basic environmental standards.

There are two more criteria that a project must meet. First, the project must not do any significant harm to any of the other environmental objectives, and second, it must meet the legal minimum safeguards.

Arguments have, however, arisen in particular over nuclear power (and also natural gas as a 'transition fuel'). Nuclear power is non-polluting and can provide 'clean' power, heat, and hydrogen from electrolysis and contributes to at least two objectives. The question is whether it could cause any harm in terms of the other environmental objectives.

On this question, there are strong views on both sides. The European Council decision to support the Finnish Presidency's wording in favour of 'renewable and climate-neutral energy sources' left the door open for nuclear energy to be classified as green in the new EU Sustainable Finance Taxonomy, which it was not originally. Indeed, the exclusion of nuclear energy on the basis of concerns about waste management may be quite invalid since all commercial nuclear activities in the EU (including spent fuel management) are already regulated to a 'Do No Significant Harm' (DNSH) standard through the laws, regulations, and procedures of

---

[4] https://ec.europa.eu/commission/presscorner/detail/en/ip_19_6793.

the EU. On the other hand, the world's scientific consensus is that nuclear energy is necessary to achieve climate change mitigation within the necessary timescale.

In March 2021, there was some resolution of this issue in a technical assessment[5] of nuclear energy with respect to the DNSH criterion of EU regulation.

A2.3 Technical Assessment of Nuclear Energy with Respect to the 'Do No Significant Harm' Criteria of Regulation (EU) 2020/852 ('Taxonomy Regulation') by the EU JRC[6]

Inclusion or exclusion of nuclear energy in the EU taxonomy was a debated subject throughout the negotiations on the EU Taxonomy Regulation. While there are indirect references in the regulation to the issue of nuclear energy (including on radioactive waste), co-legislators ultimately left the assessment of nuclear energy to the Commission as part of its work on the delegated acts establishing the technical screening criteria.

As the in-house science and knowledge service of the Commission with extensive technical expertise on nuclear energy and technology, the JRC was invited to carry out such analysis and to draft a technical assessment report on the DNSH aspects of nuclear energy including aspects related to the long-term management of high-level radioactive waste and spent nuclear fuel, consistent with the specifications of Articles 17 and 19 of the Taxonomy Regulation. This report is the result of that JRC analysis.

---

[5] *World Nuclear News* 29 March 2021 https://www.world-nuclear-news.org/Articles/EU-Taxonomy-can-move-forward-with-nuclear-JRC-find.

[6] The Joint Research Centre (JRC) is the European Commission's science and knowledge service. It aims to provide evidence-based scientific support to the European policymaking process. The text in this section is from a 387 page *JRC Science for Policy Report* published by the European Commission in 2021 under the title of the Chapter heading above. Available at https://publications.jrc.ec.europa.eu/repository/handle/JRC125953.

- The detailed assessment of the impacts of nuclear energy in its various lifecycle phases shows that all non-radiological effects and potential impact indicators are dominated by the mining and milling phase, except the GHG emissions, where Nuclear Power Plant (NPP) operation gives the largest contribution.
- The analyses did not reveal any science-based evidence that nuclear energy does more harm to human health or to the environment than other electricity production technologies already included in the Taxonomy as activities supporting climate change mitigation.
- The comparison of impacts of various electricity generation technologies (e.g. oil, gas, renewables and nuclear energy) on human health and the environment, based on recent lifecycle analyses (LCAs) presented in Chapter 3.2 of Part A, shows that the impacts of nuclear energy are mostly comparable with hydropower and the renewables, with regard to non-radiological effects.
- Management of radioactive waste and its safe and secure disposal is a necessary step in the lifecycle of all applications of nuclear science and technology (nuclear energy, research, industry, education, medical, and other). Radioactive waste is therefore generated in practically every country, the largest contribution coming from the nuclear energy lifecycle in countries operating nuclear power plants. Presently, there is broad scientific and technical consensus that disposal of high-level, long-lived radioactive waste in deep geologic formations is, at the state of today's knowledge, considered as an appropriate and safe means of isolating it from the biosphere for very long time scales.
- Measures to ensure that radioactive waste does not harm the public and the environment include a combination of technical solutions and an appropriate administrative, legal and regulatory framework. Although there remain contrasting views, it is generally acknowledged that the necessary technologies for geological disposal are now available and can be deployed when public and political conditions are favourable. No long-term operational experience is presently available as technologies and solutions are still in demonstration and testing phase, moving towards the first stage of operational implementation. Finland, Sweden, and France are in an advanced stage of implementation of their national deep geological disposal facilities, which are expected to start operation within the present decade.

The comparison of environmental impacts of various electricity generation technologies on human health and the environment, leads to the following main findings:

- Average lifecycle GHG emissions determined for electricity production from nuclear energy are comparable to the values characteristic to hydropower and wind.
- Nuclear energy has very low NOx (nitrous oxides), SO_2 (sulphur dioxide), PM (particulate matter) and NMVOC (non-methane volatile organic compounds) emissions. The values are comparable to or better than the corresponding emissions from the solar PV and wind energy chains.
- Land occupation of nuclear energy generation is about the same as for an equivalent capacity gas-fired power plant but significantly smaller than wind or solar PV.

Resolution of Taxonomy Issues Relating to Natural Gas and Nuclear, 1 January 2022

On 1 January 2022, the European Commission issued a press release headed 'EU Taxonomy: Commission begins expert consultations on Complementary Delegated Act covering certain nuclear and gas activities':

> Taking account of scientific advice and current technological progress, as well as varying transition challenges across Member States, the Commission considers there is a role for natural gas and nuclear as a means to facilitate the transition towards a predominantly renewable-based future. Within the Taxonomy framework, this would mean classifying these energy sources under clear and tight conditions (for example, gas must come from renewable sources or have low emissions by 2035), in particular as they contribute to the transition to climate neutrality.

The Commission notes that 'the activities covered in this complementary Delegated Act would accelerate the phase out of more harmful sources, such as coal, and in moving us towards a lower-carbon energy mix'.

The EU Taxonomy or 'unified classification system' *bridges the gap between international goals and investment practice and signals the types of activities that are consistent with the low-carbon transition, adaptation and other environmental objectives*. It puts environmental data in context and helps investors understand which companies are contributing to the low-carbon transition. In the security selection process undertaken by investors, the Taxonomy should enable them to identify companies whose carbon footprints and, very importantly, *transition measures* are actively contributing to environmental sustainability and are compatible with EU objectives. The EU's performance standards under the Taxonomy, or technical screening criteria as they are known, enable identification of an eligible economic activity which truly sustainable and thus 'taxonomy-aligned'.

A2.4 Reform of the EU Electricity Market Design

Clean Energy Package for All Europeans – 2019 (https://energy.ec.europa.eu/topics/energy-strategy/clean-energy-all-europeans-package_en).

In May 2019, the Council of ministers of the EU formally adopted four new pieces of EU legislation that redesign the EU electricity market to make it fit for the future. This concludes the remaining elements of the Clean Energy for All Europeans package and represents a major step towards completing the Energy Union, delivering on the priorities of the Juncker Commission. The four new pieces of legislation are in the areas of efficient energy use, more renewables, better governance of the energy union, more rights for consumers, and a smarter and more efficient energy market.

A2.4.1 *The EU Green Bond Standard*[7]

The ICMA Green Bond Principles (ICMA GBPs), which we covered above, were produced well before the TEG issued its report on the question of 'what is green'. Under these principles (ICMA GBPs), the

[7]EU Technical Expert Group (TEG) on Sustainable Finance: *Usability Guide: EU Green Bond Standard* published in March 2020 https://ec.europa.eu/info/sites/info/files/business_economy_euro/banking_and_finance/documents/200309-sustainable-finance-teg-green-bond-standard-usability-guide_en.pdf.

cornerstone is that the proceeds of a green bond issue can only be for green projects, which should be appropriately described in the legal documentation for the security. All designated green projects should provide clear environmental benefits, which will be assessed and, where feasible, quantified by the issuer.

One objective of the EU Green Taxonomy was to provide definitions of what is a green project and what is not – something that has been lacking up to now. The benefit of having the EU Green Bond Standard (EUGBS) is to help achieve the aims of the Action Plan to reorient capital flows towards sustainable investment. EU green bonds are a capital market instrument that is fully aligned with that aim. The EU GBS should not only aid in overcoming the biggest barriers to further grow this market but also create a strong foundation to accelerate the flow of capital towards the EU's environmental objectives. Indeed, the TEG Report goes on to say that 'green bonds have provided bond investors with the capacity to become involved in corporate environmental strategies. It has also enabled bond markets to become a powerful force in green finance, notably including climate mitigation finance'.

Once the taxonomy had been developed by ICMA's Principles Community, it became much easier to propose an EU GBS, building on ICMA's existing GBPs. It is therefore an example of an industry association leading by developing standards, but given that it cannot impose these on its members nor create a taxonomy, it is appropriate for a governmental body to enhance the standards and at some point to make them mandatory for a 'green bond'. While currently voluntary (like TCFD), it is not likely to remain voluntary for any entity wishing to issue a green bond.

The new EU GBS can be applied by any issuer in any country, including those outside of the EU. There are four key requirements under the framework:

1. The funds raised by the bond should be allocated fully to projects aligned with the EU Taxonomy.
2. There must be full transparency on how bond proceeds are allocated through detailed reporting requirements.
3. All EU green bonds must be checked by an external reviewer to ensure compliance with the regulation and that funded projects are aligned with the Taxonomy. Specific, limited flexibility is foreseen here for sovereign issuers.

4. External reviewers providing services to issuers of EU green bonds must be registered with and supervised by the European Securities Markets Authority. This will ensure the quality and reliability of their services and reviews to protect investors and ensure market integrity. Specific, limited flexibility is foreseen here for sovereign issuers.

A2.4.2 *The EU/EC as an Exemplar on Sustainability (Including the ETS)*

The EU has been ahead of central banks and the BIS in taking action on disclosure of 'non-financial' information. But as of the present date, the TCFD framework is essentially the one that it has adopted for future disclosure, at least on climate change. A key early document (and accompanying legislation) was the *Single Market Act—Twelve levers to boost growth and strengthen confidence*, published by the Commission in 2011, which identified the need to improve financial and non-financial companies' *disclosure of social and environmental information* by presenting a legislative proposal in this field. Thus, the European Commission sees a positive connection between non-financial information disclosure and environmentally sound growth.

The European Parliament also acknowledged the importance of businesses divulging information on sustainability such as social and environmental matters with a view to identifying sustainability risks and increasing both institutional investor and retail consumer trust. Indeed, the European Parliament acknowledged that the disclosure of non-financial (sustainability) information was vital for managing the transition towards a sustainable global economy by combining long-term profitability with social justice and environmental protection – the profits, people, planet (PPP) perspective.

The concept of each country setting its own Nationally Determined Contribution (NDC) to the goals of the Paris Agreement is a key starting point for understanding the role of the finance industry and its corporate customers in enabling a government to achieve its Paris Agreement Commitments. It is based on the fact that at the present time, the countries of the world combined are emitting 55 billion tons (CO2e) of GHG each year resulting in the concentration of long-lived GHGs in the atmosphere increasing year by year. In order to achieve the 1.5°C or well under than

2°C degree target to which all governments have committed, there is a need to achieve 'net-zero' emissions and hence no increase in concentration. There is a limited number of years available to reach net-zero emissions if the world chooses to limit the temperature rise to 'well under 2°C'. At a maximum, this is 30 years, i.e. by 2050.

Transitioning the financial system in financial firms to maximise the likelihood of a country's financial system being climate aligned requires understanding the current financial market structure and how it must change towards meeting society's new needs. Given that the world in 2020 was estimated to be 1.25–1.4°C warmer than in pre-industrial times, there is only a small, *remaining carbon budget.* To meet the goal of limiting global warming to 1.5°C, the whole world now needs to reduce emissions by 7% per year every single year in the next decade between 2020 and 2030! However, that 7% is a global average. Some countries would need to decarbonise at perhaps twice that rate yet at present global emissions are still rising.

Earlier we looked at why it is critical to the success of the Paris Agreement process that the public sector and the private sector (market approach) work together to achieve the Paris goals. Figure A2.2 from the Paris-based *Institute for Climate Economics* (IC4E)[8] is a useful summary of the main reasons why public intervention is necessary on the corporate front (numbers 1, 2, and 3) and why the same public intervention is necessary in the case of the financial sector (numbers 4, 5, and 6).

The European Commission has seen the need for legislation and regulation to minimise climate damage and sees the finance industry as the key player in achieving its goal of having no net GHG emissions across the EU by 2050 and decoupling economic growth from resource use. It sees the finance industry as being key because of its linkages to the highly polluting sectors of the economy, i.e. the lending, investing, capital-raising, and insuring services, that facilitate the growth of an economy. This viewpoint would also apply to most other governmental and regulatory agencies worldwide.

As the finance sector currently facilitates the growth of high-GHG-emitting companies ('financed emissions' when a bank itself provides financing or 'facilitated emissions' when a bank or investment bank underwrites a bond issue), the question the Commission always asks, as a representative of society's concerns in this area, is how it can put pressure

---

[8] https://www.i4ce.org/home/. Adapted from I4CE.

Figure A2.2. The limitations of the market approach to green its activities.

on financiers, and thereby on the highly GHG-emitting corporate sectors as well as those sectors that are destroying natural capital, in order to achieve its 2050 goal. The Commission cannot overcome all the limitations in the graphic above, but its regulatory intervention can certainly contribute. While there are many other issues that are pertinent to its objectives, it believes that, as shown in the following, a combination of the first five points plus (No. 6) can make a substantial impact:

1. a higher level of carbon tax or a revised cap and trade system;
2. improvement in non-financial data disclosure (ESG disclosures);
3. providing incentives for the finance industry to provide additional financing for green projects;
4. the provision of *risk mitigation services* by the European Investment Bank;
5. regulations on infrastructure project design, construction and operation;
6. self-regulation for example through trade associations such as ICMA and through stock exchanges.

No. 1 Above Refers to the EU Emissions Trading System (ETS)

If we look now at the 'effective carbon tax rate' in the EU, it has been much lower than the actual rate ($126 per metric ton) in Sweden or the IMFs estimate of the rate necessary to achieve the desired objective. The level of the tax in the EU is determined through the EU emissions trading system (EU ETS) which they call 'a cornerstone of the EU's policy to combat climate change and it is a key tool for reducing greenhouse gas emissions cost-effectively'. It is the world's first major carbon market and remains the largest. For some years the price was very low and indeed it declined to around €5 as recently as 2016. It then rose to nearly €30 a decade ago before falling again to below €20 in March and April 2020 as economic activity fell across the union as a result of the COVID-19 outbreak. By January 2021, it had risen again to over €30 for the first time in over a decade and hit nearly €100 twice in 2022 and then again in February 2023.

It is not a straightforward carbon tax but a system whereby companies are given EU Emission Allowances (EUAs). Some companies have allowances they don't need, while others have a need for more allowances. Companies are then permitted to trade allowances.

In Germany, the majority of MPs agreed in 2020 to setting the price for one tonne of CO_2 at €25 as of 1 January 2021 for companies that put petrol, diesel, heating oil, and natural gas into circulation. Unlike in the past, they will now have to buy emission allowances. The price will increase to €55 per tonne by 2025 resulting in a litre of petrol becoming 7 cents more expensive and natural gas 0.6 cents per kilowatt-hour. China also recently introduced a carbon trading system.

A2.5 We Cover the EU Corporate Sustainability Reporting Directive (CSRD) and European Sustainability Reporting Standards (ESRS) in Section 7.2

A2.5.1 *EU Corporate Sustainability Due Diligence Directive*[9] *(CSDDD)*

Undertaking 'due diligence' means making enquiries in respect of a company to uncover material information in a careful manner by putting in the 'requisite effort' to be thorough in that investigation.

In December 2022, the Council and the European Parliament today reached a *provisional* deal on the corporate sustainability due diligence

[9]Much of the information in this chapter is based on the European Commission document on this proposed Directive. Available at https://eur-lex.europa.eu/legal-content/EN/TXT/?uri=CELEX%3A52022PC0071.

directive (CSDDD), which aims to enhance the protection of the environment and human rights in the EU and globally. The due diligence directive will set obligations for large companies regarding actual and potential adverse impacts on human rights and the environment, with respect to their own operations, those of their subsidiaries, and those carried out by their business partners.

In February 2022, the European Commission issued draft regulations on both human rights and environmental due diligence. Requiring due diligence is an attempt to ensure that in these two important areas for achieving sustainability, directors of companies do not simply cast a 'blind eye' at what is happening in their company and in this case also in their whole value chain. This proposed Directive is part of an attempt by the Commission to improve corporate governance in respect of sustainability. EU companies and, importantly, non-EU companies doing more than a minimum amount of business in the EU will be required to assess both environmental impacts and human rights matters not only in their own operations but equally in their supply chains. They will be required to take action to either prevent, or if not to mitigate, any environmental or human rights harms. In respect of both the social and the environmental categories, this is potentially a huge step-up in requirements and is likely to lead to more civil cases of the type we considered above. The damages involved could be considerable and are thus highly relevant to financial firms which are considered to have provided the supply chain input (cash) to enable 'financed emissions' or their underwriting skills to provide 'facilitation of bond issues'

The need for this Directive seems to have arisen from the fact that voluntary approaches in both areas – environmental and human rights (E&S) – have failed to stop abuses or to limit GHG emissions. The Directive complements NFRD (and thus CSRD when it comes into effect). It also complements the Taxonomy. There will thus be a substantive requirement to perform due diligence to identify, prevent, mitigate and account for external harm resulting from adverse human rights and environmental impacts in a company's own operations, its subsidiaries and their value chains.

In December 2022, the Council and the European Parliament reached a *provisional* deal on the CSDDD, which aims to enhance the protection of the environment and human rights in the EU and globally. The due diligence directive will set obligations for large companies regarding actual and potential adverse impacts on human rights and the environment, with respect to their own operations, those of their subsidiaries, and those carried out by their business partners.

CSDDD will enhance information flow by mandating companies to set up additional risk management systems to those which most already have in place to report on their findings in the areas relevant to this new due diligence system, whereas up until now or at least when CSDDD is implemented, due diligence systems have been optional and are little more than a by-product from an obligation (i.e. not easily enforceable) to report under the NFRD.

The Directive could make litigation against corporates and possibly also financial firms more likely to succeed than in the past. According to Lara Wolters, a Dutch MEP:

> This new law on corporate due diligence will set the standard for responsible business conduct in Europe and beyond. We refuse to accept that deforestation or forced labour are part of global supply chains. Companies will have to avoid and address harm done to people and planet in their supply chains. The new rules will give victims a legal right to access support and to seek reparations, and will ensure fairness, a level playing field and legal clarity for all businesses, workers and consumers.

The purpose of this new directive is to improve environmental and social governance by making the absence of effective due diligence, either on environmental or social issues, grounds for a claim for reparation. The word 'reparations' is derived from the word 'repair, i.e. to return something to how it was in the past before it was damaged. In the case of the climate problem, all that can be done is to try to compensate for damage to property.

The Directive is designed to help deliver the Green Deal, which requires the EU to:

- reach net-zero GHG emissions by 2050;
- decouple economic growth from resource use;
- leave no person and no place behind.

However, achieving this will require a rapid doubling or tripling of the flow of capital into existing types of sustainable investment and new innovative energy technologies. The OECD estimates that further investments of $630 billion a year worldwide for the next decade will be needed to have even a 66% chance of limiting the temperature increase on the earth's surface to below 2 degrees.

The Directive is also designed to help achieve the *United Nations Sustainable Development Goals (UN SDGs)*, which we cover in Appendix A6.1, in both the environmental and social fields in the European Union. Achieving these objectives:

> requires implementing comprehensive mitigation processes for adverse human rights and environmental impacts in their value chains,
>
> integrating sustainability into corporate governance and management systems, and
>
> framing business decisions in terms of human rights, climate and environmental impact, as well as in terms of the company's resilience in the longer term.

The Commission notes that 'Identifying these adverse impacts in value chains will become easier if more companies exercise due diligence and thus more data is available on human rights and environmental adverse impacts'. This is in line with the objective of trying to ensure that companies do take the required effort to undertake due diligence not only on their own operations but those of suppliers and, we presume, users of their products.

A key reason for introducing this Directive is that attempts to achieve the same outcome by voluntary means have been unsuccessful as noted here:

> Voluntary action does not appear to have resulted in large scale improvement across sectors and, as a consequence, negative externalities from EU production and consumption are being observed both inside and outside the Union. Certain EU companies have been associated with adverse human rights and environmental impacts, including in their value chains. Adverse impacts include, in particular, human rights issues such as forced labour, child labour, inadequate workplace health and safety, exploitation of workers, and environmental impacts such as greenhouse gas emissions, pollution, or biodiversity loss and ecosystem degradation.

The Commission believes that the Directive will:

(1) improve corporate governance practices to better integrate risk management and mitigation processes of human rights and environmental risks and impacts, including those stemming from value chains, into corporate strategies;
(2) avoid fragmentation of due diligence requirements in the single market and create legal certainty for businesses and stakeholders as regards expected behaviour and liability;
(3) increase corporate accountability for adverse impacts, and ensure coherence for companies regarding obligations under existing and proposed EU initiatives on responsible business conduct;
(4) improve access to remedies for those affected by adverse human rights and environmental impacts of corporate behaviour. This latter objective on access to remedies is one that may have considerable.

> The Directive introduces directors' duties to set-up and oversee the implementation of due diligence and to integrate it into the corporate strategy. In addition, when fulfilling their duty to act in the best interest of the company, directors must take into account the consequences of their decisions for sustainability matters, including human rights, climate change and environmental consequences. In general, Member States shall ensure that national provisions on breach of directors' duties also apply to these provisions, *e.g. by adding personal liability of directors for non-compliance with sustainability.* Furthermore, Group 1 companies (large companies as defined in this Proposal) need to adopt a plan to ensure that their business strategy is compatible with limiting global warming to 1.5°C in line with the Paris Agreement.

The following is a case study of a company which, according to an NGO, failed to properly manage climate risk, which means that it is breaching its legal duties

Case Study: ClientEarth v Shell

ClientEarth is an NGO which takes up cases and defines itself as a non-profit organisation that uses the law to create systemic change that protects the Earth for – and with – its inhabitants. Its approach in the Shell case is somewhat unexpected. It has a small ownership of Shell shares and is taking action against Shell as a shareholder on the grounds that its behaviour is harming shareholders in general. It alleges that Shell's directors are mismanaging the risks of climate change through an inadequate plan for the energy transition and that this leaves Shell's long-term commercial viability and competitiveness in danger due to the UK government's Net Zero Strategy.

ClientEarth's[10] lawyer argues that the Board's failure to properly manage climate risk means that it is breaching its legal duties. The Board has failed to adopt and implement a climate strategy that truly aligns with the Paris Agreement goal to keep global temperature rises to below 1.5°C by 2050. We believe the Board is breaching its duties under sections 172 and 174 of the UK Companies Act, which legally requires it to act in a way that promotes the company's success, and to exercise reasonable care, skill and diligence.

This will be the first time ever that a company's board has been challenged on its failure to properly prepare for the net zero transition.

ClientEarth is pursuing shareholder litigation to compel Shell's Board to act in the best long-term interests of the company by strengthening Shell's climate plans. Its current strategy and insufficient targets put the enduring commercial success of the company and employees' jobs at risk and is not good for people or the planet. We want to make sure the Board's 'wait and see' approach to the energy transition does not come at the expense of long-term viability for the company's stakeholders, including shareholders and employees. Putting in place sufficient targets to reduce its emissions over the next 3, 5, and 10 years in order to meet net zero will secure the company's long-term value while protecting investors' capital and the climate.

It is taking 'derivative action' against Shell in the UK. A derivative action is a claim brought by a shareholder of the company, ultimately on behalf of the company, in this case to argue alleged breaches of duty by the Board. That means the shareholder bringing the claim is effectively seeking to step into the company's shoes, to pursue the Board for wrongs allegedly committed against the company.

Institutional investors are increasingly calling on all companies to align their business plans with the goals of the Paris Agreement. ClientEarth believes Shell is not responding to this objective of

[10]This section based on material on ClientEarth's website at https://www.clientearth.org/latest/latest-updates/news/we-re-taking-legal-action-against-shell-s-board-for-mismanaging-climate-risk/.

shareholders. As an example of this, in 2021, Shell was ordered by a Dutch court to reduce its overall emissions – including those from the fossil fuel products it sells – by a net 45% by the end of 2030. However, its Board has since rebuffed parts of the verdict, indicating that it is unreasonable and essentially incompatible with its business. The company has appealed the Court's decision.

Scientists say that to avoid the worst impacts of climate change, we need to halve the world's GHG emissions by 2030. Currently, more than 70% of the world's emissions come from fossil fuels, and fossil fuel producers must focus on cutting their production every year if we are to avoid the worst climate impacts. On the other hand, Shell publicly maintains that its strategy is consistent with the 1.5°C temperature goal of the Paris Agreement and has set a target to become a net-zero emission energy business by 2050, in step with society. However, its interim targets to get there seem quite inadequate and have indeed been challenged in court, as we note in the case study above.

A2.5.2 *The Just Transition – Inter-Generational and North/South Issues: Who Bears the Cost?*

Simply applying a tax to all uses of carbon by consumers such as on gas heating, petrol, and beef would have a disproportionate effect on the cost of living of many people, increasing it beyond the price rises already seen in 2022, which have impacted those on low incomes much more than those on high incomes. Clearly, a carbon tax would increase the financial stress on those on lower incomes. Equally, the ensuing transition from fossil energy to clean energy would hit the employment prospects of many of those working in, for example, the coal mining industry and some types of farming. Not only individuals would be affected but also regions of Europe and particular towns which have focused on industries based on fossil fuel.

An example of a previous industrial transition can be seen in the City of Glasgow in Scotland (the city in which the main author of this book was born). When he was growing up, the choice of employment was mainly in the iron and steel companies, coal mining, shipbuilding, and dock work on the River Clyde. Long before COP 26 in Glasgow in 2021, all those industries had disappeared. They were eventually replaced after decades of transition by jobs in the service industries, such as investment and commercial bank operations centres, call centres, research and development, and consulting. Service industries now employ around 800,000 people, while manufacturing employs only 52,000. That transition to a restructured local economy was much too slow. The objective of the EU Just Transition is to ensure that in the future, those individuals, towns, and regions impacted by the energy transition should not suffer the long drawn-out transition seen in previous eras in cities such as Glasgow. This was a city built on coal, steel, and shipbuilding. Now, there are none of these industries, but over many decades, the city has managed to transition to a service industry and SME city.

Many of those involved in the debate on carbon taxes have emphasised the need for a 'Just Transition',[11] that is, one that does not result in low-income groups, regions, or towns suffering disproportionately from a rise in their costs and a potential loss of income due to measures taken to mitigate climate change. Such mitigation measures would include retraining, investment grants, expansion of universities, etc.

Imposition of a carbon tax which hits all consumers without corresponding measures to mitigate its impact on individuals and communities could lead to an outcome such as the 'gilet jaune' protests in France when the President of France initiated higher taxes on fossil fuels on the grounds of it being one of the potentially most effective climate-change measures. The protestors said that it hit lower-income people and those living in the countryside disproportionately.

The European Union is a region that is very focused on the need for a just transition, as without measures to help those disadvantaged by the transition, social solidarity could be severely weakened. Frans Timmermans, executive vice-president of the European Commission, has said, 'We must show solidarity with the most affected regions in Europe, such as coal mining regions and others, to make sure the Green Deal gets everyone's full support and has a chance to become a reality'. The Just Transition Mechanism (JTM) in the European Union is a key tool to ensure that the transition towards a climate-neutral economy happens in a fair way, leaving no one behind. It provides targeted support to help mobilise at least €150 billion over the period 2021–2027 in the most affected regions to alleviate the socio-economic impact of the transition.

---

[11] See, for example, https://climatejusticealliance.org/just-transition/ and https://www.gov.scot/groups/just-transition-commission/.

A2.6 Inter-Generational Issues

A Just Transition is also relevant to inter-generational equity (fairness) within a country. The question is whether it is fair, and hence in line with the Just Transition, or not, to put more than a 'fair' share of the burden on to future generations than on the current generation? In 2021, Germany's highest court agreed that it was unfair to push so much of the burden on to young people in what may become a landmark case. Among the plaintiffs in court was Sophie Backsen, 22, daughter of a farming family on the North Sea island of Pellworm, who fears that rising sea levels would engulf her low-lying island, leaving her with no inheritance. In the case, the judge ruled that it would be wrong to use up large parts of the remaining CO_2 budget by requiring only small reductions now if that simultaneously meant that following generations were left with a radical reduction burden and their lives were exposed to extensive limits to freedom.

A2.6.1 *The North–South Divide*

The argument here is not about fairness (climate justice) between different regions or cities in one country or different age groups in a country but between countries. In particular, it is about which countries have been responsible for emitting GHGs in the largest amounts since the industrial revolution which set off the worldwide use of fossil fuels. The usual date for the starting point is taken to be 1750. Clearly, a large country will have much greater cumulative emissions than a small country.

The problem arises from the fact that all countries in the world suffer from these cumulative emissions, and thus developing countries with low cumulative emissions (and annual emissions per capita) suffer as much, and possibly more, than developed countries. Some will even disappear under the sea.

A2.6.2 *Sustainable Finance Disclosure Regulation: Financial Firms*

Sustainable Finance Disclosure Regulation (SFDR) is one of the mechanisms whereby the European Commission is attempting to reorient capital towards sustainable businesses. It is designed to try to ensure that those providing capital to corporates are financing them in a sustainable manner and avoiding 'greenwashing'. It also requires managers to explain how they take adverse impacts into account. The regulation is very focused on adverse sustainability factors. This refers not only to climate issues but also to social and employee issues which the Commission considers to be material to society as a whole.

A2.7 European Insurance and Occupational Pensions Authority

The European Insurance and Occupational Pensions Authority (EIOPA) is an independent advisory body to the European Commission, the European Parliament, and the Council of the European Union. It is one of the EU agencies carrying out specific legal, technical, or scientific tasks and giving evidence-based advice. In this way, it helps shape informed policies and laws at EU and national levels. EIOPA is one of three European Supervisory Authorities. The other two are the European Banking Authority (EBA) and the European Securities and Markets Authority (ESMA).

EIOPA published its 'Opinion'[12] on the supervision of the use of climate change risk scenarios in Own Risk and Solvency Assessment (ORSA). The Opinion sets out supervisory expectations on the integration of the use of climate change scenarios by insurance undertakings in their ORSA. Given that undertakings will be impacted by climate-change-related physical and transition risks, EIOPA believes it is important to encourage a forward-looking management of these risks, also in the long term. Currently, only a small minority of undertakings assess climate change risk using scenario analysis in the ORSA. Moreover, where undertakings perform a quantitative analysis of climate change risk, most assessments take a short-term perspective.

A2.8 The EU Innovation Fund

The EU is concerned not only with regulation but with providing funding for projects that otherwise might not proceed or would proceed more slowly. With revenue of more than €38 billion until 2030 from the EU Emissions Trading System (ETS), the Innovation Fund aims to create the right financial incentives for companies and public authorities to invest in

[12] https://www.eiopa.europa.eu/document-library/other-documents/application-guidance-climate-change-materiality-assessments-and_en.

the next generation of low-carbon technologies and give EU companies a first-mover advantage to become global technology leaders. The following text is from an EU press release[13]:

> Today, the EU is investing over €1.8 billion in 17 large-scale innovative clean-tech projects with a third round of awards under the Innovation Fund. Grants will be disbursed from the Innovation Fund to help bring breakthrough technologies to the market in energy-intensive industries, hydrogen, renewable energy, carbon capture and storage infrastructure, and manufacturing of key components for energy storage and renewables. The selected projects are located in Bulgaria, Finland, France, Germany, Iceland, the Netherlands, Norway, Poland and Sweden.
>
> Executive Vice-President Frans Timmermans said: 'Today's grants support innovative businesses across Europe to develop the cutting-edge technologies we need to drive the green transition. The Innovation Fund is an important tool to scale up innovations in renewable hydrogen and other solutions for European industry. Compared to the first disbursement round, the funds available have increased by 60%, enabling us to double the number of projects supported. This is a big boost for the decarbonisation of energy-intensive industry in the European Union.'
>
> The selected projects cover a wide range of sectors contributing to the EU's decarbonisation efforts such as production, distribution and use of green hydrogen, waste-to-hydrogen, offshore wind, manufacturing of photovoltaic (PV) modules, battery storage and recycling, carbon capture and storage, sustainable aviation fuels, and advanced biofuels. Together, they have a potential to save 136 million tonnes of CO_2eq over their first 10 years of operation.
>
> In addition, up to 20 projects that are promising but not yet sufficiently mature for a grant will be pre-selected for project development assistance by the European Investment Bank. These will be announced in the fourth quarter of 2022.

---

[13]Press Release. European Commission 13 July 2022 https://ec.europa.eu/commission/presscorner/detail/en/IP_22_4402.

A2.9 A New EIB Risk Sharing Initiative[14]

The European Investment Bank (EIB) is launching an initiative to support wind turbine parts manufacturers in Europe. It is doing so by extending a €500 million counter-guarantee to Deutsche Bank AG in Germany. According to the announcement, this will allow Deutsche Bank to create a portfolio of up to €1 billion in bank guarantees for new investments in wind farms in the EU.

The agreement is the first under the EIB's December 2023 plan to make €5 billion available to strengthen the provision of commercial bank guarantees to companies across the European wind industry. The initiative is part of an EU wind energy package presented by the European Commission in October 2023 to maintain a healthy and competitive wind energy supply chain across the Union. It includes counter-guarantees to the sector's main lenders, including Deutsche Bank.

'A rapid build-up of wind energy is essential for the decarbonization of the economy', Nicola Beer, EIB vice president, said in the announcement and continued:

> Together with Deutsche Bank, we are promoting the deployment of renewable energy in Europe, bringing down the price of sustainable energy. This project will also help secure and create jobs in a sustainable and competitive European industry.

The EIB counter-guarantees under the €5 billion initiative are designed to share the credit risks commercial banks face when dealing with wind industry players. It will facilitate access to prepayment and performance guarantees, benefiting both banks and the industry. In this first project, Deutsche Bank will use the EIB's backing to provide bank guarantees to wind energy producers. This will allow manufacturers to receive advance payments and provide performance guarantees when taking on new wind projects.

The guarantee scheme also allows manufacturers to pay their suppliers in advance for the delivery of wind farms and the associated

---

[14]This is an article published in EnergyWatch on 2 August 2024 *EnergyWatch is a subscription publication* https://energywatch.com/EnergyNews/Renewables/article17321467.ece.

components of the wind value chain, which includes turbines, grid connection infrastructure, cables, and substations. It is estimated that the €1 billion EIB–Deutsche Bank guarantee facility will trigger private investment of up to €8 billion.

A2.10 Carbon Border Adjustment Mechanism (CBAM)

Some countries have reported that they have reduced their carbon emissions and claimed that this shows the success of their regulatory system. However, in practice, what has happened in many instances is that countries outside Europe have taken market share from European countries in industries with high emission levels. The CO_2 emissions have simply moved from a European country to another country outside Europe. If European legislation then requires European companies to make large reductions in their CO_2 emissions or pay a high carbon tax, but countries outside Europe do not, then European companies are being disadvantaged.

CBAM is designed to prevent 'carbon leakage' and ensure that non-EU producers pay a comparable carbon price to EU manufacturers. The particular sectors which are carbon-intensive are steel, aluminium, cement, fertilisers, electricity, and hydrogen. Companies importing such products to Europe will have to purchase CBAM certificates based on the emissions 'embedded' in their products.

A2.11 EU Omnibus Simplification Package

In February 2025, the Commission published the EU Omnibus Simplification Package, which scales back or delays key sustainability reporting legislation. At the start of its new five-year term, the Commission focused on EU industrial competitiveness as top priority. As a result, in November 2024, EU leaders adopted the Budapest Declaration, which lays out a new European Competitiveness Deal (see under the following section on Clean Industrial Deal). The Omnibus proposal seeks to reduce reporting burdens by 25% for large companies and 35% for SMEs.

The Commission outline five activities aimed at increasing competitiveness:

Simplification,
Lowering barriers to the single market,

Financing competitiveness,
Promoting skills and quality jobs,
Better coordination of EU and national policies.

The key pieces of legislation impacted by the Omnibus Package are:

Corporate Sustainability Reporting Directive (CSRD),
Corporate Sustainability Due Diligence Directive (CSDDD),
EU Taxonomy,
Carbon Border Adjustment Mechanism (CBAM).

A2.12 The Clean Industrial Deal[15]

In February 2025, the EU published details of the Clean Industrial Deal (CID). It describes the CID as **'a plan for EU competitiveness and decarbonisation'**. It goes on to explain the CID thus:

> Faced with high energy costs and fierce global competition, European industries need urgent support. The Clean Industrial Deal outlines concrete actions to turn decarbonisation into a driver of growth for European industries. This includes lowering energy prices, creating quality jobs and the right conditions for companies to thrive.
>
> The Deal presents measures to boost every stage of production, with a focus on:
>
> - energy-intensive industries such as steel, metals, and chemicals, that urgently need support to decarbonise, switch to clean energy, and tackle high costs, unfair global competition, and complex regulations.
> - the clean-tech sector which is at the heart of future competitiveness and necessary for industrial transformation, circularity, and decarbonisation.

The two foci are clearly complementary in that it is the clean-tech sector (renewable energy, nuclear energy, energy storage, carbon capture

[15] https://commission.europa.eu/topics/eu-competitiveness/clean-industrial-deal_en.

and storage, clean hydrogen, clean heat, etc.) that will enable the energy-intensive sectors to decarbonise.

It is hoped that the CID will boost demand for clean products. It is expected that the Industrial Decarbonisation Accelerator Act will increase demand in the EU for EU-made clean products. It will achieve this by introducing sustainability, resilience, and 'made in Europe' criteria in **public and private procurements**. In addition, the Commission will review the public procurement framework to introduce **sustainability, resilience, and European preference criteria in public procurement for strategic sectors**.

In relation to how to finance the necessary actions that are required under the CID, the Commission believes that the CID will mobilise over €100 billion to support EU-made clean manufacturing. The Commission will:

- adopt a new CID state aid framework to accelerate the approval of state aid to roll out renewable energy, decarbonise industry, and ensure sufficient manufacturing capacity of clean tech;
- strengthen the Innovation Fund and propose an industrial decarbonisation bank, aiming for €100 billion in funding, based on available funds in the Innovation Fund, additional revenues resulting from parts of the ETS as well as the revision of InvestEU;
- launch a dedicated call under Horizon Europe to stimulate research and innovation in these areas;
- amend the InvestEU Regulation to increase the amount of financial guarantees that InvestEU can provide to support investments. This will in turn mobilise up to €50 billion for the deployment of clean tech, clean mobility, and waste reduction.

Appendix 3

Pexapark on Cannibalisation Risk

This term and others are well explained in a publication by Pexapark on cannibalisation.[1] An edited excerpt from this excellent publication is provided as follows:

> What is cannibalisation?
>
> Because of the marginal pricing method, when renewable energy injects to the grid, spot prices tend to go down. If a low-marginal-cost technology such as wind and solar is available, the system operator will give priority dispatch to that technology and will need less of the more expensive energy generated by other (typically fossil-fuel-based) technologies such as gas. Therefore, the average price for this particular time slot will fall. Cannibalisation is often referred to as 'profile risk'.
>
> In the long run, this poses a serious threat to the viability of renewables operating on the free market. When a lot of renewable energy is injected at the same time, which is particularly material for solar energy because

[1]Pexapark *Cannibalisation: Behind the Renewables Silent Risk* https://pexapark.com/cannibalization-renewables/. *About Pexapark*: We empower renewable energy players to thrive in post-subsidy markets. Keeping renewable investments attractive is our contribution to a more sustainable world. Through the phasing out of subsidy schemes, renewable energy investors become exposed to the electricity market and face entirely new challenges. How to replace guaranteed subsidy revenue with Power Purchase Agreements (PPAs)? How to sell and hedge renewable energy and actively manage energy market risks?

there's a stable generation profile shape, the realised price they will get will fall. Nowadays, renewables increasingly participate in the electricity markets on a merchant basis. The amount of renewables in the grid is set to increase quite substantially, posing a significant risk for their long-term remuneration.

Baseload Price: It is the average spot price over a selected timeframe. Remember that spot prices usually change on a half-hourly (hourly) basis. So, when we refer to baseload prices, we either refer to them on a daily, monthly, or yearly basis, accumulating all the different prices throughout the day.

Baseload Production/Power: This usually refers to the ability of a power plant to meet the average daily demand continuously. If needed, a natural gas power plant can be on 24/7. Therefore, it is able to provide baseload production, which is the opposite of what VRE can provide, which is intermittent production.

Capture Price: Capture prices are the average prices that each generator will manage to realise over a period of time; hence, they are also referred to as `realised prices'. This is where seasonality of production and the ability to provide baseload power come into play. Suppose a solar power plant produces during specific hours of the day, but its production lowers the spot price. In that case, the capture price could be lower than a natural gas plant will achieve because the latter will be able to sell during peak hours.

Cannibalisation is the phenomenon of capture prices for renewables dropping as solar peak production is at times when prices are low. With more and more solar having the same generation profile producing at these hours, prices would drop further, and the capture price would be lower.

Capture prices are calculated as an asset's volume-weighted average price, i.e. the price at which it sold, based on the volume and the time.

Capture Factors: The capture factor illustrates the ratio between baseload prices and capture prices, i.e. capture price divided by the baseload price.

Bibliography

Actuaries Institute. (2023). Limitations and assumptions of commonly used climate-change scenarios in financial services. https://actuaries.org.uk/media/qeydewmk/the-emperor-s-new-climate-scenarios.pdf

BloombergNEF. (2025). *Energy Transition Investment Trends 2025*. https://about.bnef.com/insights/finance/energy-transition-investment-trends/

Carbon Brief. (2018). How integrated assessment models are used to study climate change. https://www.carbonbrief.org/qa-how-integrated-assessment-models-are-used-to-study-climate-change

ClientEarth. (n.d.). Our groundbreaking case against Shell's Board of Directors comes to an end. https://www.clientearth.org/latest/latest-updates/news/we-re-taking-legal-action-against-shell-s-board-for-mismanaging-climate-risk/

Energy Transitions Commission. (2021). *Making Clean Electrification Possible: 30 Years to Electrify the Global Economy*. https://www.energy-transitions.org/wp-content/uploads/2022/07/ETC-Global-Power-Report-Final.pdf

EnergyTag. (n.d.). Carbon accounting & tracking for 24/7 clean grids. https://energytag.org/

EnergyWatch. (2024). https://energywatch.com/EnergyNews/Renewables/article17321467.ece

EU Technical Expert Group on Sustainable Finance. (2020). *Usability Guide: EU Green Bond Standard*. https://ec.europa.eu/info/sites/info/files/business_economy_euro/banking_and_finance/documents/200309-sustainable-finance-teg-green-bond-standard-usability-guide_en.pdf

European Commission. (2020). Report on climate-related disclosures (TEG). https://ec.europa.eu/info/sites/info/files/business_economy_euro/banking_and_finance/documents/200309-sustainable-finance-teg-final-report-taxonomy_en.pdf

European Commission. (2022). *Directive of the European Parliament and of the Council on Corporate Sustainability Due Diligence and Amending Directive (EU) 2019/1937 COM/2022/71 Final.* https://eur-lex.europa.eu/legal-content/EN/TXT/?uri=CELEX%3A52022PC0071

European Commission. (2022, July 13). Innovation Fund: EU invests €1.8 billion in clean tech projects [Press release]. https://ec.europa.eu/commission/presscorner/detail/en/IP_22_4402

European Parliament. (2015). *Energy Storage: Which Market Designs and Regulatory Incentives Are Needed?* https://www.europarl.europa.eu/RegData/etudes/STUD/2015/563469/IPOL_STU(2015)563469_EN.pdf

Green Giraffe. (2019, April). Recent trends in offshore wind finance. https://green-giraffe.eu/wp-content/uploads/2019/04/190411_green_giraffe_bilbao_-_trends_in_ow_finance_final.pdf

Hori, M. (2010). Application of nuclear produced hydrogen for energy and industrial use. In: Nuclear Energy Agency (NEA), *Nuclear Production of Hydrogen*. Paris: OECD Publishing.

IJGlobal. (2013). European refinancing deal of the year 2013: Castor UGS. https://www.ijglobal.com/articles/127334/european-refinancing-deal-of-the-year-2013-castor-ugs#comments

IJGlobal. (n.d.). The Earth moved: Castor gas storage. https://www.ijglobal.com/articles/135619/the-earth-moved-castor-gas-storage?home=1

Intergovernmental Panel on Climate Change. (2014). *Fifth Assessment Report Summary for Policymakers*. https://www.ipcc.ch/site/assets/uploads/2018/02/SYR_AR5_FINAL_full.pdf

International Energy Agency. (2024, September). *Integrating Solar and Wind.* https://www.iea.org/reports/integrating-solar-and-wind

International Energy Agency. (n.d.). Fossil vs biogenic CO2 emissions. IEA Bioenergy. https://www.ieabioenergy.com/iea-publications/faq/woody biomass/biogenic-co2/

Kraemer, S. (2020). Morocco pioneers PV with thermal storage at 800 MW Midelt CSP project. *Solar PACES.* https://www.solarpaces.org/morocco-pioneers-pv-to-thermal-storage-at-800-mw-midelt-csp-project/

LDES Council. (n.d.). *A Path Towards Full Grid Decarbonisation with 24/7 Clean Power Purchase Agreements*. https://www.ldescouncil.com/assets/2205_ldes-report_247-ppas.pdf

Luo, Z., Wang, J., Dooner, M., & Clarke, J. (2020). Cost-economic analysis of hydrogen for China's fuel cell transportation field. *Energies, 13*(24), 6522. https://www.mdpi.com/1996-1073/13/24/6522

NASA Earth Observatory. (n.d.). Climate and Earth's energy budget. https://earthobservatory.nasa.gov/features/EnergyBalance/page1.php

National Academies Press. (2005). *Radiative Forcing of Climate Change*. https://www.nap.edu/read/11175/chapter/1#vi

NOAA. (n.d.). Climate forcing. https://www.climate.gov/maps-data/primer/climate-forcing

Nordea. (2023). *How Sustainability Risks Affect Investments*. https://www.nordea.com/en/doc/final-sustainability-risks-and-how-they-affect-investments-q2-2023.pdf

Partnership for Carbon Accounting Financials. (2022). *The Global GHG Accounting and Reporting Standard Part A: Financed Emissions* (2nd ed.). https://carbonaccountingfinancials.com/files/downloads/PCAF-Global-GHG-Standard.pdf

Pexapark. (n.d.). Cannibalisation: Behind the renewables silent risk. https://pexapark.com/cannibalization-renewables/

S&P Global. (2019, April 10). *ESG Evaluation*. https://www.spglobal.com/_assets/documents/ratings/ratingsdirect_environmentalsocialandgovernance-evaluationanalyticalapproach_1176513_apr-10-2019.pdf

Taleb, N. N. (2008). *The Black Swan: The Impact of the Highly Improbable*. Penguin Books.

Task Force on Climate-related Financial Disclosures. (2017). *Recommendations of the Task Force on Climate-Related Financial Disclosures*. https://assets.bbhub.io/company/sites/60/2021/10/FINAL-2017-TCFD-Report.pdf

Trade Finance Global. (2019). Sustainable, green finance and ESG trends in 2019. https://www.tradefinanceglobal.com/posts/sustainable-finance/

United Nations. 24/7 carbon-free energy compact. https://www.un.org/en/energy-compacts/page/compact-247-carbon-free-energy

University of Reading. (2025). *MSc Climate Change and Artificial Intelligence*. https://www.reading.ac.uk/ready-to-study/study/2025/meteorology-and-climate-pg/msc-climate-change-and-artificial-intelligence

US Department of Energy, National Renewable Energy Laboratory. (2020). Hydrogen in future integrated energy system. https://www.nrel.gov/news/program/2020/study-shows-abundant-opportunities-for-hydrogen-in-a-future-integrated-energy-system.html

US Department of Transportation, Federal Highway Administration. Center for Innovative Finance Support. https://www.fhwa.dot.gov/ipd/p3/toolkit/toolkit_webinars/financing_highways/

US Energy Information Administration. (n.d.). Biomass explained. https://www.eia.gov/energyexplained/biomass/

World Economic Forum. (2019, January). 3 ways to be a good leader in the Fourth Industrial Revolution. https://www.weforum.org/agenda/2019/01/the-fourth-industrial-revolution-needs-new-forms-of-leadership/

World Energy Council. (n.d.). Balancing the energy trilemma. https://www.worldenergy.org/assets/downloads/Trilemma-what-is-the-energy-trilemma.jpg

Index

www.ingramcontent.com/pod-product-compliance
Lightning Source LLC
LaVergne TN
LVHW051919051126
851858LV00009B/138
* 9 7 8 1 8 0 0 6 1 9 0 1 2 *